Bernhard Welz

Atomic Absorption Spectroscopy

verlag chemie

Bernhard Welz

Atomic Absorption Spectroscopy

Verlag Chemie · Weinheim · New York · 1976

First published in German under the title "Atom-Absorptions-Spektroskopie" by Verlag Chemie in 1972.
This English version was translated from the completely revised 2nd German edition 1975.

Translated by Christopher Skegg

Dr. Bernhard Welz
c/o Bodenseewerk
Perkin-Elmer & Co GmbH
Postfach 1120
D-7770 Überlingen

This book contains 73 figures and 46 tables

CIP-Kurztitelaufnahme der Deutschen Bibliothek

Welz, Bernhard
Atomic absorption spectroscopy.—1. Aufl.—Weinheim, New York: Verlag Chemie, 1976.
Dt. Ausg. u. d. T.: Welz, Bernhard: Atom-Absorptions-Spektroskopie.
ISBN 3-527-25680-6

Printed by: Zechnersche Buchdruckerei, D-6720 Speyer
Bookbinder: Georg Kränkl, D-6148 Heppenheim
Printed in West Germany

Preface

It was very convenient that the translation of my book into the English language was undertaken just as I was completing the second German edition. Therefore, all the latest developments and publications could be incorporated directly into the English edition.

Usually, years go by between the publication of the original book and the completion of a translation which, therefore, typically does not represent the latest state. Here, however, the translation could be published about a year after the original German edition. This is of special importance for the rapidly growing field of furnace atomic absorption which was hardly known a few years ago when the first edition of my book was published. In the meantime it has found worldwide acceptance among analysts.

So, to all my friends and colleagues who have been involved in the translation and completion of this book, I would like to express my thanks for the time that they have spent and for all the effort that they have put into it so that it could be published so early. Last not least, I want to express my pleasure that my book on Atomic Absorption Spectroscopy has been accepted for translation into English. I hope that it will prove a stimulus to atomic absorption spectroscopy and will help analysts and spectroscopists in their daily work.

Meersburg, March 1976 Bernhard Welz

Preface to the Second German Edition

Within the last year or two, atomic absorption spectroscopy has received tremendous impetus through flameless atomization in graphite furnaces. The reason for this is the increase in sensitivity by two to three orders of magnitude; it has thereby been possible to keep abreast of the increasing demands made by all branches of analysis without having to resort to time-consuming enrichment procedures. Furthermore, atomic absorption spectroscopy has been used in new fields that previously could only be served by more complicated analytical techniques.

These new developments have received comprehensive treatment in the second edition; the chapter on atomization has been logically divided into two sections—atomization in flames and flameless atomization. The totally different mechanisms occurring within the graphite furnace have been explained as fully as possible. Nevertheless, this has not always been easy since investigations on the processes occurring in the graphite tube during the thermal pre-treatment and atomization stages have only recently begun.

As well as in this special section, flameless atomic absorption has been taken into consideration in all other chapters, and especially the chapters on technique, the individual elements, and

specific applications have been correspondingly revised and extended. It must nevertheless be emphasized that the publications in this sector are still very limited and, in part, contradictory. Therefore, a relatively large number of personal opinions and experiences have been worked into the text. These may require a certain amount of revision with the course of time.
New publications up to about the end of 1974 have been taken into account in all chapters and the bibliography has been extended by more than 60%. Worth mentioning in this context are the developments in electrodeless discharge lamps and background correctors. The latter have gained particular importance in the train of flameless atomic absorption. Under the methods for sample introduction, next to the DELVES system and its variants, the hydride system has earned particular attention. In this latter system, elements that form covalent hydrides are conducted in gaseous form to the flame or to a heated tube.
Finally, the problems that occur with the direct analysis of solid samples, especially in connection with atomization in graphite tubes, have been treated in a separate section. Further, a number of new digestion methods have been described in greater detail in the section on applications since they will almost certainly facilitate analyses by atomic absorption spectroscopy.
All in all, this second, fully revised and extended edition has been brought up to date, and not only with respect to flameless atomic absorption spectroscopy. Mr. E. Klebsattel, who drew a number of new diagrams and revised the entire figures, has again made a substantial contribution. I also received important suggestions from Dr. E. Witte, who has been intensely involved with the problems of atomization in graphite tubes. I would also like to thank those observant readers who drew my attention to errors in the first edition. Despite all efforts, it is unlikely that the second edition will be completely free of mistakes and every notice of error or recommendation for improvement will be gratefully received.

Meersburg, May 1975 Bernhard Welz

Preface to the First German Edition

During the decade from 1960, atomic absorption spectroscopy rapidly developed into a universally applicable, highly selective and sensitive analytical method that nowadays finds use in all branches of analysis. The number of publications dealing with this method is correspondingly large.
To obtain a maximum density of information within a limited space, topics for which other comprehensive sources are readily available to the reader have only been briefly broached. This principle has been applied consistently throughout the chapter on specific applications, since detailed information on all the possibilities of application would expand this book beyond all

resonable bounds; the various examples have only been briefly discussed and a comprehensive bibliographical survey is provided in lieu of a more detailed treatise. Furthermore, comprehensive compendia of methods containing proven procedures are nowadays available. The introductory chapter on theory only advances the physical-spectroscopical principles of atomic absorption spectroscopy as far as is required for an understanding of the fundaments. Numerous textbooks of physics or physical chemistry are available for further information.

These measures are designed to increase the readability of the book by not going to too great length on the one hand, but without foregoing important information on the other; the close interweaving of theory and practical application serves the same purpose.

Since atomic absorption spectroscopy is a relatively young analytical method, it has not been possible to include the latest knowledge in all sections of this book. While flame atomic absorption has achieved a relative stability, flameless atomic absorption has undergone drastic development in the year since the completion of the manuscript. The end of this development is nowhere yet in sight. It will certainly be opportune to devote more space to this technique in the future and possibly to extend the chapter on interferences to include the latest findings. While the interferences known from the flame are hardly observed in the graphite furnace, considerable difficulties can occur if the element under study forms a volatile compound with a component of the matrix. It has further been shown that non-specific light losses due to scattering on solid particles in the light beam are relatively common. Thus a deuterium background corrector is of decisive importance for this technique—the more so since with the ballistic signals generated by flameless atomic absorption, the non-specific signals are very difficult to reproduce and thus virtually impossible to eliminate through subsequent measurement.

Finally, initial results have shown that plasmas may possibly gain great significance in atomic absorption spectroscopy. Flame-like plasmas of suitable temperature have been described which, because of their fully inert atmospheres, could be of especial use for the determination of refractory elements.

To close, I do not wish to miss the opportunity of thanking all those who, directly or indirectly, have contributed to the success of this book, even when they are not mentioned by name. Dr. H. Stenz willingly read through the greater part of the manuscript, and in numerous discussions contributed much to a clarification of the theoretical and physical aspects. I would like to express especial thanks to Mr. E. Klebsattel for careful and conscientious production of the numerous figures.

Meersburg, February 1972 Bernhard Welz

Contents

1. Introduction

Atomic Absorption Spectroscopy is the observation of the absorption of light energy by atoms.

1.1. History

The phenomenon of light absorption had already been investigated at the beginning of the eighteenth century, principally on crystals and liquids. It was observed that the original radiation intensity is resolved into three components: into reflected, transmitted and absorbed radiation. Using BOUGER's work [147], LAMBERT [723] found in 1760 that the amount of light passing through a layer of uniform thickness of a homogeneous medium is dependent upon the thickness d of this layer and that the ratio of the intensity of the transmitted light I_t to the intensity of the incident light I_0 is independent of the radiation intensity.

$$I_t = I_0 \cdot e^{-\varkappa' d} \tag{1.1}$$

The proportionality factor $\varkappa'$ is a measure of the power of attenuation of the layer and is termed the absorption coefficient.

If the irradiated medium is not a single substance but a solution of an absorbing substance in a non-absorbing medium, the absorption coefficient is proportional to the concentration c.

$$\varkappa' = k' \cdot c \tag{1.2}$$

LAMBERT's law underwent a thorough examination by BEER [95] and is nowadays mainly used in the form

$$A \equiv \log \frac{I_0}{I_t} = k \cdot c \cdot d\,. \tag{1.3}$$

It states that the absorbance A (or alternatively, the logarithm of the reciprocal transmission) is proportional to the thickness through which the light is transmitted and to the concentration of the absorbing substance.

The history of absorption spectroscopy is closely connected with the observation of sunlight [1248]. In 1802 WOLLASTON discovered the black lines in the sun's spectrum which were later investigated in detail by FRAUNHOFER. In 1820 BREWSTER expressed the view that these Fraunhofer lines were caused by absorption processes in the sun's atmosphere. The underlying principles of this absorption were established by KIRCHHOFF and BUNSEN during their systematic examination of the line reversal in the spectra of alkali and alkaline-earth metals [648–651]. They conclusively demonstrated that the typical yellow line emitted by sodium salts in a flame is identical to the black D line of the sun's spectrum. The classical experimental arrangement is shown in figure 1.

The relationship between emission and absorption was formulated by KIRCHHOFF in his law, which is generally valid and states that any material which can emit light at a given wavelength will also absorb light at that wavelength.

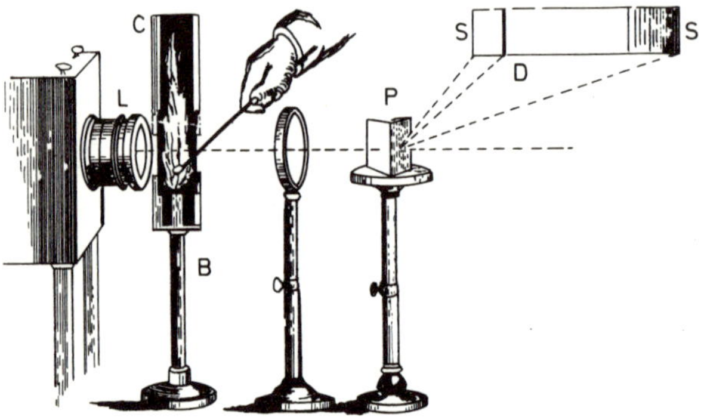

Figure 1. Experimental set-up of KIRCHHOFF and BUNSEN for investigating the line inversion in the sodium spectrum (according to [1248]). Light from a continuous source is focused by the lens **L** through the flame of a Bunsen burner **B** into which sodium chloride is introduced with a spatula. The light beam is dispersed by the prism **P** and observed on the screen **S**. The sodium D line appears as a black discontinuity in the otherwise continuous spectrum.

PLANCK (1900) established the quantum law of absorption and emission of light, according to which an atom can only absorb light of a specific wavelength (frequency), i.e. it can only take up and release definite amounts of energy ε.

$$\varepsilon = h\nu = \frac{hc}{\lambda} \tag{1.4}$$

Characteristic values of ε and ν or λ exist for each atomic species.
If continuous radiation falls, for example, on sodium or mercury vapour contained in a quartz vessel, spectral dispersion of the transmitted light produces the entire continuous spectrum of the light source with the exception of a narrow line seen as a black discontinuity. This "black" line is characteristic of the vaporized material and lies in the visible range at 589.2 nm for sodium and in the ultraviolet range at 253.6 nm for mercury.
If the irradiated quartz vessel is observed at right angles to the beam path of the continuous light source, the atoms present in the vessel will emit only a single wavelength which is found to be precisely that wavelength which appears as a black line in the spectrum of the transmitted light.
It has not only been shown that the same wavelength as is emitted is absorbed, but also that the same quantity of light is involved. This phenomenon is therefore termed "resonance fluorescence". On the basis of this and many other observations, RUTHERFORD proposed his atomic model in 1911, the fundamental principle of which is that atoms do not exist in random energy states, but only in certain fixed states which differ from each other by integral quantum numbers.
Upon absorbing a quantum of energy, an atom is transformed into a particular, energy-enriched state "containing" the light energy which has been taken up. Later (more precisely, after

a period of 10^{-9} to 10^{-7} s) the atom can re-emit this energy—as a resonance fluorescence quantum—and thus return to the ground state.
Although KIRCHHOFF had already recognized the principle of atomic absorption in 1860 and the theoretical basis was steadily extended during the following decades, the practical significance of this method was not recognized.
Since the work of KIRCHHOFF, the principle of atomic absorption was merely used by astronomers to determine the concentration of metals in the atmospheres of stars. Chemical analyses were only carried out very sporadically by this method; the determination of mercury vapour did however acquire a certain degree of importance [1354]. The actual year of birth of atomic absorption spectroscopy was 1955. In that year, publications authored independently by WALSH [1282] and by ALKEMADE and MILATZ [33] [34] appeared, in these publications atomic absorption spectroscopy was recommended as a generally applicable method of analysis.
In the following years it was principally WALSH and his co-workers at C.S.I.R.O.*) who developed atomic absorption into a quantitative analytical method of high sensitivity and selectivity. WALSH receives the credit for developing not only the theoretical basis, but also the practical applications and the instrumental principles.

1.2. Atomic Spectra

It has already been mentioned that only well defined amounts of energy can be absorbed and re-emitted by an atom, and that after absorbing a quantum of energy the atom is transformed into a particular, energy-enriched or "excited" state. Upon returning to the lower energy state, the atom usually releases the absorbed energy in the form of radiation.
Three different modes of radiative energy absorption or release can be distinguished:
If the atoms are excited thermally or electrically, then absorbed energy is released as an emission spectrum. If the excitation is by light energy, the atoms only absorb exactly defined amounts of energy (i.e. light of definite frequency) and an absorption spectrum can be observed. The energy absorbed in this manner is released in the form of a fluorescence spectrum.
If an atom could exist in only *one* excited state, then only one emission, absorption or fluorescence line would appear. As can be demonstrated experimentally, however, this is not the case. Thus the absorption spectrum of an atom is seen to consist of a sequence of numerous lines following one another, evidently according to a set plan. Towards the short wavelength end, the lines appear closer and closer together and decrease regularly in intensity until a "convergence limit" is reached; that is, a value beyond which there are no more lines and only continuous absorption occurs.
In the absorption spectrum of sodium, for example, it has been possible up to now to observe and measure 57 absorption lines (figure 2b).
The emission spectrum (figure 2b) which is produced on excitation by a glow discharge, a flame, or an electric arc has many more lines and does not appear to exhibit the pronounced regularity of the absorption spectrum.

---
* C.S.I.R.O.—Commonwealth Scientific and Industrial Research Organization, Australia.

An atom can exist in many different states and many spectral transitions in absorption and emission are therefore possible. This situation is most clearly represented by an energy level diagram.

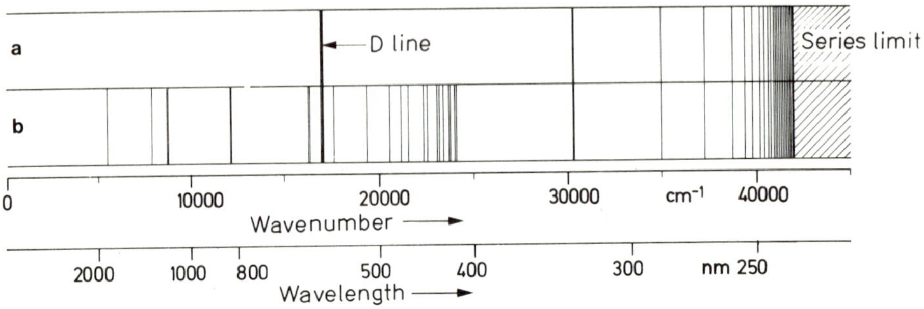

Figure 2. a) Absorption and b) emission spectrum of sodium.—At the temperatures normally used in absorption measurements the majority (~99.9%) of the atoms formed are in the ground state. For the most part only transitions that start from the ground state can occur (main series) in absorption. On thermal or electrical excitation of the sodium atoms all possible excitation states can be reached from which the emission of the main and all secondary series is possible. The overlapping of these series leads to the observed multiplicity of lines in the emission spectrum.

Each spectral line can be regarded as the difference between two atomic states or terms $\tilde{\nu}$:

$$\frac{1}{\lambda} = \tilde{\nu}_k - \tilde{\nu}_j \tag{1.5}$$

The differences between two such terms are the energy differences between the excited atomic states multiplied by $1/h \cdot c$.

$$\tilde{\nu}_k - \tilde{\nu}_j = \frac{1}{h \cdot c}(E_k - E_j) \tag{1.6}$$

The various states in an atom under study can be described with the aid of three quantum numbers; the principal quantum number n, the secondary quantum number L and the "inner" quantum number J. In table 1 the possible terms in an alkali metal spectrum are given as an example.

In order to analyse a spectrum it is necessary to know which of these terms can be combined through optical transitions in absorption (or emission). The information is provided by selection rules, which in the case of alkali metal absorption spectra, for example, require an increase of the secondary quantum number L by one unit. The principal quantum number n, on the other hand, can alter by any amount.

The energy level diagram for sodium is shown in figure 3 (page 6). The lowest term here is $3s\,^2S_{\frac{1}{2}}$ because the photoelectron is situated in the 3s orbital above the filled K ($n=1$) and L ($n=2$) shells.

Table 1. Possible terms in alkali metal spectra*)

	$l=0$ (s)	$l=1$ (p)	$l=2$ (d)	$l=3$ (f)
$n=1$	1s $^2S_{\frac{1}{2}}$			
$n=2$	2s $^2S_{\frac{1}{2}}$	2p $^2P_{\frac{1}{2}}$; $^2P_{\frac{3}{2}}$		
$n=3$	3s $^2S_{\frac{1}{2}}$	3p $^2P_{\frac{1}{2}}$; $^2P_{\frac{3}{2}}$	3d $^2D_{\frac{3}{2}}$; $^2D_{\frac{5}{2}}$	
$n=4$	4s $^2S_{\frac{1}{2}}$	4p $^2P_{\frac{1}{2}}$; $^2P_{\frac{3}{2}}$	4d $^2D_{\frac{3}{2}}$; $^2D_{\frac{5}{2}}$	4f $^2F_{\frac{5}{2}}$; $^2F_{\frac{7}{2}}$
n	*n*s $^2S_{\frac{1}{2}}$	*n*p $^2P_{\frac{1}{2}}$; $^2P_{\frac{3}{2}}$	*n*d $^2D_{\frac{3}{2}}$; $^2D_{\frac{5}{2}}$	*n*f $^2F_{\frac{5}{2}}$; $^2F_{\frac{7}{2}}$

At the temperature of normal sodium vapour in usual flames, nearly all the atoms are in the ground state (see section 1.4), i.e. the photoelectron corresponds to the 3s $^2S_{\frac{1}{2}}$ term. For this reason and as a result of the selection rules, all lines occurring in the absorption spectrum arise from transitions from this term to P terms.

$$3s\,^2S_{\frac{1}{2}} \dashrightarrow 3p\,^2P_{\frac{1}{2},\frac{3}{2}} \quad (589.593\text{ nm}/588.996\text{ nm})$$
$$3s\,^2S_{\frac{1}{2}} \dashrightarrow 4p\,^2P_{\frac{1}{2},\frac{3}{2}} \quad (330.294\text{ nm}/330.234\text{ nm})$$
$$3s\,^2S_{\frac{1}{2}} \dashrightarrow np\,^2P_{\frac{1}{2},\frac{3}{2}}$$

If, on the other hand, sodium vapour is excited thermally (e.g. in a hot flame) or electrically, terms with arbitrary n and L can be attained and the emission spectrum can contain all lines which are possible starting from these terms in compliance with various selection rules. A complete emission series can be ascribed to each secondary quantum number. Since these emission series overlap, the regularity of the individual series is obscured.
Hence it is obvious that the emission spectrum at these temperatures has many more lines than the absorption spectrum. This applies particularly to elements having several photoelectrons. While the situation is still fairly simple for sodium, atoms having complicated structures give rise to highly complex emission spectra.
Atomic fluorescence is the reversal of the absorption process; i.e. the fluorescence spectrum is produced by release of radiation energy taken up during the absorption process.
Fluorescence spectra are generally simple. They correspond to the transition of atoms excited by absorption of light energy to atoms in lower-lying excited states or in the ground state. Fluorescence lines generally have the same or lower frequencies (longer wavelengths) than the corresponding absorption lines. Relevant details will be discussed in a later section.

1.3. Selection of the Lines

From the energy level diagram of sodium (figure 3), it is clear that in absorption practically only those lines can occur which originate from the ground term of the neutral atom. These lines are referred to as resonance lines since they can come into resonance with light of suitable energy.

* Details will be found in standard textbooks of physical chemistry.

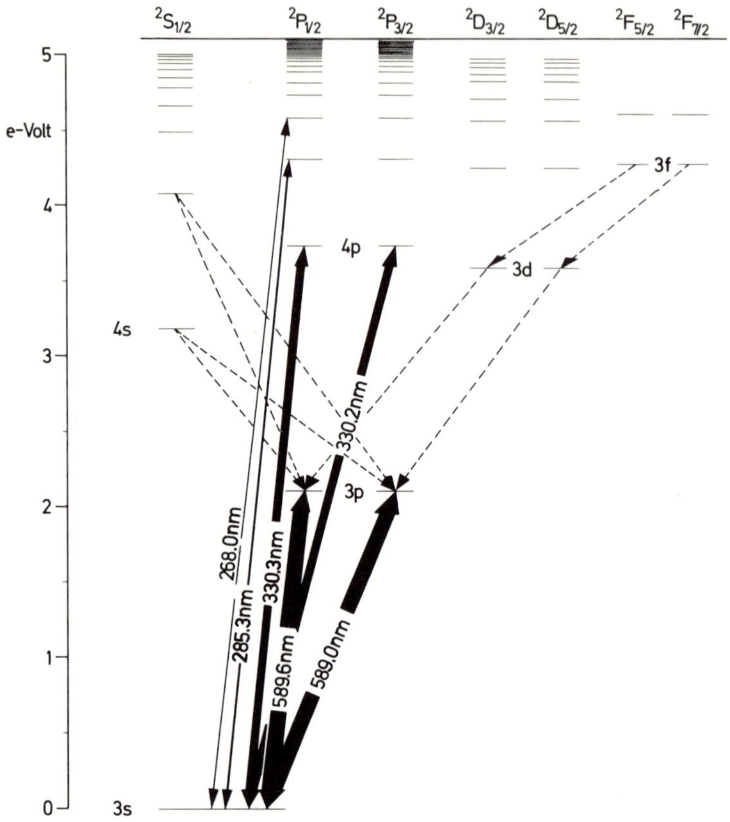

Figure 3. Energy level diagram for sodium.—The continuous lines with double arrows represent transitions for the main series and occur both in absorption and emission. The broken lines are for the transition of secondary series and at the temperatures of interest occur only in emission.

The most intense and easily excited lines should be the "last lines", corresponding to the transition from the ground state to the first excited state. The name "last line" comes from the fact that the transition to the first excited state requires the least energy, so that these lines are the absorption lines of lowest frequency and therefore appear at the red end of the absorption spectrum.

Such a clear-cut situation, however, is encountered only with alkali and alkaline-earth metals owing to their very simple atomic structures. With various other elements, lines corresponding to a transition of the atom from the ground state into a higher excited state can also be more intense.

The atoms of the transition elements frequently exhibit a fairly large number of low-lying excited states, better described as ground states of slightly raised energy, whose populations depend upon the temperature. This problem will be discussed in detail in a later chapter. Which lines have the greater intensity can only be decided by experiment.

ALLAN [37] [38], DAVID [270], and MOSSOTTI and FASSEL [885] examined the absorption spectra of a large number of transition elements and lanthanides by means of an emission spectrograph in front of which was mounted a flame containing high concentrations of the respective metals. In this way the intensities of the absorption lines could be measured and the most sensitive lines determined.

In principle all elements can be determined by atomic absorption since the atoms of any element can be excited and are therefore capable of absorption. At present the limitations of the method lie practically only in the field of instrumentation.

Thus measurements below 200 nm in the so-called vacuum UV range become difficult owing to incipient absorption by oxygen and especially by the hot flame gases. This range, for example, contains resonance lines of selenium (189.1 nm and 196.1 nm) and arsenic (189.0 nm and 193.7 nm), which can still be determined by a good atomic absorption spectrophotometer, and also the resonance lines of all gases and the typical non-metals. With somewhat modified instruments and a shielded flame [654] [669] or with a heated graphite furnace [779] [780] even such elements as iodine (183.0 nm), sulphur (180.7 nm) and phosphorus (178.3 nm and 178.8 nm) can still be satisfactorily determined.

Apart from the remaining non-metals, all metals and metalloids can be determined by atomic absorption spectroscopy. The elements cerium and thorium are exceptions which up to the present have evaded analysis by this method. Occasional references to direct determinations of cerium at 522.4 nm and 569.7 nm respectively are doubtful and the detection limits quoted are not very encouraging [582].

Artificial elements and a number of strongly radioactive substances have not yet been examined for obvious reasons.

1.4. Thermal Excitation

The differences between the thermal and the optical excitation process should first be appreciated in order that the principal differences in the observed emission, absorption, or fluorescence spectra be understood.

The ratio of the number of atoms N_j in an excited state j to the number of atoms N_0 in the ground state when the temperature is not too low in given by:

$$\frac{N_j}{N_0} = \frac{P_j}{P_0} \cdot e^{-E_j/kT}, \tag{1.7}$$

where P_j and P_0 are the statistical weights of the excited and ground states, E_j the energy of excitation, k the BOLTZMANN constant, and T the absolute temperature. Since the wavelength of the resonance line is inversely proportional to its energy of excitation (see equation 1.4), the number of excited atoms increases exponentially with increasing wavelength (see figure 4).

In equation (1.7) the exponent is inversely proportional to the absolute temperature, which means that the increase in the relative number N_j/N_0 of excited atoms with increasing temperature is exponential. WALSH [1282] has calculated the ratio N_j/N_0 for a number of elements at various temperatures (table 2). It is found that N_j is always small compared to N_0; i.e. the number of atoms in the excited state can be ignored relative to the number of atoms in the ground state at temperatures below 3000 K and wavelengths less than 500 nm.

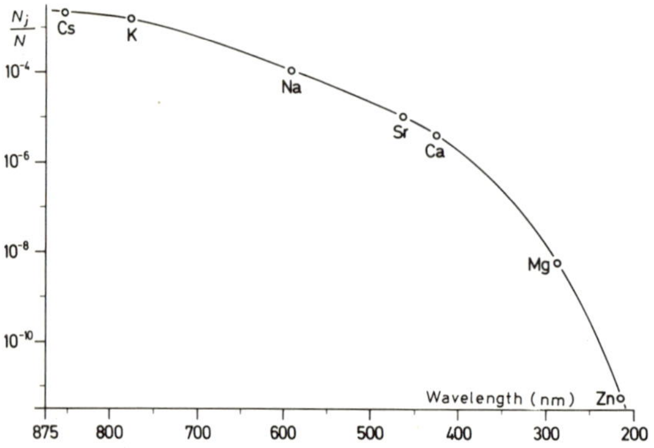

Figure 4. For a given temperature the number of excited atoms N_j increases exponentially with increasing wavelength. The values in this curve were calculated for 2500 K.

For absorption, it can therefore be assumed that the number of atoms in the ground state is virtually identical to the total number of atoms formed. This means that the number of atoms in the ground state is independent of the energy of excitation E_j and of the temperature, provided that the latter factor does not influence the total number N of atoms present.
This influence of temperature on the total number of atoms, i.e. on the effectiveness of atomization, will be discussed at length in a later chapter. It has nothing to do with the statement in equation (1.7), however. If the total number N of atoms formed alters, then N_0 and N_j are both altered to the same extent. In practice this leads to the same percentage attenuation of both the emission and the absorption lines, the ratio N_j/N_0 remaining unaffected.

Table 2. Temperature and wavelength dependence of the ratio N_j/N_0—according to WALSH [1282]

Element	Resonance line nm	$\frac{P_j}{P_0}$	N_j/N_0 2000 K	3000 K	4000 K
Zn	213.9	3	$7 \cdot 10^{-15}$	$6 \cdot 10^{-10}$	$1 \cdot 10^{-7}$
Ca	422.7	3	$1 \cdot 10^{-7}$	$4 \cdot 10^{-5}$	$6 \cdot 10^{-4}$
Na	589.0	2	$1 \cdot 10^{-5}$	$6 \cdot 10^{-4}$	$4 \cdot 10^{-3}$
Cs	852.1	2	$4 \cdot 10^{-4}$	$7 \cdot 10^{-3}$	$3 \cdot 10^{-2}$

1.5. Absorption Coefficient

Free atoms in the ground state are able to absorb light energy of exactly defined frequency (light quantum $h\nu$) with concomitant transformation into an excited state. The amount of energy absorbed per unit time and volume is proportional to the number N of free atoms per unit volume, the light energy $h \cdot \nu_{jk}$, and the spectral radiation intensity S_ν at the resonance frequency:

$$E_{abs} = B_{jk} \cdot N S_\nu \cdot h \nu_{jk} \,. \tag{1.8}$$

The proportionality factor B_{jk} is the EINSTEIN probability coefficient of absorption for the transition $j \to k$. The product $B_{jk} S_\nu$ is an expression for the fraction of all the atoms present in the ground state that can absorb a photon of energy $h \cdot \nu_{jk}$ per unit time.
In unit time a radiation unit of $c \cdot S_\nu$ (c=speed of light) or $c \cdot S_\nu / h_\nu$ photons respectively pass through a unit volume. The fraction of the photons that is absorbed by atoms in the ground state is proportional to the total number N of free atoms and to the "effective cross-section" of an atom, the so-called absorption coefficient $\varkappa_{jk}$. The total amount of energy absorbed per unit volume can then be expressed as the product of the number of absorbed photons and their energy.

$$E_{abs} = \varkappa_{jk} \cdot N \cdot c \cdot S_\nu \tag{1.9}$$

By equating the energy in equations (1.8) and (1.9), the absorption coefficient can be expressed as

$$\varkappa_{jk} = \frac{h \cdot \nu}{c} \cdot B_{jk} \,. \tag{1.10}$$

An atom can also be regarded as an oscillating electrical dipole where the electrons circling round the nucleus represent the oscillators in equilibrium with the radiation. In the electromagnetic field of a light beam these can be excited to a motion of higher frequency.
According to the laws of electrodynamics, the total amount of energy absorbed by such a harmonic oscillator per unit time can be expressed as

$$E_{abs} = f \frac{\pi e^2}{m} \cdot S_\nu \,, \tag{1.11}$$

where e is the charge and m the mass of an electron. f is a dimensionless factor, the so-called oscillator strength, which merely represents the effective number of classical electron oscillators corresponding to transition $j \to k$ for the absorption effect of an atom. Expressed simply, f is the average number of electrons per atom which can be excited by the incident radiation ν_{jk}. By equating the energies in equations (1.8) and (1.11), and considering that equation (1.11) is valid for 1 atom, an expression for the absorption coefficient is obtained

$$\varkappa_{jk} = \frac{\pi e^2}{mc} \cdot f_{jk} \,. \tag{1.12}$$

The absorption coefficient $\varkappa_{jk}$ has the dimensions of an area (as expected for an effective cross-section) and is a measure of the quantity of light at frequency ν which can be absorbed by an atom.

In practice it is more convenient to use the absorption coefficient K (dimension: length^{-1}) referring to the unit of volume rather than the absorption coefficient $\varkappa$ referring to an atom.

$$K_{jk} = N \cdot \varkappa_{jk} \tag{1.13}$$

Equation (1.12) then takes the form:

$$K_{jk} = \frac{\pi e^2}{mc} \cdot N \cdot f_{jk} \tag{1.14}$$

K_{jk}, the probability of light absorption per unit volume, is now a measurable quantity. The other side of the equation is made up of an unequivocally calculable constant and two unknowns, namely the total number of atoms per unit volume and the oscillator strength f. If N can be measured, f can be calculated.

On the other hand, if f is known, the absolute number N of free atoms can be calculated. This procedure originally proposed by WALSH [1282] has been successfully used by various authors for the calculation of f [777] [778] [1074] [1269] [1270].

Thus in absorption, the excitation energy is provided by a light quantum, no temperature factor is present. The absorption coefficient is determined by the product of the total number of atoms present per unit volume and the oscillator strength of the resonance line.

1.6. Line Width

In the previous section frequent mention was made of the frequency ν_{jk} of the resonance line and also the radiation intensity S_{jk} at the resonance frequency. Nothing, however, was said about the dimensions. Rather it was assumed that the spectral radiation intensity S_ν is constant in a finite frequency interval Δ_ν and that the frequency ν_{jk} lies within this interval. The expression for the absorption coefficient in equation (1.12) represents, for a line of finite width, an integral over the line width (or the frequency interval):

$$\varkappa_{jk} = \int \varkappa_\nu \cdot d\nu \tag{1.15}$$

WALSH showed that owing to the probability distribution over each energy level, the natural width of a resonance line is in the order of 10^{-5} nm. In practice this natural line width is influenced by a variety of factors, above all by the disordered thermal motion of the atoms and by various collisions of the atoms.

The former influence, disordered thermal motion of atoms, causes the line profile to have the form of a MAXWELL distribution of the atomic velocities which can be expressed as a GAUSS function. This so-called DOPPLER effect is given by

$$D = \frac{\nu}{c}\sqrt{\frac{2RT}{M}}. \tag{1.16}$$

The DOPPLER line width D is proportional to the square root of the absolute temperature T, and inversely proportional to the square root of the atomic weight M.
The latter effect, collision broadening, comes from the fact that the energy levels which are decisive for the emission or absorption of light quanta are somewhat indefinite due to the collison of atoms. For this reason, light quanta of slightly differing frequency are emitted and also absorbed. Depending upon the type of particles involved in the collision, the effect is also slightly different. If the interaction involves electrically charged particles it is referred to as the STARK effect, while collisions with uncharged particles lead to the VAN DER WAALS effect; collisions between atoms of the same kind give rise to the resonance (or HOLTSMARK) broadening effect. Since it is very difficult in practice to differentiate between the three effects, they are frequently designated collectively as LORENZ broadening.

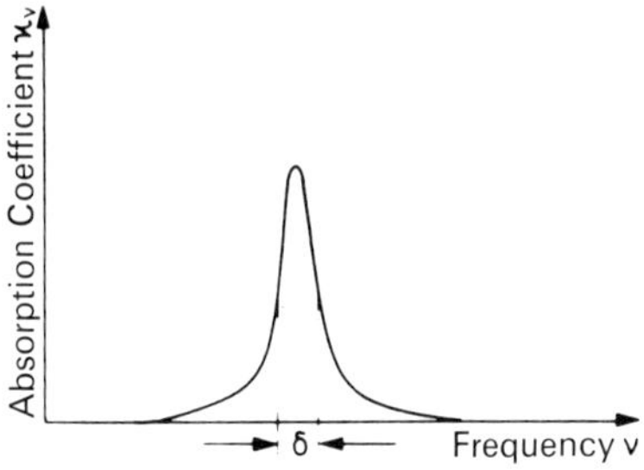

Figure 5. Profile of a resonance line.—The width at half height δ is frequently used as a measure for the line width.

Finally it should be pointed out that DOPPLER and LORENZ broadening often occur simultaneously, and both profiles then combine to give the VOIGT profile which can only be expressed by relatively complicated functions. In figure 5 the profile of a resonance line is shown schematically. The significance of the peak width at half height, or half-width, of emission and absorption lines for the absorption process will be discussed in more detail in the next chapter.
The half-width of resonance lines increases with increasing pressure and with increasing temperature. Various authors [664] [665] [778] [953] [999] [1160] have measured or calculated the half-widths of the resonance lines of a large number of elements. The values are in the order of 0.0005 nm to 0.005 nm.

1.7. Measuring the Absorption

Since atoms are only able to absorb light within a very narrow frequency interval, certain demands must be placed upon the light source. Although continuous light sources afford a high total illumination intensity, the illumination intensity in the interval of interest of about 0.0005 nm to 0.005 nm is nevertheless too weak. For this reason WALSH recommended that the light source used for absorption measurements should emit the spectrum of the element to be determined. With such an arrangement the required resonance line merely has to be separated from other spectral lines by means of a monochromator (see figure 6).

Up to now we have considered only the absorption of radiation per unit volume, a quantity which is difficult to measure in practice. If, on the other hand, the usual quantity employed

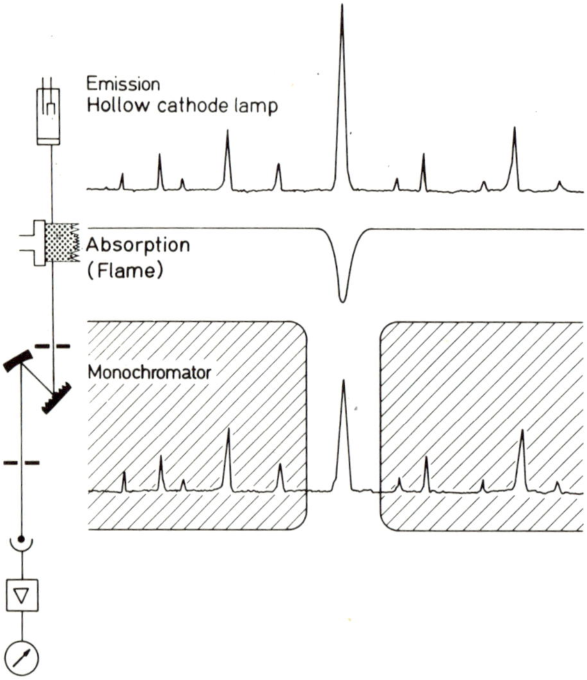

Figure 6. Measurements in absorption.—The spectrum of the element under study is emitted from a hollow cathode lamp. In the flame a portion of the resonance line, corresponding to the concentration of this element, is absorbed. Lines that do not occur in absorption are not attenuated. After dispersion of the light in a monochromator the resonance line is separated by the exit slit and all other lines are masked. The detector 'sees' only the resonance line, whose attenuation is then displayed.

in normal absorption measurements, the flux density J, is used for measurements in atomic absorption the BEER-LAMBERT law (equations 1.1 to 1.3) can be used in the form

$$J_{\nu,l} = J_{\nu,0} \cdot e^{-\varkappa_\nu N l}, \tag{1.17}$$

where $J_{\nu,0}$ and $J_{\nu,l}$ are the beam flux densities prior to and after passing through the absorbing layer of length l respectively. $\varkappa_\nu$ is the spectral atomic absorption coefficient and N the total number of free atoms.

After rearrangement, equation (1.17) is derived from analogy to equation (1.3) in the form

$$A \equiv \log \frac{J_{\nu,0}}{J_{\nu,l}} = 2.303 \cdot \varkappa_\nu N l, \tag{1.18}$$

where A, the absorbance, is directly proportional to N, the total number of free atoms present. The total number N of free atoms present admittedly cannot be determined from equation (1.18), but this is not necessary for routine measurements. Like many other spectroscopic methods, atomic absorption spectroscopy is a relative method since a linear relationship exists between the concentration of free atoms in the beam path and their absorbance*).

At higher absorbance values a deviation from the BEER-LAMBERT law is observed relatively frequently in practice. This deviation can be seen as a curvature of the calibration plot (absorbance versus concentration) towards the concentration axis. Various authors [11] [1006] [1073] have examined whether this effect is due to resonance line broadening. It could be shown, however, that this is not the case [30]. Non-linearity of the calibration curve is occasionally caused by a hyperfine structure of the spectral line. Most spectral lines have such a fine structure due to isotope shifts and nuclear spin, which are, however, normally blanketed by the above broadening effects. Only with very light or very heavy elements do the individual components cause line separation. In these cases non-linearity due to the hyperfine structure has indeed been demonstrated [409] [1348].

Most frequently the non-linearity of the calibration curve is caused by light which cannot be absorbed [862], this can be stray light in the accepted sense, but more usually it is "background radiation" from the light source [1147]. For this reason this phenomenon is fully explained in the next chapter in connection with light sources for atomic absorption spectroscopy.

It has already been mentioned that quantitative absorption of incident light by atoms can only take place if the half-width of the emission line from the radiation source is considerably smaller than the half-width of the absorption line. If this is not the case, the resonance line will be flanked on both sides by non-absorbable residual radiation which acts as radiation background. Such broadening of the resonance lines emitted from spectral light sources occurs, for example, with a high gas pressure in the light source. The self-absorption that usually takes place at the same time is a further source of interference in absorption measurements.

In addition to these effects, the total width of the emission lines also has a considerable influence on the occurrence of spectral interferences in atomic absorption spectroscopy. FASSEL [342] in particular has drawn attention to this fact and has reported effective line widths (including the oscillations) in the order of 0.1 nm. This problem will be further described in the context of spectral interferences.

1.8. Instrumentation

The general construction of an atomic absorption spectrophotometer is simple and is shown schematically in figure 7. The most important components are the light source **1**, which emits the spectrum of the element of interest; an "absorption cell" in which the atoms of the sample to be analysed are formed by thermal dissociation—usually in a flame **2**; a monochromator **3** for the spectral dispersion of the light with an exit slit for selction of the resonance line;

---

* Occasionally the measurement is quoted not in absorbance but in *% absorption* (*% Absn*). The conversion formula is

$$A = \log \frac{100}{100 - \% Absn} \quad \text{or} \quad A = 2 - \log\,(100 - \% Absn)$$

where $100 - \%$ *absorption* $= \%$ *transmission*.

a detector **4** permitting measurement of radiation intensity followed by an amplifier **5** and a display instrument for the readout of the absorption values.

Such a system, as shown in figure 7a, has a crucial disadvantage. Upon closer scrutiny, an atomic absorption spectrophotometer built on these lines is seen to be nothing other than an emission flame photometer with a light source irradiating the flame. In atomic absorption, however, the flame should ideally be an absorption cell in which the sample is atomized and only atoms in the ground state are produced. In contrast, the flame used in flame emission photometry should produce as many excited atoms as possible, thus ensuring maximum feasible intensity of the emission spectrum produced by the element to be analysed.

Although the number of excited atoms at normal flame temperatures is always considerably lower than the number of atoms in the ground state (see section 1.4), many practical examples are known in which the emission of the flame interferes with absorption measurements in this type of instrument. This is hardly surprising when it is considered that a "true sample" usually contains a multitude of elements which all display their own emission spectra in the flame (compare [176]).

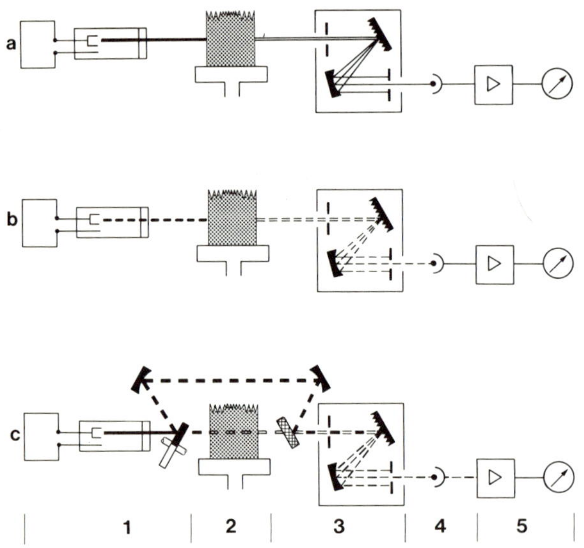

Figure 7. Schematic construction of an atomic absorption spectrophotometer.—**1** Light source, **2** Flame, **3** Monochromator, **4** Detector, **5** Amplifier, electronics and display. a) Single-beam continuous light (DC) instrument; b) Single-beam alternating light (AC) instrument; the alternating light can be produced either by the use of pulsed direct current to the lamp, or mechanically by a rotating chopper situated between the lamp and the flame. c) Double-beam alternating light (AC) instrument; the light from the hollow cathode lamp is sent alternately through the flame and around the flame by a rotating mirror-chopper. The two beams are recombined by a semitransparent mirror behind the flame.

If it proves impossible to separate the resonance line of the element under study from an emission line of another element, i.e. to eliminate the interfering line by means of the exit slit of the monochromator, then "spectral interferences" take place [7] [176] [605]. In such a case the quantity measured is not the ratio $J_{\nu,0}/J_{\nu,l}$, from which the absorbance is calculated (see equation 1.18), but rather $J_{\nu,0}/(J_{\nu,l}+J_{\nu',E})$, where $J_{\nu',E}$ is the intensity of an emission line of frequency ν' which is not removed by the monochromator. The absorbance obtained from this ratio is always smaller than that expected. In highly unfavourable cases $(J_{\nu,l}+J_{\nu',E})$ can be even greater than $J_{\nu,0}$, resulting in an apparently negative absorbance.

To eliminate this interference due to flame emission, virtually all atomic absorption spectrophotometers nowadays operate with a chopped or alternating light (AC) system as shown in figure 7b instead of direct light (DC) as depicted in figure 7a. The light beam is modulated either mechanically or electrically at a fixed frequency and the amplifier electronics are tuned to the same frequency (selective amplifier). In this AC system only the light from the source having the modulation frequency is amplified, while the emission from the flame, which is not modulated, is neglected. Accordingly, spectral interferences are practically impossible in such a system.

A further refinement of this principle is the double-beam AC system shown in figure 7c [604]. Here, the beam from the light source is divided by a rotating mirror-chopper into a sample beam (crossing through the flame, $J_{\nu,t}$) and a reference beam (bypassing the flame, $J_{\nu,0}$) and then recombined behind the flame. The electronics of this system are designed to yield the ratio of the two beams with the measurement beam as the denominator and the reference beam as the numerator. Since both beams come from the same light source, pass through the same monochromator, are received by the same detector and amplified by the same electronics, any variations in the lamp emission, detector sensitivity, or amplification appear in both the numerator and the denominator and therefore cancel out. The stability of this system is thus clearly better than that of a single-beam system [1273], and will only be influenced by variations in the burner system, which nevertheless can be dominant in individual cases as will be shown later.

A further variation is the dual-channel system in which two light sources are used to irradiate the same absorption cell (e.g. a flame) and are then handled independently by two monochromators, two detectors, and two electronic units. As well as permitting the simultaneous determination of two elements, this system affords a means of working with an internal standard. Hence adverse influences of the nebulizer-burner system on the stability of the analyses can be further reduced. The dual-channel system is a special form of the (not yet commercially available) multichannel system which will be discussed elsewhere.

In the following chapters the principal components of an atomic absorption spectrophotometer and their significance for analyses will be discussed in full. A tabular survey of the most important commercially available instruments will be found at the close of this book.

Figure 8. Atomic Absorption Spectrophotometer, Series 400 (Bodenseewerk Perkin-Elmer & Co. GmbH), a modern double-beam instrument.

2. Light Sources

Very many of the advantages of atomic absorption spectroscopy can be directly or indirectly traced to the narrow half-widths of the resonance lines, i.e. the absorption of an element takes place within a very limited spectral range. This advantage becomes very noticeable if the light sources used for excitation emit the spectrum of the element under study in lines that are at least as narrow. As light sources, spectral lamps (hollow cathode lamps, low pressure gas discharge lamps, etc.) or flames, into which high metal concentrations are sprayed, can be employed. BUTLER and BRINK [194] have published a review on light sources for atomic absorption and atomic fluorescence in which the various lamp types and their function are described.

2.1. Hollow Cathode Lamps

In modern practice, hollow cathode lamps are used principally for excitation because they generally meet the requirements for performance and operating convenience. The technology of this type of lamp is not new since it was first described by PASCHEN [955] in 1916. After the introduction of atomic absorption, WALSH and co-workers [589] [1075] modified and simplified the construction. With the introduction of the Intensitron® hollow cathode lamps in recent years, it has finally been possible to optimize the performance with respect to intensity and purity of the radiation.

A hollow cathode lamp consists of a glass cylinder filled with an inert gas (neon or argon) under a pressure of a few Torr into which an anode and a cathode have been fused (see figure 9). The cathode is in the form of a hollow cylinder and is either made of the metal under study or filled with it. The anode is in the form of a thick wire, usually of tungsten or nickel.

Figure 9. Schematic construction of a hollow cathode lamp.

If a voltage of 100–400 V is applied across the cathode and anode, a discharge can be observed. The filling gas is ionized and the gas ions, following the voltage gradient, strike the metal surface of the cathode and are discharged there. Simultaneously, metal atoms are displaced from the surface of the cathode and excited. These excited atoms then emit the emission spectrum of the cathode material. Since the majority of the emission is radiated from inside the hollow cylinder, it is relatively well focussed.

The manufacture of hollow cathode lamps is not without its problems. The nature and pressure of the filling gas, the selection of the cathode material, and the applied voltage and current, for example, play an important part in the intensity and purity of the lamp spectrum and more importantly in the half-width and form of the emitted lines. Since the quality of commercially available hollow cathode lamps is usually excellent and several of the manufacturers give very detailed information on the optimum operation, discussing these aspects in detail is irrelevent.

HUMAN [540] found that for hollow cathode lamps whose gas pressures were approaching zero, temperature had no noticeable effect on the line width. This means that the lines emitted from hollow cathode lamps generally have an appreciably smaller half-width than the absorption lines in the flame since the latter are strongly broadened at atomspheric pressure and higher temperatures. Therefore, hollow cathode lamps represent ideal light sources for atomic absorption spectroscopy.

Earlier hollow cathode lamps, however, possessed a number of disadvantages. The metal atoms displaced from inside the cathode by the gas ions formed a cloud of free atoms in the ground state in front of the cathode opening, especially when the cathode itself was relatively hot. This lead to a reduction of the intensity of the emission owing to self-absorption and an alteration of the line profile because of resonance broadening (see section 1.6). This self-absorption could be particularly observed when higher currents were applied to metals of low vapour pressure [613] [1280].

The phenomenon of self-absorption or self-reversal, which is depicted schematically in figure 10, is worth further consideration; it occurs not only with spectral lamps of the most varying types, but may also be observed more often with emission and fluorescence measurements at higher atom concentrations.

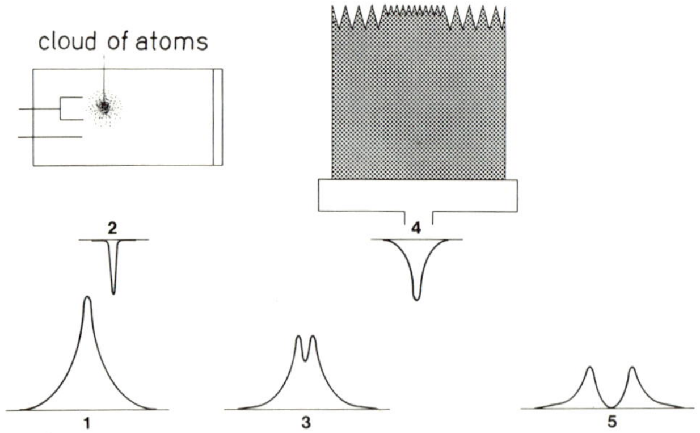

Figure 10. Self-absorption in hollow cathode lamps.—**1** an emission line is emitted from the hollow cathode lamp. A portion of the resonance line is absorbed by the cloud of atoms **2** in front of the cathode opening, and especially from the middle of the line so that 'self-reversal' **3** takes place. Because of the line broadening that usually takes place at the same time, only a part of the resonance line can be absorbed in the flame **4**. Non-absorbable residual radiation **5** remains, leading to a non-linear calibration plot.

Normally, by raising the applied current so that the number of metal atoms excited by gas ions is increased, the emission intensity of a hollow cathode lamp can be increased. With easily volatilized metals, the higher current and the resulting higher temperature of the cathode

lead to increased vaporization of the cathode material, which, at the same time, increases the number of collisions with neutral and charged particles of higher temperature, hence leading to a clear broadening of the line. The cloud of neutral atoms in the ground state formed simultaneously at the cathode opening is now in a position, owing to its relatively low temperature and the narrow half-width of the absorption line, to absorb light selectively from the emission maximum while relatively little is absorbed from the flanks. This leads to a line profile with a very unsuitable frequency distribution for atomic absorption. At the position of greatest probability of absorption of the emission line there is a minimum, while on the flanks where the absorption probability falls off sharply there are two maxima.
This alteration of the line profile caused by broadening and self-reversal means that only part of the resonance lines emitted from the light source can be absorbed by the atoms in the flame. The absorption curve does not then approach 0% transmission asymptotically, but to a value corresponding to the residual radiation. In practice this means a deviation from the BEER-LAMBERT law and thereby a non-linear calibration plot.

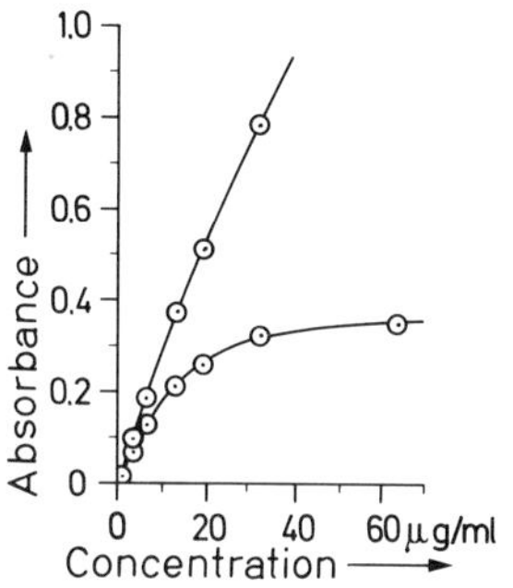

Figure 11. Calibration curves for nickel, measured with new and old hollow cathode lamps respectively. The older lamp (lower curve) emits an intense hydrogen spectrum which leads to strong curvature.

It would be appropriate at this point to mention several further causes of non-linearity of the calibration plot which are also due to the lamp. All these effects have the same result; non-absorbable radiation passes through the monochromator and forms an additive member to $J_{\nu,t}$, which for higher concentrations cannot approach zero but retains a finite value.
As principal causes of the non-absorbable radiation, which passes through the monochromator within the transmitted spectral range, the following come into consideration: Non-specific background radiation, emission lines from the filling gas or the cathode support material, if this is not identical to the cathode, and emission lines from the metal itself corresponding to higher transitions.
For a long time, obtaining high purity cathode materials free of hydrogen was a problem. With older lamps, hydrogen was often released during operation, which then under the influence of the glow discharge emitted an intense continuous spectrum. The portion of this continuum to the total radiation increased on occasions to over 50% [851], which greatly reduced the usability of such lamps (see figure 11).
If the utmost consideration is not taken of the spectra of these foreign elements during selection of filling gases and support materials for cathodes of easily volatile metals, it can easily occur that these materials have an emission line directly neighbouring the resonance line of the metal under study, which leads to effects similar to those of the hydrogen continuum.

It should be noted, however, that in practice only two gases (neon and argon) come into consideration as filling gases for hollow cathode lamps and that not very many metals are suitable as supports for the cathode, so that on occasions the fact that lesser emission lines of the filling gas or support material lie within the set slit width of the monochromator [810] cannot be avoided.

Finally, there are elements that have so many lines in their emission spectra that the resonance line cannot be separated from other non-absorbable lines [856]. This problem will be discussed further in connection with the spectral bandwidths of the monochromators used in atomic absorption spectroscopy. When the reasons for the frequently observed deviation from the BEER-LAMBERT law became known and could be largely ascribed to hollow cathode lamps, considerable activity took place in the development of new lamps which has led to considerable improvements and to the elimination or reduction of many difficulties.

In 1959, WHITE [1316] showed that after several hundred operating hours, a normal cylindrical hollow cathode had changed its geometry, owing to metal transport from the hot inner side to the surface (see figure 12). A cathode consisting of a hollow sphere with a relatively small opening was produced.

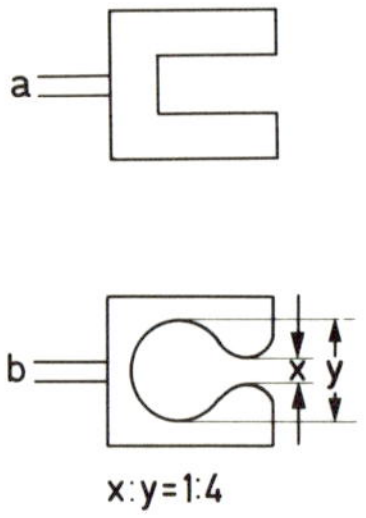

Figure 12. After a long period of operation of a cylindrical hollow cathode lamp **a**, the cathode alters its form owing to metal transport to the rim **b**. When the ratio of the diameter x of the opening to the diameter y of the hollow sphere is 1:4 virtually no further vaporization of metal from the cathode takes place.

WHITE showed that with a ratio of the diameter of the opening to the internal diameter of the hollow sphere of 1:4 together with careful determination of the optimum gas pressure, virtually no more metal atoms could escape from inside the cathode and therefore no self-absorption or loss of metal could take place. This finding was verified by VOLLMER [1277] and has been used commercially since about 1967 [812].

Use of the WHITE cathode form, especially for easily melted and vaporized elements, together with complete electrical insulation of the cathode and anode with ceramic materials and mica discs (see figure 13) has limited the glow discharge practically to the inside of the cathode. As well as a considerably increased radiation density, a much "purer" spectrum is obtained. This means that non-absorbable foreign lines and ion lines of the element under analysis are both strongly reduced in their intensity in comparison to the resonance lines. Through this and the absence of self-absorption, many of the above described difficulties which lead to strong curvature of the calibration plot have been eliminated.

For a number of elements, particularly stable emission can be obtained if the metal is in the molten form in the cathode; a procedure only having continuous success in combination with the WHITE cathode form [813] [1275] [1276]. A further improvement in the radiation intensity can be obtained when the hollow cathode lamps are fed with short pulses of correspondingly higher current strength instead of direct or alternating current [1007]. With this procedure an improved signal-to-noise ratio can be achieved from commercially available hollow cathode lamps without adversely influencing their lifetime through the higher pulsed current strength.

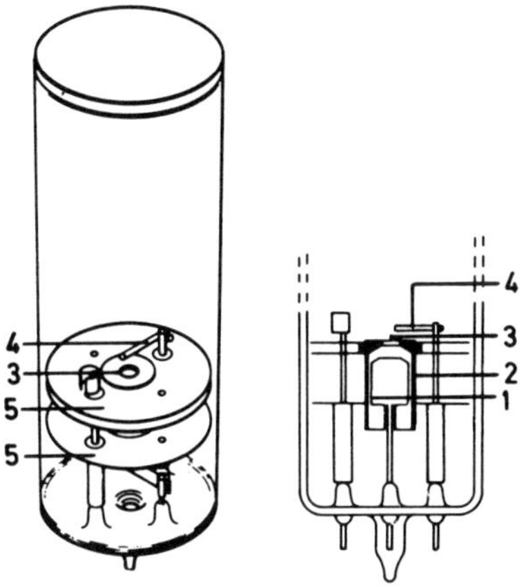

Figure 13. Schematic of an Intensitron®*) hollow cathode lamp.—**1** Hollow cathode, **2** insulation of ceramic, **3** restricted cathode opening, **4** anode, **5** mica discs.

2.2. Multielement Lamps

In principle an atomic absorption analysis requires an individual light source for every element to be determined, this will always be found a certain disadvantage. For this reason, endeavours were made to produce multielement hollow cathode lamps at an early stage in the development of atomic absorption.

WALSH constructed a lamp with several cathodes of different metals within the same glass cylinder. The greatest difficulty with these lamps was correct adjustment in the spectrophotometer. MASSMANN [828] and BUTLER and STRASHEIM [197] constructed multielement lamps whose cathodes consisted of rings of various metals pressed together. The rings were arranged in order of their volatilities.

The usual conception used nowadays for the construction of multielement lamps was developed by PERKIN-ELMER [1104] [1140]. Here, various metals in powder form are mixed in determined ratios, pressed and sintered. Using this method, virtually any desired metal combinations can be manufactured [364] [501] [811].

If a multielement lamp is to be used in atomic absorption, however, various aspects must be taken into consideration. The first general requirement is that the intensity of the spectrum of each element in a multielement lamp is not markedly weaker than the comparable single-element lamp. Under no circumstances may the enormous advantage of atomic absorption spectroscopy, namely the selectivity, be lost by the use of mulielement lamps through the introduction of spectral interferences caused by unsuitable element combinations.

*) Intensitron® is a registered trade mark of the Perkin Elmer Corporation, Norwalk, Conn., USA.

As previously explained, spectral interferences in atomic absorption spectroscopy are eliminated, in principle, by modulating the light source and tuning the amplifier to the same modulation frequency. Spectral interferences occur in emission spectral analyses if two emission lines of different elements cannot be separated in the monochromator and fall simultaneously on the detector.

By the use of unsuitable metal combinations in multielement lamps, resonance lines of various elements contained in the cathode can lie so close together that they cannot be separated in the monochromator. Since these lines are emitted from the same lamp, they have the same modulation frequency and are amplified to the same degree by the detector. In practice it is no longer possible to assign a measured absorption to a determined element [574].

Great care must be exercised in the manufacture of multielement lamps to exclude spectral interences, and it must always be axcertained when multielement lamps are purchased whether they have been specially manufactured for atomic absorption spectroscopy.

While combination lamps with two or three elements can usually be used without hesitation, lamps with four or more elements are not recommended for all applications. With these multielement lamps the radiation intensity of the individual resonance lines is, in part, considerably lower than with the single-element lamps. This results in a somewhat poorer signal-to-noise ratio which can influence both the precision and the detection limit. Furthermore, the calibration plots occasionally exhibit stronger curvature owing to the multitude of lines of these lamps, so that the linear working range is shorter. Nevertheless, for a very large number of routine analyses, multielement lamps bring a clear advantage.

2.3. Demountable Hollow Cathode Lamps

Various authors have described demountable hollow cathode lamps [431] [685] [686] [870] [1060] [1177]. Their use in routine work will not find great interest since their operation is not simple and requires experience of high vacuum techniques. For various research purposes, on the other hand, they may have certain importance. Further developments in this direction do not, however, come within the scope of this book.

2.4. High Intensity Lamps

To eliminate self-absorption and to increase the radiation intensity, SULLIVAN and WALSH [1193] constructed a special type of hollow cathode in which the glow discharge on the cathode was used principally to atomize as much metal as possible. The cloud of atoms, which in normal hollow cathode lamps is responsible for the self-absorption, was then excited by a second discharge produced in front of the cathode by an additional electrode isolated from the first discharge. The authors reported a hundred-fold increase in the intensity of these lamps compared to that of normal hollow cathode lamps. The half-width and profile of the resonance lines were said not to have altered and, above all, practically no more self-absorption was to have been observed. Accordingly, the first calibration curves published by SULLIVAN and WALSH showed good linearity up to high absorbance values. The findings of SULLIVAN were verified through work by CARTWRIGHT [213–215], in which attention was also drawn to the considerable improvements for such difficultly vaporized elements as Si, Ti, and V. For these elements, only lamps of low light intensity, which mostly gave unsatisfactory results, were available.

Although these high intensity lamps were initially greeted with great interest, it was soon discovered that they did not offer just advantages in routine operation. In particular, the second power supply for the additional discharge and the long warm-up period necessary for stabilization of the emission were found to be a nuisance. The higher price resulting from the complicated construction also bore no relation to the often observed short lifetime.

When it later became evident that the same results could be obtained from normal hollow cathode lamps through considerably simpler means, such as the WHITE cathode form and isolation of the space behind the cathode opening [812], the general interest in high intensity lamps disappeared rapidly. Today, high intensity lamps are practically only used for research purposes. In connection with atomic fluorescence spectroscopy, on the other hand, they may again increase in importance since the obtainable sensitivity is directly dependent on the illumination intensity in this technique.

2.5. Vapour Discharge Lamps

Very volatile metals such as mercury, thallium, zinc and the alkali metals were often previously determined using low pressure vapour discharge lamps since these were offered commercially at low prices and gave a high radiation density. Various work has been published in which these lamps are compared with hollow cathode lamps [810] [1074] [1151]. With the greatly improved quality of hollow cathode lamps, the vapour discharge lamps are increasingly losing importance for atomic absorption analyses. Only in the cases of rubidium and cesium are they better than hollow cathode lamps. The disadvantage of these lamps under normal working conditions is that they emit strongly broadened lines owing to self-absorption of high internal atom concentration and are therefore little suited to atomic absorption measurements [1240]. Vapour discharge lamps must be operated with highly reduced currents to avoid excessive self-absorption in atomic absorption spectroscopy. This can lead to an increasing instability of the lamp, however. All in all, low pressure gas discharge lamps are not always so problem-free in use as modern hollow cathode lamps.

2.6. Electrodeless Discharge Lamps

For a number of years, electrodeless discharge tubes excited by radio frequency have found increasing interest, especially in atomic fluorescence spectroscopy [25] [62] [169] [252] [263] [824]. For atomic fluorescence, line profiles and the spectral zone around the resonance lines are of less importance since the light emitted from the lamp does not fall on the detector. The light intensity is of principal interest since the sensitivity over a wide range is directly proportional to the radiation intensity.

The biggest advantage of electrodeless discharge lamps is the increased intensity of the radiation by several orders of magnitude compared to conventional hollow cathode lamps. Furthermore, they can be manufactured rather cheaply. They consist of a sealed quartz tube of several centimeters in length and about 5–10 mm in diameter, filled with a few milligrams of the element of interest (as pure metal, halide, or metal with added iodine) under an argon pressure of a few Torr [263] [1355]. The tube is mounted within the coil of a high frequency generator (e.g. 2400 MHz) and excited by an output of a few Watts up to 200 Watts.

At the present, however, not all of the problems connected with the manufacture of electrodeless discharge lamps appears to have been solved [25] [27] [516]. WEST and WINEFORDNER give detailed instructions for the construction of good tubes [253] [255] [817] [1378].

Various factors, such as degassing and cleaning of the quartz tubes to be used, the tube dimensions, the filling pressure of the inert gas and, especially, the quantity and chemical nature of the filling material, all play an important part [105] [381] [567] [857].

HOARE [516] observed that small tubes, in particular, were only usable for a relatively short time. Good success could be obtained with tubes that were fused into a vacuum jacket for better thermal insulation. Particularly high stability could be achieved if the lamps were thermostatted [172]. HAARSMA and co-workers have recently published a critical review on the manufacture and operation of electrodeless lamps, it serves as an excellent guide for self-manufacture.

For a long time, various views prevailed on the advantages of electrodeless discharge lamps in atomic absorption spectroscopy since here higher radiation intensity does not influence the sensitivity; only the signal-to-noise ratio can occasionally be improved, leading to higher precision and to more suitable detection limits. From an early date a number of authors have, nevertheless, employed these lamp types in atomic absorption with success [458] [1355]. Electrodeless discharge lamps are of great advantage especially for advancing into the vacuum UV range [654] [669] [670].

For the elements which can be determined in this range, no lamps, or only unsatisfactory other light sources are available. Furthermore, the high radiation intensity is particularly important since higher energy losses owing to the lower transparency of air, the flame and lenses, besides poorer reflection of mirrors, are to be expected.

For a long period of time, the greatest disadvantages of electrodeless discharge lamps for use in atomic absorption were their relatively short life and the long warm-up periods usually required, during which they frequently drifted. In 1973 a lamp specially designed for atomic absorption spectroscopy, in which the known disadvantages had largely been over-

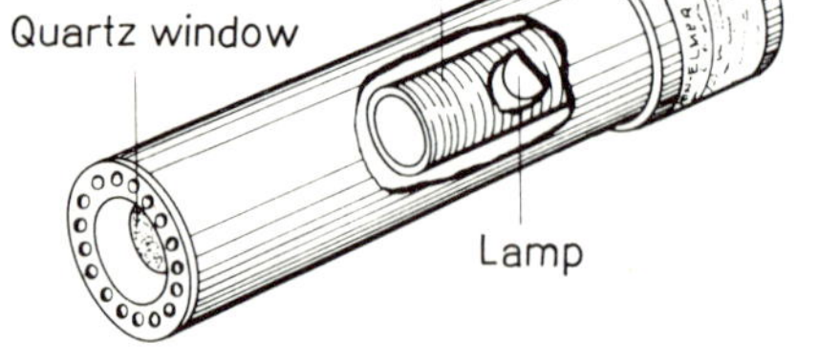

Figure 14. Schematic of an electrodeless discharge lamp.—The lamp consists of a quartz sphere into which the element is fused under a filling gas pressure of a few Torr. Excitation is by a high frequency field.

come, appeared on the market for the first time [74]. The principal difference from conventional lamps is that the element to be tested is filled into a quartz tube, which is firmly joined to the high frequency coil and mounted into an insulated jacket (see figure 14). These lamps are very simple to operate, have a relatively short warm-up period and very good stability. A further advantage is that the lamps are operated with only 27 MHz so that the power supply unit is much simpler. BARNETT [74] reported a considerable improvement in the signal-to-noise ratio and thereby to the detection limit, and also in the linearity of calibration plots, particularly for arsenic and selenium.

2.7. Flames as Light Sources

In 1955, ALKEMADE and MILATZ [34] used a flame, into which high metal salt concentrations were sprayed, as a light source for atomic absorption spectroscopy; various authors have followed this example [1134]. A flame as the primary light source is advantageous since it is cheap, universal and very flexible. Particularly for multielement analyses [197] [1178], it offers the advantage of admitting practically every desired element combination. The disadvantage of these light sources is that they are somewhat unstable and do not exhibit a particularly intense emission. Furthermore, the half-widths of the emission and absorption lines are the
same, which ... is the use of an electric arc, such as the KRANZ
arc [702], as ... ilized arcs are said to have
a high emissi ... results obtained with them
can be comp...

2.8. Contin...

Light sourc... ntness (hydrogen or deuterium
lamps, high ... en-iodine lamps [346]) would
appear at fi ... ons. They show good stability,
make mult... ly when many elements have
to be anal... s are more than offset by two
disadvanta... ion only takes place within a
very small...
Over a wi... onance line, a continuous light
source ha... al light source, even when the
total inte... urthermore, such a light source
places ver... omator. With the best monochro-
mators of... photometers, a loss in sensitivity
by a fact... This fact was recognized when
the proce... eason is that there are practically
no monc... s corresponding to the half-width
of an ab... ral bandwidths have to be used,

a considerably higher portion of non-absorbable radiation to either side of the resonance line will always fall on the detector. This leads to the above described reduction in sensitivity and to strong curvature of the calibration plot. Additionally, there is an increased risk of introducing spectral interferences when continuous sources are employed since the light source itself is no longer selective.

Despite these general disadvantages, FASSEL [348] has made a thorough study of these light sources and has found that under given prerequisites they can be quite useful.

3. Atomization

The emission spectrum of the element under study produced in the light source is passed through an "absorption cell" in which a portion of the incident light is absorbed by atoms produced by thermal dissociation. Accordingly, the most important function of this absorption cell is to produce metal atoms in the ground state from the ions or molecules present in the sample. This is without doubt the most difficult and critical process within the whole atomic absorption procedure. The success or failure of an analysis is virtually dependent upon the effectiveness of the atomization; the sensitivity of the determination is directly proportional to the degree of atomization of the element under study in the sample. Finally, all known chemical interferences in atomic absorption spectroscopy are nothing more than influences on the atomization, i.e. the total number of atoms that can be formed.
The longest practiced, and therefore most widely propagated, method for the dissociation of a sample into atoms in atomic absorption spectroscopy is the spraying of a solution into a flame. Next to this, flameless atomization procedures have gained considerable importance in recent years, and especially atomization in a graphite tube heated by electrical resistance. Since the mechanisms of these procedures are completely different, atomization in a flame and flameless atomization are treated separately.

3.1. Atomization in Flames

To atomize a sample in a flame, it is usually sprayed into the flame in form of a solution by means of a pneumatic nebulizer. Methods have also been described in which solid samples are introduced directly or as suspensions into the flame. A special case of solid sample introduction is the boat technique with its variants in which the solid (or dried) sample is introduced into the flame in a tantalum boat or nickel crucible. A further special case is the introduction of gaseous samples into the flame, such as with hydride systems.
By using a nebulizer to spray the sample into the flame, a steady, time independent signal is produced, whose height is proportional to the concentration of the element under study, and which exists as long as sample solution is aspirated and sprayed. For other types of sample introduction, such as the boat technique or the hydride system, time dependent signals are produced whose heights (or areas) are proportional to the absolute quantity of the element under study.

3.1.1. Thermal dissociation

If a dissolved sample (e.g. a solution of a salt in water) is sprayed into a flame, the solution droplets are initially dried, i.e. the solvent is evaporated. The speed of evaporation is dependent upon the size of the droplets and the nature of the solvent. The solid particles (e.g. salt crystals) formed can now undergo various alterations under the influence of the flame temperature; organic substances are burnt, inorganic components react with each other or with the flame gases. Hydrated chlorides, which can be formed by many elements in hydrochloric acid solution, can eliminate either hydrogen chloride to form oxides, or water to form free chlorides. Similarly, oxides can be formed from carbonates and sulphates, and pyrophosphates from phosphates. All further processes, such as melting and vaporization (or sublimation, etc.), are frequently dependent upon these transition reactions; one only has to remember the thermal stability of refractory oxides (compare [36]). Many of the causes of chemical interferences are also to be found here. The above mentioned thermal transition reactions are frequently dependent on the other reaction partners contained in the sample. From pure hydrochloric acid solutions, calcium ions form a dehydrated chloride which is very easily vaporized; if, on the other hand, sulphate or phosphate ions are added to the solution, an oxide or a pyrophosphate is formed which melts and vaporizes considerably more slowly or only at higher temperatures. Hence it is already clear that temperature will play an important role in later considerations on chemical interferences.

Finally, if gaseous molecules are present after the particles have vaporized, the actual thermal dissociation process of molecules into atoms begins. This is an equilibrium reaction which can often be very complicated owing to the large number of similar reactions taking place parallel to it or joining in.

In a flame (and this should be exclusively considered) there are a variety of combustion products besides the atoms under study, which are, in some cases, in considerable excess (CO_2, CO, C, H_2O, O_2, O, H_2, H, OH, NO, N_2). The cleavage products of the solvent and other possible substances in the sample must also be taken into consideration. Atoms, radicals, ions, and so on are not capable of existence under normal conditions (excepting the inert gases), and are only formed in the flame by temperature-dependent equilibrium reactions. If the same cleavage product is produced by two equilibrium reactions, both reactions influence each other according to the law of mass action through the partial pressure of the reaction products. This will be of interest for our observations if the anionic constituents of the metallic compound under analysis are also produced simultaneously in high concentrations (high partial pressure) as a result of other equilibrium reactions in the flame. The thermal dissociation of the element under study can then be strongly influenced. In principle, the flame can be regarded as a "solvent" in which a trace of metal atoms is to be found [580]. Besides the suppression of the dissociation equilibrium

$$\mathrm{MeX} \leftrightharpoons \mathrm{Me} + \mathrm{X}$$

through a parallel equilibrium reaction in which X is also produced, the total number of metal atoms Me formed can also be influenced by a subsequent reaction such as ionization (compare [1065])

$$\mathrm{Me} \leftrightharpoons \mathrm{Me}^{\oplus} + \mathrm{e}^{\ominus}$$

or through the formation of a compound with another reaction partner

$$\mathrm{Me} + \mathrm{Y} \rightleftharpoons \mathrm{MeY}\,.$$

A schematic representation of the most important processes possible in a flame is shown in figure 15.

3.1.2. The ideal absorption cell

When it is known which processes of chemical and physical nature take place (or can take place) on the way from solution droplets to free atoms, then one can specify one's idea of an ideal absorption cell, without regard to its feasibility. In short, it will be a device in which the processes for the formation of free atoms take place rapidly and quantitatively, no parallel reactions hinder the atomization, and no subsequent reactions can take place.

If these requirements are considered with respect to a flame, the time required for the conversion of solution droplets into gaseous molecules will certainly be dependent on the size of the droplets and the temperature of the flame. The atomization itself is also dependent upon the flame temperature and possibly upon the partial pressure of components of the flame (combustion products). This can bc influenced by the choice of suitable flame gases. For

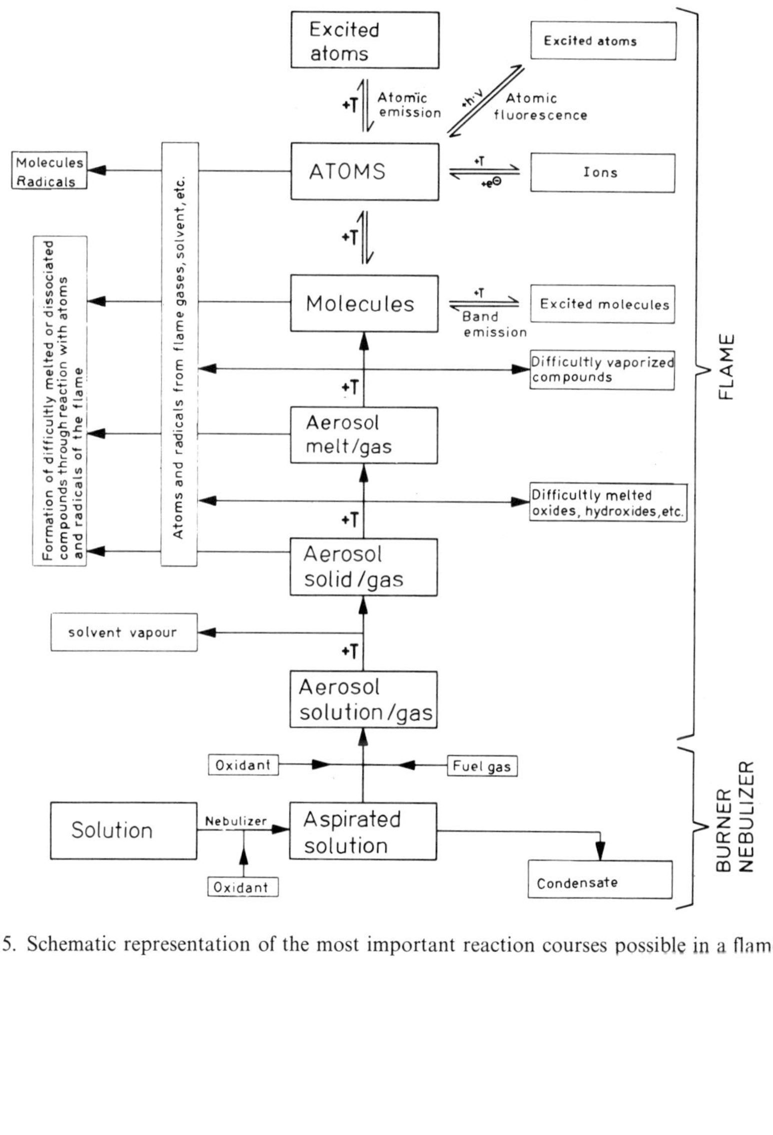

Figure 15. Schematic representation of the most important reaction courses possible in a flame.

example, the partial pressure of atomic oxygen in an oxygen/acetylene flame is around 200 times higher than in an air/hydrogen flame. Subsequent chemical reactions can also be influenced in the same way, i.e. by choice of suitable gas mixtures. Ionization, on the other hand, is temperature dependent, in the same manner as the atomization itself. Accordingly, the temperature must be so chosen that the atomization takes place rapidly, but that ionization does not occur. Finally, it is also possible to influence the whole atomization process by increasing the residence period of the sample in the flame, i.e. in flames of lower combustion speed the atomization can take place more completely. This becomes particularly clear if it is realized that even in flames of relatively low combustion speed, such as the air/acetylene flame, a distance of 1 cm is passed through in less than 10 ms.

It should also be considered that in absorption measurements the light beam passes through the flame. In other words, the flame must act as an absorption cell and be optically transparent. This means that the flame itself must not exhibit any absorption and that the atomization should take place quantitatively, if possible. Incomplete atomization means that not only fewer atoms are formed, but also that solid or liquid particles, which can lead to light scattering, are still in the flame and thereby in the beam path. Even undissociated molecules can lead to non-specific light losses because they absorb radiation (usually over a greater spectral range).

After these rather theoretical considerations, the customary means of atomization, with their advantages and disadvantages, will now be discussed, especially with respect to the quantitative atomization of the sample.

3.1.3. Turbulent and laminar burners

The direct nebulizing turbulent burners, known from emission flame photometry, consist of two concentric tubes through which fuel gas and oxidant are fed separately to the jet (see figure 16). The fuel gas and oxidant are actually first mixed in the flame and thus considerable (optical and acoustic) turbulence takes place. The sample solution is aspirated through a central vertical capillary in the burner by the drop in pressure of the oxidant at the jet and is directly nebulized into the flame. These burners are often referred to as total consumption burners because all the solution nebulized passes into the flame.

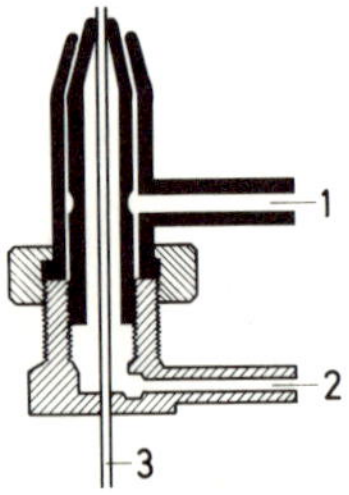

Figure 16. Turbulent burner with direct nebulizer.—**1** Fuel gas line, **2** oxidant feed, **3** aspiration capillary for the sample solution.

The most important advantages of the turbulent burner are its simple construction, which enables it to be cheaply manufactured, its safety (a flashback of the flame is impossible), and the possibility of using virtually all imaginable gas mixtures—however, the corresponding combustion speeds must be taken into account in the design.

In the laminar burner with mixing chamber (often referred to as mixing chamber or premix burners), which found use in Australia for the first applications of atomic absorption spectroscopy [237] [1328], the fuel gas and oxidant are thoroughly mixed before they are ignited. The sample solution is sprayed into the mixing chamber with a pneumatic nebulizer and forms an aerosol with the flammable gas mixture. The design has been repeatedly improved with respect to its resistance and stability [190] [1139]. In figure 17, a modern version of this burner type is depicted. The flame usually forms a long, rectangular cell with the light beam passing through the complete length.

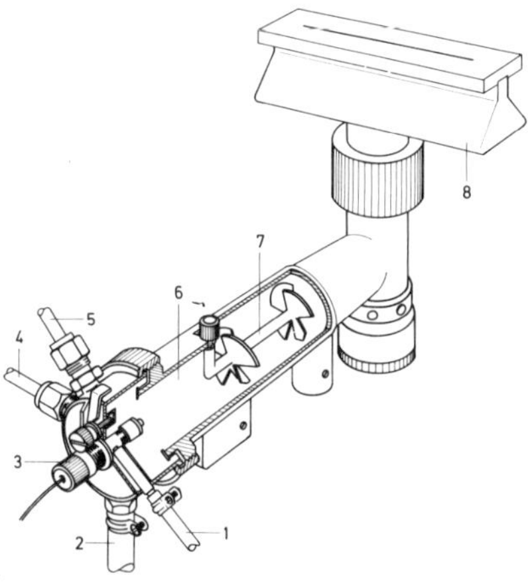

Figure 17. PERKIN-ELMER mixing chamber burner.—**1** Oxidant feed for nebulizer, **2** drain hose, **3** nebulizer, **4** oxidant feed, controllable, **5** fuel gas line, **6** mixing chamber, **7** spoiler, **8** interchangeable burner head.

The fact that the relatively large volume of flammable gas in the mixing chamber is a potential source of danger is regarded as the greatest disadvantage of the mixing chamber burner. Modern burner designs are largely protected against flashback of the flame. Mixing chamber burners also have the disadvantage that only flames of relatively low combustion speeds can be used; this excludes, for example, the use of gas mixtures employing pure oxygen.

In the following section, both burner designs will be examined with respect to their atomizing properties: Both burner types operate with a pneumatic nebulizer system and have comparable aspiration rates, which lie between 2 and 10 ml/min. In general, the direct nebulizers have a slightly lower rate of aspiration. The rate of aspiration of the direct nebulizer is determined by the pressure and flow of the oxidant. An alteration in the pressure and flow of the fuel gas also causes an alteration in the aspiration rate. This is a disadvantage of the turbulent burner since independent optimization of the nebulizer and flame is not possible. Mixing chamber burners, on the other hand, usually have independently adjustable nebulizer systems which operate with constant oxidant pressure.

Pneumatic nebulizers produce a sample aerosol with a wide variety of droplet sizes. There is a maximum droplet volume up to which a given flame is just able to vaporize a given solvent within a determined time. Owing to the varying droplet sizes, the vaporization takes place at various positions in the flame; very small droplets are dried shortly after entering

the flame, while large droplets are dried much later. This is naturally valid for all other processes in the flame, such as melting and vaporization, as shown in figure 15.
Since both burner types operate with comparable nebulizer systems, the fineness of the aerosol produced and the maximum droplet size are also comparable. While in the turbulent burner the sample is nebulized directly into the flame, in the mixing chamber burners the aerosol must travel a determined distance and pass contact surfaces (spoilers) before it reaches the flame. The larger droplets condense on the spoilers and run out through a drain in the mixing chamber. Although a large part of the sample solution is wasted in this way (only 5–20% of the solution reaches the flame), the maximum droplet size is considerably reduced. KERBYSON [639] carried out interesting comparative measurements in which the aerosol produced by a direct nebulizer was first examined directly and then with a mixing chamber inserted. The maximum droplet volume measured at 5 cm above the burner (no flame) was found to be approximately 7 nl for the direct nebulizer and about 0.05 nl for the mixing chamber burner.
Furthermore, the exit speed of the gases from the jet of a turbulent burner is greater than from a laminar burner by a factor of about 50. It is then to be expected that the larger droplets, which pass through the flame of a turbulent burner within a few milliseconds, do not have the opportunity to be quantitatively atomized. GIBSON [412] found that with turbulent burners even water droplets could pass through the whole flame without being completely vaporized. This finding has been verified by other authors [282] [1344].
If the concentration of atoms in the tubulent flame of a direct nebulizer is observed at any given height in the flame, only relatively small alterations can be detected; i.e. the atomization takes place throughout a wide area of the flame. In the same way, solvent droplets, salt crystals, etc., are to be found at every height over the whole length of the flame. The turbulent flame from a direct nebulizer has little structure. In the laminar flame of a mixing chamber burner, on the other hand, a quite clearly formed zone structure is to be found, since the reduced droplet size makes optimum atomization within a defined flame zone possible [241] [460] [1018].
Various authors have compared the advantages and disadvantages of both burner designs [7] [682] [1145], especially with respect to the occurrence of chemical interferences [14] [1384]. The most common cause for the occurrence of chemical interferences can be traced to the influence of other chemical components in the sample on the degree of conversion of the aerosol droplets into free atoms of the element to be determined [25]. Other workers found that physical factors, in particular, influence the rates of vaporization and atomization. Decreasing droplet size of the aerosol, higher flame temperatures and longer residence in the flame reduce or eliminate a large number of interferences [511] [683] [1374]. This has also been verified by the discovery that interferences occur more weakly in higher flame zones than directly over the burner slot [35] [391]. Precise investigations on known interferences, such as the influence of phosphate [1155] and similar anions on calcium, have shown that they can be almost completely eliminated by the choice of suitable flames and burners [344] [368] [884] [976].
As mentioned at the beginning, the requirements placed on the ideal absorption cell are that it should be optically transparent and that the sample should be quantitatively atomized if possible. The latter requirement is well fulfilled by a laminar burner with a mixing chamber,

since it produces a considerably finer aerosol and has a slower combustion speed than a turbulent burner. The temperatures achieved by both types are comparable. This will be discussed in detail in the next section. The transparency of a laminar burner is practically ideal, as demonstrated by means of schlieren photographs [140] [1145], while a turbulent flame is optically non-uniform, even when the light scattering on non-vaporized particles is disregarded.
The optical noise of a flame results from the varying optical properties of the flame. These fluctuations are caused by convection of the gases and by turbulence. The influence of flame noise on the precision is considerably greater in atomic absorption spectroscopy than in flame emission photometry or atomic fluorescence spectroscopy, since in the latter two the flame is either not irradiated or the beam passing through the flame is not taken into consideration.

In a large number of investigations on various flames and burners, LEBEDEV [737] found that mixing chamber burners, especially with multi-slot burner heads, produced minimum optical noise. In such flames, completely stable middle zones can be observed. Turbulent burners, on the other hand, display the strongest optical noise, especially when organic solvents are sprayed.

3.1.4. Flames

All the flames usable for atomic absorption spectroscopy are compiled in table 3.
The flames using pure oxygen as the oxidant can only be employed with direct nebulizing turbulent burners because of the high combustion speeds. Owing to the high combustion speed and the insufficient atomization, these flames have lost importance, especially after the introduction of the nitrous oxide/acetylene flame, which has a comparable temperature but a considerably lower combustion speed.
The first practical applications of atomic absorption spectroscopy by WALSH and co-workers were carried out using air/coal gas flames. The low combustion speed of this flame gives a good absorption cell, but the low flame temperature is insufficient for complete atomization of many elements [7] [413]. As explained previously, the flame temperature has no noticeable influence on the absorption characteristics. For this reason the temperature of the flame should be raised as far as possible to avoid chemical interferences while still maintaining the good optical characteristics but without introducing ionization too strongly. The air/coal gas flame, like the air/propane flame, is seldom used nowadays.
The flame most well known and used in atomic absorption spectroscopy is the air/acetylene flame. For many elements it offers a suitable environment and a temperature sufficient for atomization. In only a few cases (alkali metals) does noticeable ionization occur. The flame is completely transparent over a wide spectral range and only shows noticeable self-absorption below 230 nm, increasing to about 65% at the wavelength of arsenic (193.7 nm) (see figure 18). Furthermore, the emission of the air/acetylene flame is very low so that for many elements the ideal conditions are given. Normally, this flame is operated stochiometrically or weakly oxidizing; however, it is variable over a wide range which further increases its applicability. A number of noble metals such as gold, iridium, palladium, platinum and rhodium, for example, may be determined with the highest sensitivity and largely free of interferences in a strongly

Table 3. Characteristics of various flames*)

Gas mixture Oxidant	Fuel gas	Combustion speed (cm/s)	Temperature °C	Range**)	References
Argon + diffusion air	Hydrogen		400	(350–1000)	[254]
Air	Natural gas	55	1840	(1700–1900)	[842]
Air	Methane	70	1875		[842]
Air	Coal gas	55	1900		[746]
Air	Propane	80	1930		[842]
Air	Hydrogen	440	2045	(2000–2050)	[587]
Air	Acetylene	160	2300	(2125–2400)	[587] [936]
Nitrous oxide	Acetylene	180	2750	(2650–2800)	[936] [1377]
Oxygen	Hydrogen	1150	2660	(2550–2700)	[773]
Oxygen	Acetylene	2480	3100	(3060–3155)	[504]
Oxygen	Cyanogen	140	4500		[67]
High frequency plasma			7800		[440]

* A summary of further flames can be found in KNISELEY [676] and WILLIS [1334].
** Compare REIF [1029] and SNELLEMANN [1159].

oxidizing air/acetylene flame. While the alkaline-earth metals are determined most favourably in a slightly reducing flame (i. e. a slight excess of fuel gas), DAVID [271] found that molybdenum exhibited the most intense absorption in a highly luminous flame (larger excess of fuel gas).

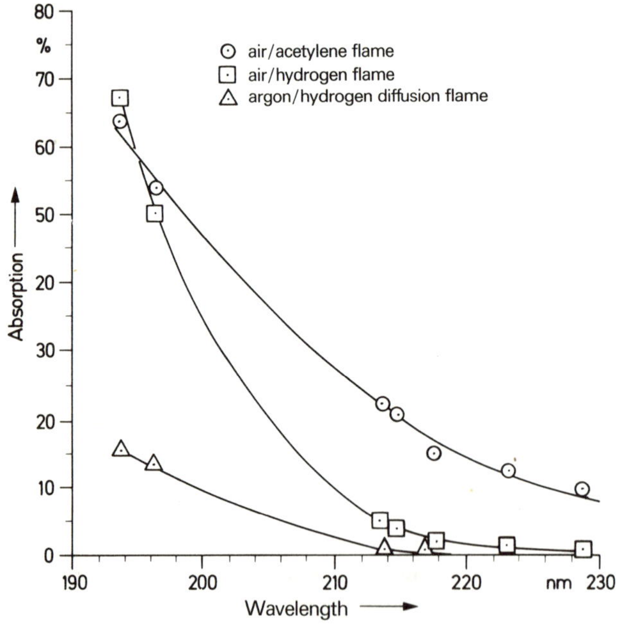

Figure 18. Self-absorption of various flames depending upon the wavelength.

Similar flame conditions have also proved to be of value for chromium and tin. The disadvantages of these highly reducing flames are their relatively high emission and a noticeable self-absorption owing to unburnt carbon. The emission leads to increased noise, especially with low-energy light sources.

The temperature of the air/acetylene flame is nevertheless insufficient to dissociate a large number of principally oxidic compounds or to prevent their formation in the flame. That the oxidation frequently first takes place in the flame was detected by RANN and HAMBLY [1018] for the case of molybdenum. They observed that in a stochiometric air/acetylene flame, directly above a primary reaction zone, there was a high concentration of molybdenum atoms in a very narrow band that reacted very rapidly with the oxygen present.

The occurrence of a large number of interferences, which are often difficult to control in the determination of molybdenum, tin and especially chromium in a fuel gas rich air/acetylene flame, indicates quite clearly that this flame is not particularly suitable for these elements. The occurrence of chemical interferences always indicates that certain chemical compounds are insufficiently dissociated in the flame or that the atoms formed react spontaneously with components in the flame. The frequent occurrence of such interferences should then be taken as an indication that the performance limit of a given flame has been reached. There are about 30 elements that cannot be determined using the air/acetylene flame. These are the elements that form very stable oxidic bonds.

For this reason therefore, attempts were made during the early years of atomic absorption spectroscopy to determine those elements which form refractory oxides in a hot oxygen/acetylene flame [187] [281] [312] [349] [677] [885], frequently using organic solvents [40] [1146], or with modified burners [47] [791], or even in an oxygen/cyanogen flame [1039]. The addition of complexing agents was also investigated [276]. It was shown that a fuel gas rich pre-mixed oxygen/acetylene flame was suitable for the dissociation of several of these elements [345], but because of its lower temperature [842] was not sufficient for all [241].

Certainly the most important development in flames was the introduction of the nitrous oxide/acetylene flame by WILLIS in 1965 [1333]. As a result of its low combustion speed, this hot flame offers a favourable chemical, thermal and optical environment for virtually all metals that give difficulties in the air/acetylene flame.

The usual nitrous oxide/acetylene flame is operated with a slight excess of fuel gas and exhibits a 2–4 mm high blue-white primary reaction zone above which is a characteristic 5–50 mm high red reducing zone. Above this is then a pale blue-violet secondary reaction zone in which the oxidation of the fuel gas takes place.

The red reduction zone is usually the one of analytical interest. The thermal dissociation of the sample into atoms takes place there and no noticeable oxidation of the metal atoms can occur.

A large number of workers investigated this flame soon after its introduction [48] [150] [689] [794] [795] [976] [977] [1152] and found that it was largely free of interferences [547] [764] [1155]. MARKS and WELCHER [823] investigated the influence of the flame composition and instrument settings on interferences in the nitrous oxide/acetylene flame. They established that the height of observation and the oxidant/fuel gas ratio had the strongest influences, and that with the correct selection of these parameters, interferences disappeared or were at least reduced. To describe the flame conditions as exactly as possible, they introduced

a parameter ϱ representing the molar oxidant/fuel gas ratio as a fraction of the stochiometric ratio 3:1 ($3\,N_2O + C_2H_2 \rightarrow 2\,CO + 3\,N_2 + H_2O$).
Various authors have also established that a larger number of free atoms are produced in the nitrous oxide/acetylene flame than in all other usual flames [350] [575] [688]. These free atoms are exactly localized and strongly dependent on the stochiometry, however, since the atomization is an equilibrium process [350]. RASMUSON and co-workers [1023] have established that for every metal which has a monoxide with a dissociation energy of less than 6.5 eV, there is a flow relationship ϱ (nitrous oxide : acetylene) at which the atomization is complete (naturally assuming complete vaporization of the sample).
Nevertheless, the nitrous oxide/acetylene flame has two disadvantages which must be taken into consideration. Firstly, many elements are more or less strongly ionized in the hot flame and thereby show reduced sensitivity. Secondly, it has a relatively strong self-emission [794]. While the ionization interferences can often be easily removed by adding an excess of another easily ionized element (see section 6.2), the emission can on occasions cause genuine difficulties [127].
The CN, CH and NH bands, which occur over a wide spectral range and are often very intense, can cause "emission noise" (see section 5.2) if they fall together with a given element. This noise can influence the precision of an analysis, or in the case of weak light sources make an analysis impossible.
Nevertheless, if these difficulties are understood, they can usually be avoided, and there is often no reason why the nitrous oxide/acetylene flame should not be used. The advantages and the success of this flame caused SLAVIN [1149] to state in a review, in 1969, that there were still a number of chemical interferences that could not be eliminated even by the use of the nitrous oxide/acetylene flame. Without doubt this flame has made a considerable contribution to the elimination of the difficulties encountered in dissociating stable oxidic compounds [1334].
Recently, a nitrous oxide/hydrogen flame was investigated independently by two reseach teams [260] [1336]. It was shown that for atomic absorption at least, this flame was inferior to the nitrous oxide/acetylene flame. LUECKE [765] has reported a nitric oxide/acetylene flame which can give an increase in sensitivity of up to 60% for the elements barium, boron, strontium and zirconium. An excellent review on flames of higher temperature and their use has recently been published by WILLIS [1334].
ALLAN [43] found that the sensitivity for tin in an air/hydrogen flame was considerably better than in an air/acetylene or a nitrous oxide/acetylene flame. Other authors have verified this finding [206] [208]. More chemical interferences are, of course, to be expected in this flame. A thorough investigation of the air/hydrogen flame [1087] has shown that for the determination of a number of other elements it has distinct advantages, for example the lower ionization of the alkali metals in comparison to the air/acetylene flame. Furthermore, it has low self-absorption in the range 230 nm to 200 nm (see figure 18). However, more chemical interferences must be taken into account and the use of organic solvents, although possible, is critical [1219].
During his work on atomic fluorescence spectroscopy, WINEFORDNER examined an argon/hydrogen/entrained air flame [1264] [1368], and DAGNALL has reported the low self-absorption of a nitrogen/hydrogen/entrained air flame [253] [254]. In both these flame types, hydrogen

serves as the fuel gas while argon or nitrogen serve to nebulize the sample into the mixing chamber. When the flame is ignited, the hydrogen, diluted with inert gas, burns in the surrounding air. This results in a flame with a unique profile, which is particularly noticeable if a three-slot burner head is used [139]. On the outside, where mixing with the surrounding air takes place, it has a temperature of about 850 °C, while in the middle, the temperature only reaches 300–500 °C depending on the flame height [254]. This type of flame is termed a diffusion flame.

KAHN [617] has investigated the applicability of the argon/hydrogen flame (this simplified name will be used in future) to atomic absorption spectroscopy and has found that for arsenic, selenium, cadmium and tin there are considerable improvements in the sensitivity and detection limits. In all probability the atomization takes place with the active participation of hydrogen [903] [1067] [1068], although the reaction mechanism is not completely clear. All authors, however, draw attention to the considerable interferences that occur in a flame of such a low temperature; these can be both chemical interferences and high non-specific light losses. This flame shows its greatest advantage at the start of the vacuum UV range because of its high transparency compared to other flame types (see figure 18). It is therefore particularly suitable for the determination of arsenic and selenium. In connection with the hydride system, the usual disadvantages are of no great importance because in this procedure the element under study is separated from the remaining matrix and conducted to the burner as a gas [799].

Also in connection with emission and atomic fluorescence measurements, WEST and co-workers have reported on separated air/acetylene [517] [518] [660] and nitrous oxide/acetylene [250] [658–662] flames. Here, the flames burn either in a quartz tube, or in an atmosphere of argon or nitrogen so that the secondary reaction zone is separated from the primary. This separation and the shielding of the flame from atmospheric oxygen gives an intermediate zone with very low background radiation [668]. This is particularly interesting, because the highest atom concentration can be observed directly above the primary reaction zone of the flame. At this position in all hydrocarbon flames, the combustion of carbon monoxide takes place supported by the entry of atmospheric air. This is evident by the occurrence of particularly intense OH bands at 309 nm and 306 nm and the broad chemiluminescence continuum of the reaction $CO + O \rightarrow CO_2 + h\nu$ directly above the primary reaction zone. Through the separation of the primary and secondary reaction zones, the high atom concentration above the primary reaction zone is maintained, while the emission connected with the secondary zone is displaced to higher regions. A reduction in the emission by more than two orders of magnitude could be measured at the position of highest atom concentration in the separated flame. This has been shown to be particularly advantageous for emission and atomic fluorescence measurements.

KIRKBRIGHT and co-workers [666] investigated the temperature relationships in normal and shielded nitrous oxide/acetylene flames and found, according to the observed height and stochiometry, a decrease of 40–180 °C in the intermediate zone of interest for atomic absorption. Despite this lower temperature, the authors found at least the same, if not better sensitivity [659], for elements that form refractory oxides; that is to say, the flame temperature is not so decisive as a high partial pressure of reducing gases and low partial pressure of oxygen. Although the first applications of separated flames did not bring any great improvements in atomic absorption analyses [652] [657], they have nevertheless recently shown interesting possibilities.

RUBEŠKA [1066] established that no further interfering CaOH emission could be observed for the determination of barium in a shielded nitrous oxide/acetylene flame, even with a large excess of calcium.

KIRKBRIGHT and co-workers [654] [669] were able to determine iodine, sulphur and phosphorus in the vacuum UV range successfully only by means of a shielded nitrous oxide/acetylene flame. The partial pressure of oxygen, which exhibits the strongest absorption in this range, is very small in a fuel gas rich, shielded nitrous oxide/acetylene flame. In principle, according to the reaction $2N_2O + C_2H_2 \rightarrow 2CO + H_2 + N_2$, no oxygen or water will be formed in such a flame. Additionally, the reaction $CN + O \rightarrow CO + \frac{1}{2}N_2$ can take place, so that operation is under strongly reducing conditions. The absorption in such a flame is also correspondingly low: At the wavelength of arsenic at 193.7 nm, a shielded nitrous oxide/acetylene flame only absorbs 5% of the radiation compared to about 65% with a non-shielded flame. Even at 178.3 nm, the resonance line of phosphorus, only 45% of the radiation is absorbed by a shielded flame (compared to nitrogen), so that it is still possible to work reasonably in this range.

Finally, a brief mention should be made of "electrical flames" or plasmas, which have been investigated over some period of time by various workers [843] [1265] [1304] [1305] [1307]. This method of atomization has found hardly any practical importance up to the present, because the equipment is very expensive and the introduction of the sample is not without its problems. In virtually all work, plasma has been used for emission measurements [300] [338] [667] because very high temperatures can be achieved. In some cases, astonishing detection limits have been reported for elements such as boron [840], or others that usually cause difficulties by forming stable monoxides [300], or elements that are normally difficult to determine by emission such as arsenic, selenium or iodine [667].

In principle, plasma should also be suitable in atomic absorption, especially for elements that form stable compounds with flame gases or oxides. A plasma produces a fully chemically inert environment and the sample is so diluted that practically no inter-elemental effects can occur when the sample remains long enough in the plasma. The only interference to be expected is an ionization interference. Inductive coupled plasma and microwave plasma are relatively similar to a flame and have a good future for the determination of refractory elements (compare [343]).

3.1.5. Burners and nebulizers

While the advantages and disadvantages of turbulent and laminar burners have already been described (section 3.1.3), a number of interesting variations and additions will be mentioned in this section.

To be able to use the various flames described in the last section, it is generally necessary to carry out certain modifications to the burner. In the case of turbulent burners, various complete burners are offered for the different gas mixtures. With mixing chamber burners, on the other hand, only the burner heads need to be adapted to the flame to be used, because they are usually interchangeable. Various slot burners are commercially available with burner slots of around 5–10 cm long and 0.5 mm (air/acetylene) to 1.5 mm (air/propane) wide. Burner heads made of titanium are of especial advantage because of their high chemical resistance. A burner head, much in use nowadays, with three parallel slots 11 cm long and 0.45 mm wide was recommended by BOLING [139]. This three-slot burner head has better flame optics [140] [737] leading to an improved signal-to-noise ratio. A number of elements such as calcium, chromium and molybdenum can be determined with improved sensitivity. The greatest advantage of this burner head, however, is that relatively high total salt concentrations can be burnt in it without the slots becoming blocked up. With normal slot burners up to about 3–5% total salt content can be used, however solids contents of up to 30% or even undiluted blood serum can often be burnt for longer periods with the three-slot burner. This burner head is suitable for the air/acetylene, and air/propane flames, and also diffusion flames such as the argon/hydrogen flame. An interesting variation is the burner head proposed by BUTLER [191] in which the slot has been replaced by five rows of holes. For a number of easily atomized elements using an air/propane flame, an improvement in the sensitivity by a factor of ten has been reported. ALDOUS and co-workers [26] used a burner in which the slot was replaced by a row of capillaries. They found that this design gave lower emission and high stability. The greatest advantage seen by the authors was the ability to use extreme gas mixtures without the risk of a flashback of the flame. SCHMIDT and SANSONI [1093] reported a burner which had cooling directly at the slot with water thermostatted at 30 °C. They found that this burner gave a considerably improved signal-to-noise ratio. This burner design showed a very low blockage tendency, particularly when highly concentrated salt solutions were being sprayed, and subsequent spraying of water produced a high degree of self cleaning.

For the hot nitrous oxide/acetylene flame, specially designed burner heads are usually required [792] [912], which are, however, also suitable for other gas mixtures. Because of the somewhat higher combustion speed, a narrower slot must be selected to prevent flashback of the flame. Various endeavours have been made to avoid the deposition of pyro-carbon observed with this flame, but with little success up to the present (compare [497]). The hotter the design and materials of such a burner head enable it to be, the more advantageous this appears for its performance.

HELL [497] and VENGHIATTIS [1267] independently attempted to improve on the low use of the aspirated solution caused by separation of the larger droplets by heating the mixing chamber. Drying the solution droplets in the flame is thereby largely saved. For a large number of elements, an increase in the sensitivity by a factor of around ten was found. However, this procedure can only be used with dilute aqueous solutions.

In 1963, FUWA and VALLEE [399] introduced a long tube burner in which the flame of a turbulent burner is directed at the opening of a tube of ceramic material mounted in the

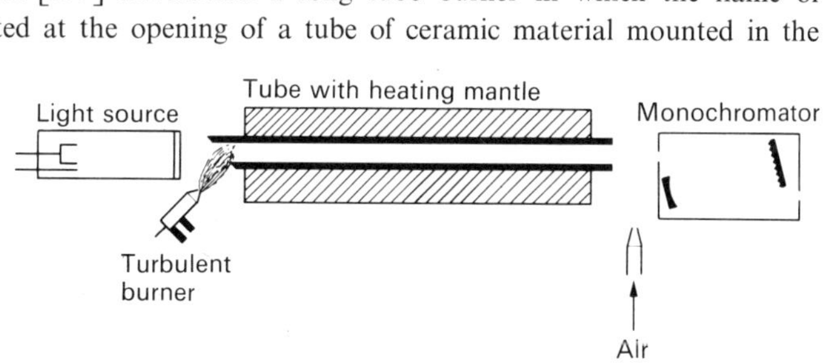

Figure 19. Long tube burner according to FUWA and VALLEE [399] in the heated form of MOLDAN [872].

beam path of an atomic absorption spectrophotometer (figure 19). In this way the atoms are forced to remain longer in the beam path, with the result that the sensitivity is increased. This is only valid however, if the atoms observed have a sufficiently long lifetime. According to equation (1.18) (page 12), the absorbance should increase proportionally to the length of the tube. To within a given degree this is also correct [685], but the length of the tube is always limited by the lifetime of the atoms [1236]. A number of difficulties such as carry-over and increased chemical interferences are certainly caused by the lower temperature in the tube [1186], but these can be largely removed by heating it [872] [1070] [1071]. Various fields of application have been treated in detail by AGAZZI [20], CHAKRABARTI [224], FUWA [398] [399], KOIRTYOHANN [685], RAMAKRISHNA [1015], RUBEŠKA [1064] [1071] and ŠTUPAR [1186] [1189].

An interesting procedure for the direct atomization of solids—usually ores and rocks—has been proposed by VENGHIATTIS [1266]. In this, the finely ground sample is thoroughly mixed with a flammable substance and the mixture is pressed into a pellet which is then ignited in the beam path of an atomic absorption spectrophotometer. The flame, so produced, has a similar temperature to an air/acetylene flame, so that about the same number of elements can be determined by this method as in an air/acetylene flame. Since the dependence of this method on the matrix is greater than for conventional atomic absorption spectroscopy, analyses must usually be carried out against pre-analysed standard rocks or ores, or the method of additions must be employed. The achievable precision is then about 5–10%.

It has often been found unsatisfactory that the conventional pneumatic nebulizer, like the direct nebulizer, produces droplets of greatly varying size and that the maximum droplet volume is considerably above the size that can be atomized in the flame. The effectiveness of this nebulizer is around 10%; i.e. about 90% of the aspirated sample solution is wasted in one way or another. This would appear to be a real starting point for an improvement in the sensitivity of atomic absorption analyses. Various authors have therefore described ultrasonic nebulizers which often produce a very even and fine aerosol. In practice, however, these nebulizers are not so easy to operate and the effective improvement in the sensitivity is not very great owing to their relatively low aspiration rates [515] [1163] [1187] [1188] [1307].

HARRISON and JULIANO [473], and also KASHIKI and OSHIMA [626], attempted to directly analyse solid samples by introducing them into the flame as suspensions. Water-insoluble tin compounds could be maintained in suspension without a stabilizer as long as required for the measurement [473], while aluminium oxide catalysts could be held in suspension in methanol with a special stirrer [626]. The results were dependent upon the physical nature of the samples, they however showed few interferences, and by standardizing against similar suspensions a good correspondence with results determined by digestion methods could be obtained.

COUDERT and VERGNAUD [240] and also GOVINDARAJU [435] used a transport screw to feed solid, pulverized samples directly into the flame or the flame gases respectively. For various industrial products and geological samples, remarkably precise results were obtained. For elements that form gaseous covalent hydrides, such as antimony, arsenic, bismuth, germanium, selenium, tellurium and tin, etc., the possibility exists of introducing the sample in gaseous form into the burner instead of spraying it as a solution. HOLAK [524] proposed this system initially for arsenic. He produced nascent hydrogen by adding zinc to the hydrochloric acid solution and collected the arsine in a trap cooled in liquid nitrogen. At the end of the reaction he warmed the trap and conducted the arsine with a stream of nitrogen into the flame. DALTON and MALANOSKI [266] simplified this procedure by conducting the arsine together with the hydrogen produced directly into an argon/hydrogen flame. Finally, FERNANDEZ and MANNING [362] [799] optimized this method and introduced a commercial system (figure 20). Later, other reducing agents were recommended by POLLOCK and WEST [993] [994], and by SCHMIDT and ROYER [1092], for the production of hydrogen to extend this method to other elements. FERNANDEZ [358] made a thorough examination of the reduction with sodium borohydride and has obtained the optimum parameters with respect to acid concentration and induction period of the reducing agent for seven elements.

3.1.6. Sampling boat systems

One of the continuous aims in atomic absorption spectroscopy has been to improve the sensitivity and the detection limits. Accordingly, many attempts have been made principally to improve the effectivness of the nebulizer-burner system. Remarkable success has thereby been achieved with a very simple means of atomization, the so-called sampling boat technique [616], which acts on the two weakest points of the nebulizer-burner system. As already explained, only about 10% of the aspirated sample solution reaches the flame, while the other 90% is wasted. A further disadvantage of spraying the sample solution into the flame

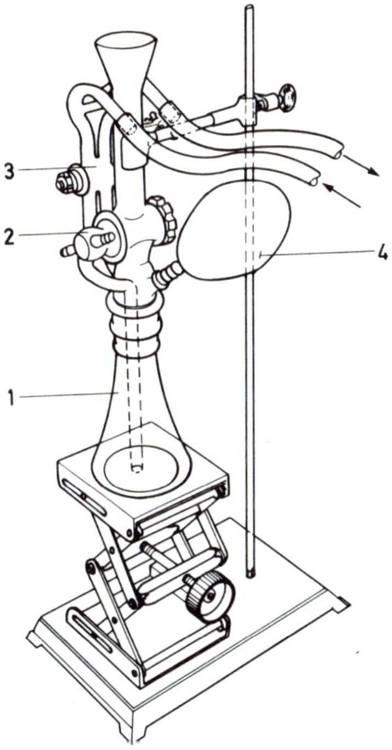

Figure 20. Hydride system. A tablet of sodium borohydride is introduced through the dosing stopcock (2) into the acidified sample solution (1). The gaseous hydride formed is collected in the balloon reservoir (4). After 15–30 seconds the three way stopcock (3) is opened and the argon stream flushes the hydride to the burner.

is that the transport of the sample to the flame is limited by the aspiration rate of the nebulizer. Both disadvantages are eliminated by the boat technique. A small amount of sample (maximum 1 ml) is introduced into a small tantalum boat, the solution is dried (for example, near the flame of an atomic absorption spectrophotometer or in a muffle furnace [527]) and then the boat is introduced into the flame of the spectrophotometer. The sample situated in the boat is then rapidly (within a few seconds) and quantitatively atomized. This results in a tall, narrow ballistic signal which can be registered on a strip chart recorder. Since the temperatures that can be achieved with the boat technique are only around 1200 °C, this procedure is only suitable for easily atomized elements such as arsenic, bismuth, cadmium, lead, mercury, selenium, silver, tellurium, thallium and zinc. For these elements, the boat technique brings an improvement in the detection limits by factors of 20–50 times (table 4).

Table 4. Absolute and relative detection limits with the boat technique and the Delves system

Element	Absolute detection limits g		Relative detection limits µg/l		
	Boat technique	Delves system	Boat technique (1 ml sample)	Delves system (0.1 ml sample)	Flame AA
Ag	2×10^{-10}	1×10^{-10}	0.2	1	2
As	2×10^{-8}	2×10^{-8}	20	200	100
Bi	3×10^{-9}	2×10^{-9}	3	20	40
Cd	1×10^{-10}	5×10^{-12}	0.1	0.05	1
Hg	2×10^{-8}	1×10^{-8}	20	100	500
Pb	1×10^{-9}	1×10^{-10}	1	1	20
Se	1×10^{-8}	1×10^{-7}	10	1000	100
Te	1×10^{-8}	3×10^{-8}	10	300	90
Tl	1×10^{-9}	1×10^{-9}	1	10	30
Zn	3×10^{-11}	5×10^{-12}	0.03	0.05	2

This method rapidly became very popular, because elements of interest for toxicology and environmental control could be determined with increased sensitivity, and biological samples such as urine and blood could be analysed directly [618]. The boat technique has also been successfully applied to the analysis of water [186], milk [724] and other foodstuffs [527], and also to geological samples [329] [766] [767]. Beaty [92] found working with the boat to be particularly simple if only 10 µl sample volume were used, since the drying stage could then be omitted and at the same time better precision and increased lifetime of the tantalum boat obtained.

A little later, Delves [292] advanced a modification of the boat technique, which he had developed specially for the determination of lead in blood. The long tantalum boat was replaced with a small round nickel cup, and to increase the sensitivity an open tube (of nickel or, better, quartz) was mounted over this. The beam path passed through the tube and the element of interest was atomized into the tube. This tube increased the residence period of the atoms in the beam path (figure 21). The blood samples were dried in the nickel cup on a hot-plate and chemically treated before introduction into the flame. Many authors have checked [359] and modified [85] [318] [357] [1056] this procedure for the determination

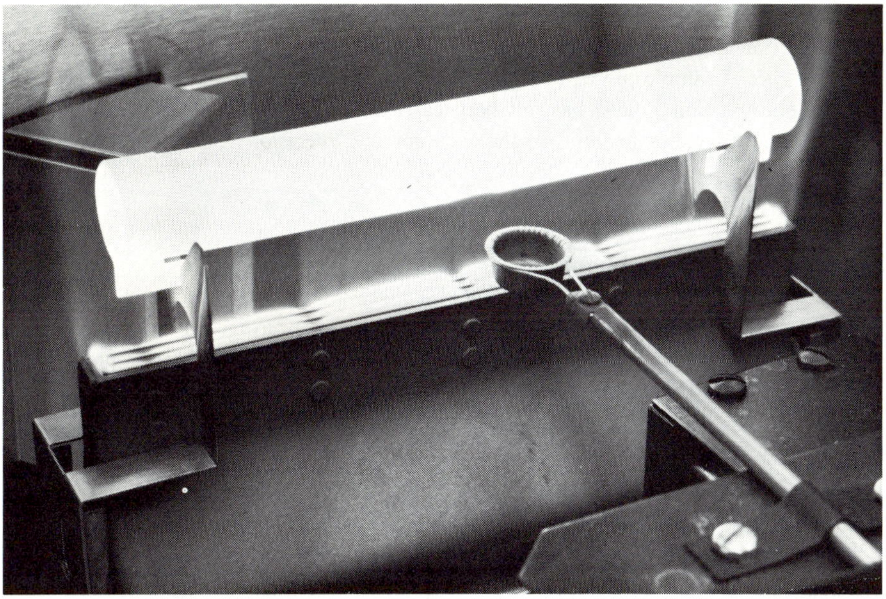

Figure 21. DELVES system. The sample is introduced into the flame of a three-slot burner in a nickel cup. The atoms pass through a hole into the quartz tube which is mounted above the burner directly in the beam path of the atomic absorption spectrophotometer.

of lead in small blood samples and found that it was especially suitable for screening tests. Later, the DELVES technique was also applied to the analysis of milk [456] and paint [502], and further, to the determination of cadmium in blood [218] [319] [592] and other biological materials [1120].

A further modification of the DELVES system has been described by CERNIK and SAYERS [219], the blood sample is absorbed on a filter paper, and after drying, a piece is punched out and placed into the nickel cup for introduction into the flame. CERNIK reported the simplicity and reliability of this method [217], and that the signal is largely independent of alterations in the cup material. Other authors have been able to verify this finding [591] [593].

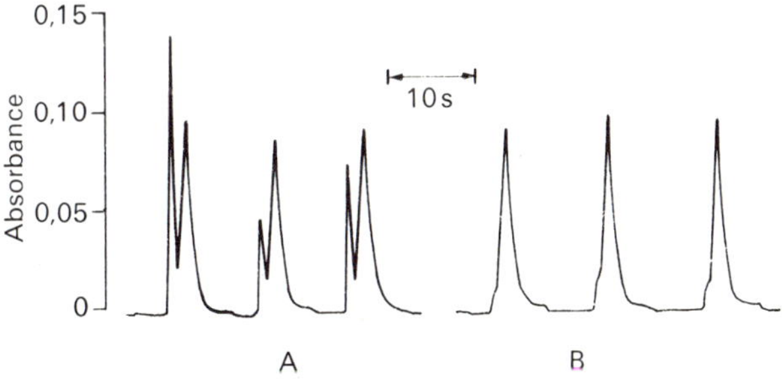

Figure 22. Determination of lead in blood with the DELVES system.—**A**: Without background correction a poorly reproduced signal appears before the lead signal. **B**: With background correction only the element-specific lead signal is obtained.

It is important to realize that with the boat technique or with the DELVES modification, considerable non-specific signals can occur during the atomization owing to the formation of smoke if special sample preparation [85] [592] or extraction of the element under study with an organic solvent [1056] has not been carried out before the determination. The use of a background corrector to improve the accuracy and precision is therefore recommended for this procedure (figure 22).

3.2. Atomization without Flames

In addition to the flame, various electrically heated furnaces have been described in recent years; a fixed sample volume is introduced into the furnace, and after thermal pre-treatment, is rapidly atomized. This results in a high, time-dependent signal whose height or area is proportional to the quantity of the element under study. Apart from graphite furnaces, which are the most frequently employed, atomization with heated metal bands, in the discharge of a hollow cathode, and with lasers have been described.
A special method has also been developed for the determination of mercury based on the fact that it is the only metallic element exhibiting distinct vapour pressure at room temperature.

In the following sections, atomization in graphite furnaces is treated in some detail since this method has found the greatest distribution of all the flameless procedures, and furthermore, the mechanism has been the longest and most thoroughly investigated. It must be pointed out, however, that the processes taking place during atomization in a graphite furnace are nowhere yet as clear as those in a flame. The following statements should not be regarded as unbreakable rules, but rather as an attempt to clarify the present state of knowledge.

3.2.1. Atomization in graphite tube furnaces

If a liquid, a dissolved, or even a solid sample is placed into a graphite tube, an attempt is first made by raising the temperature, either stepwise or uniformly, to remove as much of the accompanying material as possible before the element under study is thermally dissociated into atoms by a final rapid increase in the temperature. The first major difference to atomization in the flame is that a portion of the sample (e. g. the solvent) is separated before the atomization of the element takes place and thus cannot have an interfering effect as in the flame.
As will be shown later, the more successful the separation of the accompanying substances in a graphite tube furnace, the freer from interferences the determination will be.
A further advantage is the inert gas atmosphere in which atomization in graphite tube furnaces takes place. This is further enhanced by the strongly reducing properties of glowing carbon. While the risk always exists that the atoms produced by thermal dissociation can form compounds with the highly reactive components of the flame (e.g. oxygen atoms, hydroxide radicals) and thereby be lost to the determination, in the graphite tube the inert, reducing atmosphere promotes the atomization. Finally, the direct introduction of fixed sample volumes into the graphite tube (e. g. with a micropipette) is a great advantage in many cases since the physical properties of the sample such as viscosity, surface tension, density, etc., become insignificant,

while in the flame they can influence the nebulization process considerably. Even solid samples can be introduced into the graphite tube, although this is not always without its problems.

The atomization processes taking place in the glowing graphite tube will now be treated in more detail. Certainly the most thorough theoretical and practical investigations have been carried out by L'vov [8], who has been occupied with this question since 1959.

For the following considerations, the question of how the sample to be analysed is distributed over the surface of the graphite is of importance. A simple calculation shows that the space requirements of the element under study is in the range 0.0001 mm^2 to 10 mm^2 according to the nature and quantity of the sample. In every case this is considerably less than the surface of the graphite tube. If it is assumed that there is an even distribution of the element over the area wetted by the sample solution, then a monoatomic or monomolecular layer must exist. Even when small heaps of atoms or molecules (crystals) are formed during the drying process, which is likely, there is still always good contact of the element with the graphite surface. This may not be the case when solid samples are introduced directly into the tube or when highly concentrated salt solutions are pipetted, but this will be fully discussed later.

If the atomization itself is observed, it can emanate from either molecules or atoms, according to the nature of the sample and the behaviour of the element under study. If it emanates from molecules, the atomization can be either a simple thermal decomposition (dissociation) of a compound, or the reduction of an oxide on the glowing graphite surface. The difference between these two mechanisms is solely the active participation of the tube material graphite on the dissociation of the sample molecules, and frequently a clear differentiation is not simple.

If atomization emanates from the metal, it can be either desorption or vaporization. Because of the small quantity and wide distribution of the sample, boiling as the atomization mechanism can be excluded by deduction.

If a monoatomic layer exists after drying of the sample, i.e. every atom is isolated from the next, the atomization will be a desorption from the graphite surface and will be determined by the adsorption isotherm. It must be directly stated, however, that this mechanism is very unlikely and that up to now no indication of such an atomization could be found. If, on the other hand, the atoms are present as small heaps (e.g. crystals), which can be assumed as probable, then the atomization is a pure vaporization. The vaporization temperature can lie considerably below the boiling point or even below the melting point. Because of the extremely small sample amount and the correspondingly low partial pressure of the atom cloud which is formed, quantitative vaporization can take place even at very low vapour pressures. According to the kinetic gas theory, the mass G which the sample looses per unit time and area when it is heated to temperature T is given by equation (3.1).

$$G = P\sqrt{\frac{M}{2\pi RT}}\ [\mathrm{g/cm^2 \cdot s}]\,, \tag{3.1}$$

where P is the saturated vapour pressure at temperature T and M is the molecular weight. Nevertheless, it should be taken into account that equation (3.1) is only valid for atomization in a vacuum and that in the presence of a gas which does not react with the sample it can be up to two orders of magnitude lower.

The vapour pressure at which an element can be atomized within 0.1 second (a requirement that will be explained later) may be quickly calculated. If it is assumed that a sample amount of 10 ng with a mean atomic weight of 50 is distributed over an area of 1 mm^2 at an atomization temperature of 3000 K, from equation (3.1) it can be calculated that the pressure P is about 0.1 Torr. For most elements, the saturated vapour pressure at 3000 K is greater than 0.1 Torr, however, with the exception of tantalum, tungsten and rhenium, and a few others that form carbides with extremely low vapour pressures such as hafnium, niobium, thorium and zirconium. The majority of the other elements can be atomized either in less than 0.1 second or at temperatures below 3000 K.

Under the given pre-requirements, an element can be atomized within about 0.1 second when the saturated vapour pressure is about 0.1 Torr. The atomization can already begin at considerably lower vapour pressures (i.e. temperatures), but the atomization time is then correspondingly longer. This consideration is important, because during the thermal pre-treatment the sample may not be heated above the point at which atomization just begins since losses of the element under study will otherwise take place. Furthermore, it can be deduced that the time it takes to reach the atomization temperature also plays an important role. With a slow rate of heating, a considerable portion of the sample will have vaporized before the actual atomization temperature is reached (i.e. the atomization takes place slowly), because the vaporization will take place within the specified time of 0.1 second only with an infinitely fast increase of the atomization temperature.

The atomization time is of the greatest importance; the maximum density of the cloud of atoms can be achieved only when the atomization time is shorter than the residence period of the atoms in the graphite tube. As will be explained in the following section dealing with the advantages of graphite tube furnaces, the residence period of the atoms in the graphite tube is only a few tenths of a second, so that an atomization time of 0.1 second is absolutely essential. L'VOV has shown that it really is possible to achieve the maximum atom cloud density and thereby the highest sensitivity. He has also pointed out that atomization in the graphite tube is around a thousand times slower than in the flame, and that this is the reason for the relative freedom of determinations in the graphite tube from interferences since refractory substances have more time to decompose.

The actual atomization process will now be considered again; i.e. the discussion on the mechanism of atomization in the graphite tube and the question whether it emanates from metal or from molecules, and how far the glowing graphite takes an active part in the atomization. For the experimental clarification of this question, a system of measurements, which has been tested in the author's laboratory for a large number of elements in the presence of various anions, can be employed. Double curves are established in which the signal height is plotted versus the temperature. In the first curve the height of the signal at the optimum atomization temperature is plotted versus the pre-treatment temperature as the variable, and in the second the signal height is plotted versus the temperature of atomization where this is variable (figure 23). The first curve shows the temperature at which a sample can be thermally pre-treated without loss of the element under study. Furthermore, it is possible to derive the temperature at which the element is quantitatively vaporized within a given time from the first curve. From the course of the curve between these two points it is sometimes possible to draw conclusions on conversion reactions in the sample. From the second curve, the temperature at which the atomization first takes place and also the optimum atomization temperature at which the maximum atom cloud is achieved can be read off.

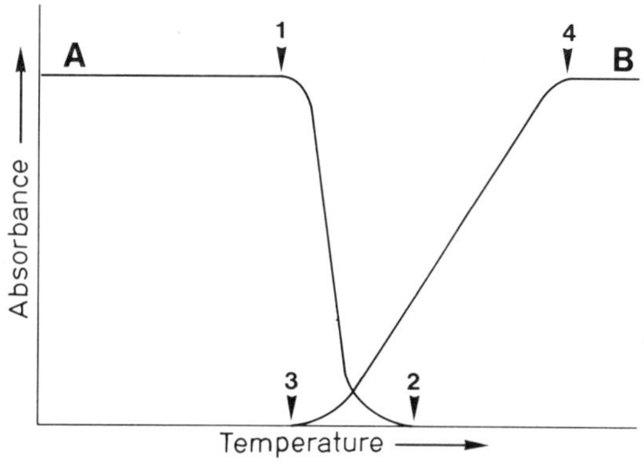

Figure 23. Decomposition/atomization curves in the graphite tube furnace.—In the decomposition curve **A**, the absorbance is measured at the optimum atomization temperature (**4**) with varying pre-treatment temperatures and plotted against these pre-treatment temperatures. **1** shows the highest temperature to which the matrix can be thermally pre-treated without loss of the element under study taking place. **2** is the lowest temperature at which the element under study can be quantitatively atomized (or volatilized). The atomization curve **B** shows the absorbance in dependence on the atomization temperature. **3** is the temperature at which the first atomization signal is observed and **4** is the optimum atomization temperature.

If physical data such as melting point, boiling point and decomposition point of the element are entered into these experimentally determined curves, it is frequently possible to draw conclusions about the atomization mechanism. This is a purely empirical method, but it has led to very useful results [1297]. In the following, a number of examples will be mentioned which give indications of differing mechanisms.

Figures 24 and 25 show the decomposition/atomization curves for gold and iron respectively. In both examples, the first losses during the thermal pre-treatment take place at 100 to 200 °C below the melting points of the respective elements, and at the same time a small atomization signal can be measured. Both curves are independent of the salt used or the acid added. This finding, which incidently is very similar for all the noble metals examined and various other elements, is a clear indication that the atomization emanates from the metal. This is also in agreement with the earlier consideration that with the very small amounts of metal to be atomized, very low partial pressures are already sufficient for atomization and this can take place below the melting point.

This however means that the sample must be reduced to metal early in the thermal pre-treatment phase, as assumed by FULLER [393].

According to this, the difference between the previously discussed mechanisms is solely the time point at which reduction or decomposition takes place. If the decomposition temperature of the salt (with or without the participation of the hot graphite) is clearly lower than the vaporization temperature, the sample will be reduced to metal during the thermal pre-treatment and the atomization is then a pure vaporization. If, on the other hand, the decomposition

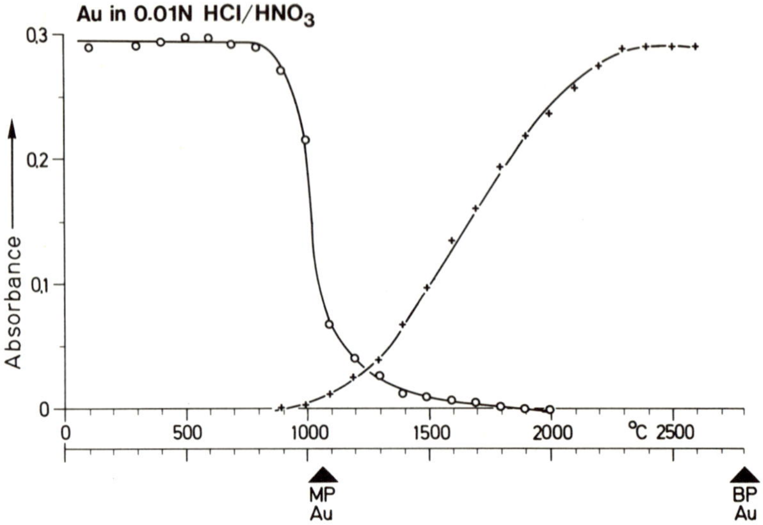

Figure 24. Decomposition/atomization curve for gold.—The first losses and the start of the atomization lie somewhat below the melting point of gold. The atomization emanates from metal.

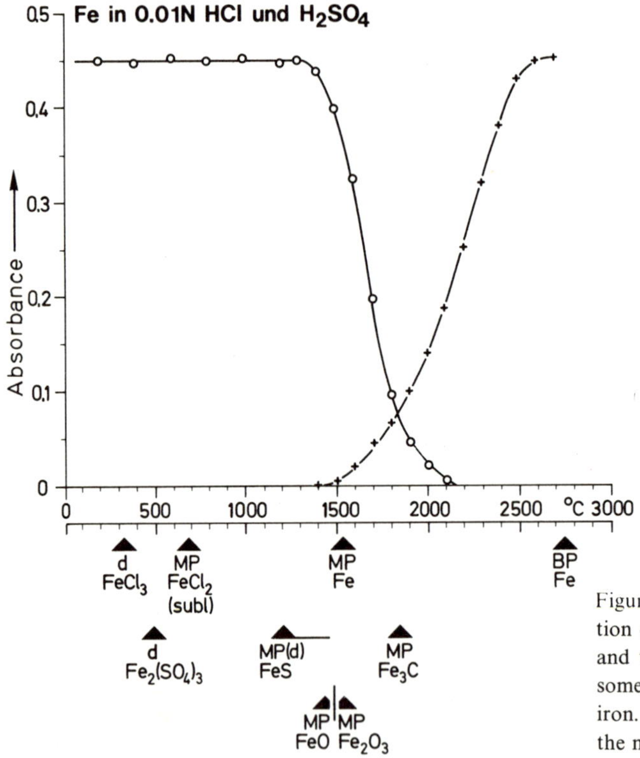

Figure 25. Decomposition/atomization curve for iron.—The first losses and the start of the atomization lie somewhat below the melting point of iron. The atomization emanates from the metal.

temperature of the compound is higher than the vaporization temperature of the metal, the decomposition temperature is simultaneously the atomization temperature.

An example of the second mechanism is the decomposition/atomization curve for antimony in sulphuric acid solution shown in figure 26. The point at which the first losses occur and at which the atomization signal can be simultaneously observed is virtually identical to the decomposition point of antimony(III) sulphide. This is obviously formed during the thermal pre-treatment by reduction of the sulphate on the hot graphite.

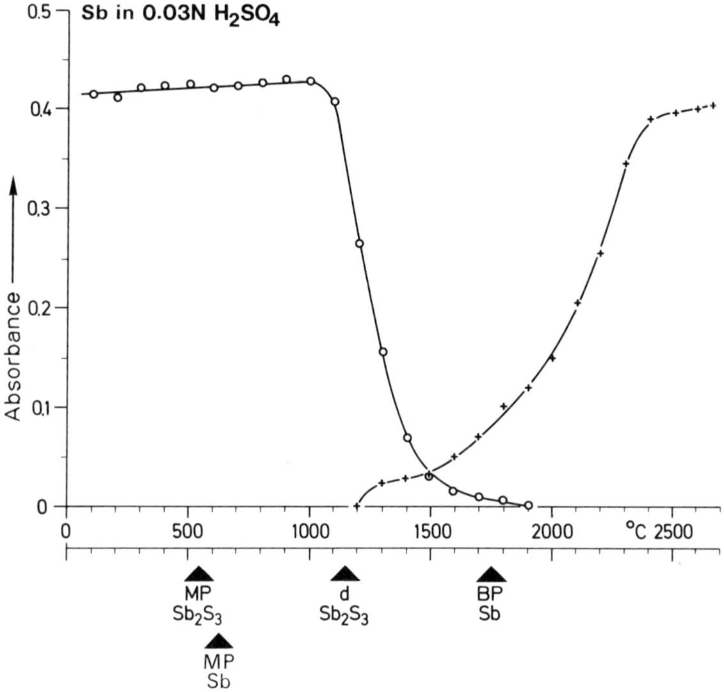

Figure 26. Decomposition/atomization curve for antimony in sulphuric acid. The first losses and the start of the atomization coincide with the decomposition point of antimony(III) sulphide.

A particularly interesting curve sequence is exhibited by the element beryllium. If this element is determined in acid solution (figure 27), even at relatively low temperatures there is a fall-off in the decomposition curve, but no atomization signal can be measured. The point at which the first losses occur coincides with the decomposition point of beryllium sulphate or the boiling point of beryllium chloride respectively. Since no atomization can be measured at the same time, the beryllium must be transported from the tube in a molecular form and hence lost to the subsequent determination.

By increasing the temperature further, the decomposition curve reaches a plateau, i.e. with further increase in the temperature, no further loss of beryllium takes place. This indicates the formation of a stable compound. The renewed fall-off in the curve and the simultaneous appearance of the first atomization signal coincide with the decomposition point of beryllium carbide. It is then clear that at the first decomposition point, beryllium sulphate forms beryllium carbide and a volatile compound, which has yet to be identified.

The curve sequence is completely different when an alkaline beryllium solution is used (i.e. the hydroxide is present). This is shown in figure 28. Beryllium oxide, which is thermally very stable, is formed during the thermal pre-treatment and can therefore be more strongly heated in the graphite tube. In this case, the atomization clearly appears to be a reduction of the oxide on the glowing graphite surface. A very similar curve sequence is also to be found for such elements as aluminium and silicon. They can only be satisfactorily determined if the oxide is formed during the thermal pre-treatment.

CAMPBELL and co-workers [205] have given the reduction mechanism

$$MeO + C \rightarrow Me + CO$$

as a requirement for many metals. They calculated the free energy for the corresponding reaction and compared the values determined thermodynamically with the temperatures at which the first atomization signal appears and found good agreement. From the above observations, however, this oxide reduction is not the only atomization mechanism coming into question.

A further indication that the active participation of the hot graphite is not in all cases necessary is shown by experiments on the direct determination of easily atomized elements in solid rock samples [768] [1295]. A few milligrams of the powdered rock are placed into the graphite

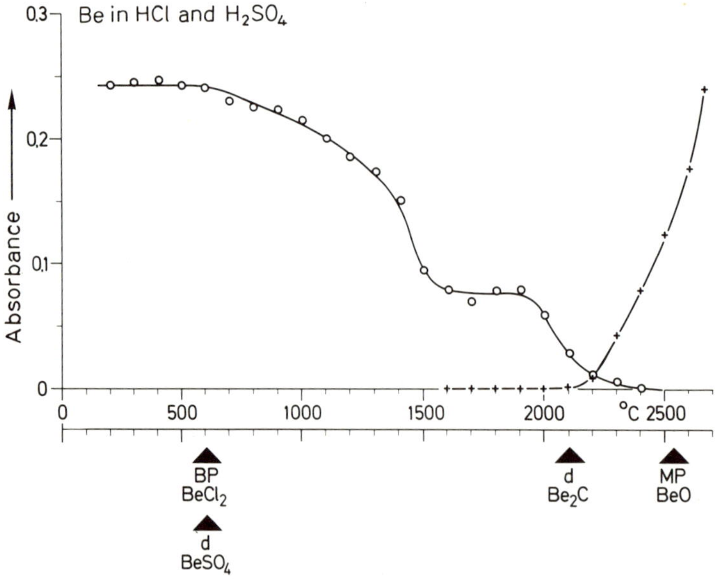

Figure 27. Decomposition/atomization curve for beryllium in acid solution.—The first losses start at the boiling point of the chloride or the decomposition point of the sulphate respectively. The first atomization signals are obtained only at the decomposition point of the carbide.

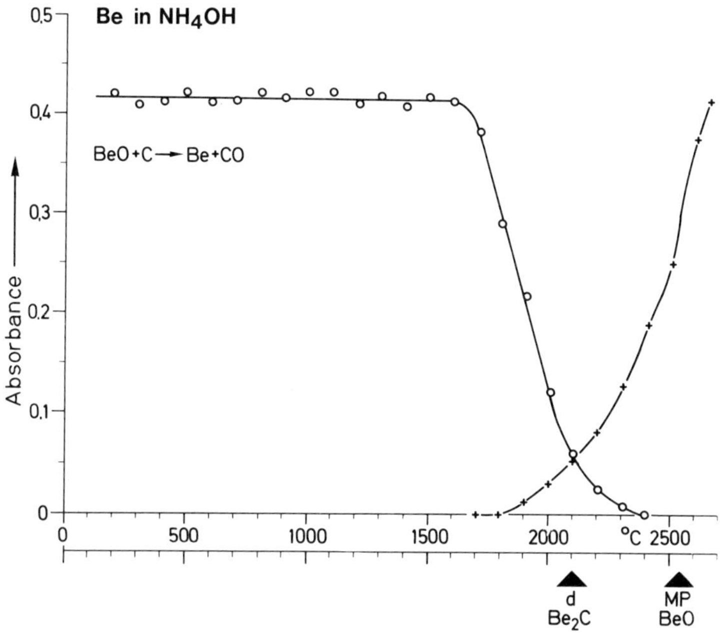

Figure 28. Decomposition/atomization curve of beryllium in ammoniacal solution.—The first losses and the start of the atomization fall together, but have no relation with the transition point of a compound. Here, the atomization is a reduction of the oxide on the hot graphite.

tube and heated to around 1400–1700 °C. In this procedure, the element under study (e.g. cadmium, lead, mercury, silver, thallium, zinc) can hardly come into direct contact with the graphite, nevertheless the first atomization signals can be recorded below the melting points of the elements. The sensitivities here, however, are considerably lower than when working with solutions.

Finally, the importance of good contact between the sample and the graphite surface must be mentioned, because it is necessary for perfect atomization and good sensitivity. Frequent mention has been made of the importance of atomizing the sample in about 0.1 second. This is only possible if it is in good contact with the glowing graphite surface. In the flame, the sample is always surrounded by the hot flame gases and therefore has the best "contact" with the temperature. In the graphite tube, the energy transfer can take place practically only from the hot graphite surface.

From this, the next requirement can immediately be drawn, i.e. that the sample should remain in good contact with the graphite tube as long as possible during heating. The longer the sample remains in contact with the graphite tube, the hotter it will be, and in the first approximation there is an exponential relationship between the rate of atomization and the temperature. A temperature increase of 100–200 °C corresponds to an increase in the rate of atomization by around one order of magnitude [8].

BRATZEL and CHAKRABARTI [157] have carried out temperature measurements on metal atom vapours in a graphite tube heated to around 3000 K and found 2080 K for indium and 2770 K for gallium. As expected, gallium, the element more difficult to atomize, had reached the higher temperature because it had been in contact with the graphite surface longer. Therefore, there is no temperature equilibrium between the graphite tube and the atom vapour, and the temperature of the inert gas is without doubt considerably lower than the temperature of the cloud of atoms.

3.2.2. Advantages of graphite tube furnaces

The height of the absorption signal measured in a graphite tube furnace is dependent upon the density of the cloud of atoms. For this reason, the statement was made in the last section that the atomization time must be considerably shorter than the residence time of the atoms in the graphite tube to enable the maximum atom cloud density to be achieved. Therefore, the factors influencing the residence period of the atoms will next be considered.
Loss of atom vapour from the graphite tube can be caused by:
1. A forced inert gas stream through the graphite tube
2. Convection near the hot walls of the graphite tube
3. Expulsion of excess atom vapour when there is insufficient space in the tube
4. Diffusion through openings in the tube and through the porous walls of the tube.

The inert gas stream through the graphite tube, which is necessary with an open furnace construction, can be shut off briefly during the atomization stage without causing disadvatages. The sensitivity can thereby be increased by a factor of ten compared to a graphite tube through which gas flows continuously [619] [1294]. With a horizontal furnace, the convection is small compared to the diffusion and can therefore be ignored in the first approximation. The expulsion of excess atom vapour from the graphite tube when insufficient space is available does not usually cause any problems. A graphite tube 5 cm long with a diameter of 8 mm offers sufficient space for the atoms which are produced from about 1 mg of solid substance (naturally the atomic weight and the temperature play a part). This means that the space filled by the element under study is many orders of magnitude smaller than the space available in the tube. The tube volume can occasionally be a problem when high matrix amounts must be atomized simultaneously with the element—a further reason for separating the matrix before the actual atomization.
The most important remaining factor is the loss of atoms from the tube due to diffusion. According to equation (3.2), the relative mass loss of atomic vapour

$$\frac{\mathrm{d}M}{M} = -\frac{8D}{l^2}\,d\tau \qquad (3.2)$$

is inversely proportional to the square of the tube length l and independent of its diameter (D=diffusion coefficient). This means that the length of the tube plays a decisive part in the residence period of the atoms in the beam path. From equation (3.2) by differentiation, the mean residence period τ of the atoms in the cell is directly proportional to the square of the length l of the graphite tube

$$\tau = \frac{l^2}{8D}\,. \qquad (3.3)$$

L'vov has calculated the diffusion of mercury from a 5 cm cell at 1500 K as an example, and found a mean residence period of 1.8 seconds. A value of 0.5 seconds was obtained from measurements, this is a much shorter residence period.
Through experiments with radioactive tracers, L'vov and Khartsyzov [781] were able to show that the diffusion of metal atoms through the walls of the graphite tube (in dependence on the temperature) could be greater than from the open tube ends. In the initial experiments, the inner surface of the graphite tube was covered with tantalum foil which immediately prevented these losses. Later, L'vov found that tubes of pyrolitic carbon (produced by pyrolizing hydrocarbons, e.g. methane, at 2000°C), or tubes that were coated with pyrolitic carbon, had the same properties. With these coated tubes, doubled sensitivity could be obtained for a number of elements because of the prevention of diffusion losses. To obtain a continuous renewal of the pyrocarbon layer, Morrow and McElhaney [883] mixed 10% methane with the inert gas and found that among other things there was a clear increase in the lifetime of the tube.
The influence of the tube dimensions on the sensitivity and detection limit in flameless atomic absorption will now be examined in somewhat greater detail. It has already been shown that the mean residence period τ is proportional to the square of the tube length and that the maximum signal height can be achieved when τ is considerably greater than the atomization time. Through evaluation from the peak height, a maximum sensitivity is achieved which cannot be increased by further lengthening of the graphite tube. With integration of the peak area, on the other hand, the sensitivity can be further increased proportionally to the square of the tube length by increasing the length of the tube, since here the sensitivity is only dependent on the residence period. The tube length cannot be increased at random, however, because with increasing length the current consumption also increases, so that tubes more than 10 cm long can no longer be used. As shown experimentally, the optimum length is between 3 cm and 5 cm [8] [882] [1294]; although, in principle the optimum tube length for every element and every temperature is different.
Not only the tube length, but also the tube diameter plays a certain role. Admittedly it does not influence the diffusion and thereby the residence period of the atoms in the tube, but since the volume increases with the square of the diameter, the cloud of atoms will be correspondingly diluted. In other words, the absolute sensitivity in a graphite tube furnace is inversely proportional to the tube diameter. Therefore, a long narrow tube is required for a good absolute sensitivity.
However, since a larger graphite tube can normally contain a larger sample volume, the relative sensitivity (i.e. the sensitivity in concentration units) is independent of the tube diameter in the first approximation. The same proportionality is valid for the tube length as for the absolute sensitivity, i.e. the requirement for a long tube.
The fact that the tube is mounted in the beam path of an atomic absorption spectrophotometer is a further detail which must on no account be forgotten when the tube dimensions are under consideration. This means that the beam from the light source should pass as unhindered as possible through the graphite tube since stronger shadowing resulting from a tube cross-section which is too narrow can lead to increased noise and other difficulties [834]. However, the tube cross-section must also not be too large since the atom cloud is then poorly utilized. The optimum tube dimensions should conform to the light beam. A parallel light beam

should be encircled as exactly as possible, while with a beam that has a focal point in the sample chamber, the tube length l and the diameter d should obey the mathematical interrelationship shown in figure 29 [1303]

$$d = b + l \cdot \tan\alpha . \tag{3.4}$$

Summarized, the greatest advantage of flameless atomic absorption in graphite tube furnaces, namely the high sensitivity, is best obtained when a 3–5 cm long tube of dense graphite with a cross-section matched to the light beam is used. Furthermore, the atomization time must be short compared to the residence time of the atoms in the tube. This is best achieved by heating the sample as rapidly as possible. Since the atomization time of around 0.1 second

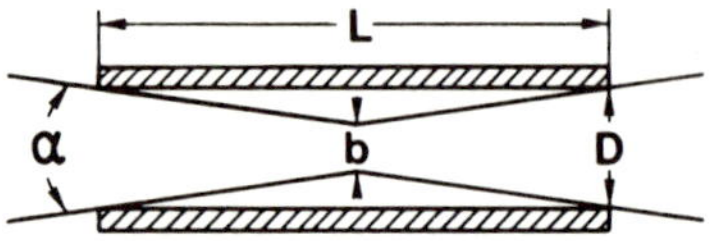

Figure 29. Optimum graphite tube dimensions.—With a beam path focussed in the sample chamber, the optimum length and diameter of a graphite tube are calculated according to $d = b + l \cdot \tan\alpha$.

is several orders of magnitude longer than with flame atomic absorption, refractory substances are more easily atomized. Furthermore, in the inert gas atmosphere, reactions in the gas phase are hardly to be expected, so that a determination largely free of interferences can be reckoned with.

Graphite tube furnaces then approach very closely to the requirements made in section 3.1.2 for the ideal absorption cell. Also, the high optical transparency, which makes working in the far UV range easier, should be mentioned as a further advantage [779] [780].

3.2.3. The various graphite tube furnaces

In the continuous attempts to avoid the disadvantages of the nebulizer-burner system, to achieve better detection limits, and to construct absorption cells offering a suitable environment for physical measurements, flameless atomization devices were designed and employed at an early stage in the development of atomic absorption spectroscopy.

In 1959, L'vov developed a graphite tube furnace [775–777] based on the work of King [641], it was 10 cm long and internally lined with tantalum foil. The sample was placed onto a graphite electrode, dried, and then introduced with the electrode through a hole into the graphite tube. The tube was heated by resistance heating and the sample was simultaneously atomized by a direct current arc. In an improved version [778], a short tube about 5 cm long made of pyrocarbon was used and the graphite tube itself acted as the counter electrode. In this way, the arc which atomizes the sample is formed the moment the electrode is introduced into the graphite tube (figure 30). The whole arrangement was mounted in a sealed argon chamber having quartz windows to allow the light to pass through, and it could be operated at increased or reduced pressures. With this arrangement, L'vov was able to obtain detection limits between 10^{-10} g and 10^{-14} g. These values are several orders of magnitude better than with flame atomic absorption.

MASSMANN constructed a simplified absorption cell (figure 31) which also consisted of a 5 cm long graphite tube but which was heated by applying a high current (500 A) at low voltage (10 V) to the ends of the tube [832] [833]. This resistance heating enabled a very fine gradation of the temperature to be made and thereby the selection of the optimum temperature conditions for the atomization of every individual element. MASSMANN's cell was not mounted in a closed argon chamber like that of L'VOV and therefore had to be continuously flushed with an inert gas stream to prevent the ingress of atmospheric oxygen. The sample (maximum 0.5 mg solid or 50 µl solution) was introduced into the graphite tube through a hole.

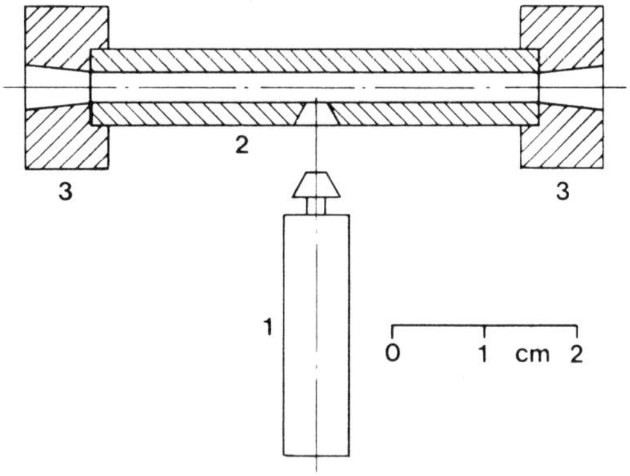

Figure 30. Flameless atomization according to L'VOV [788].—**1** Movable graphite electrode, **2** graphite tube, **3** contacts.

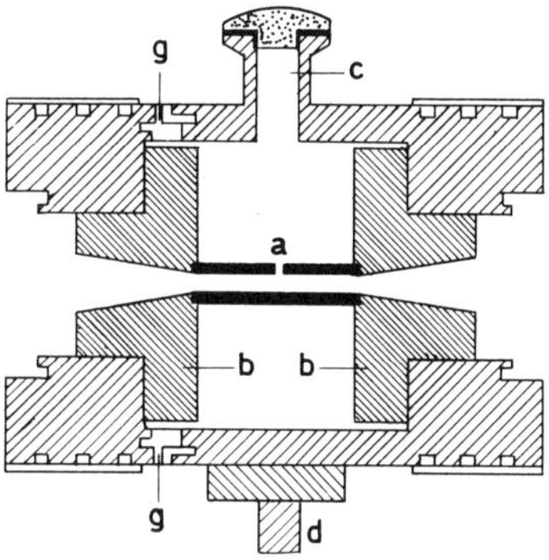

Figure 31. Argon chamber with graphite cell for flameless atomic absorption after MASSMANN [832].—**a** Graphite tube, **b** steel flanges, **c** opening for sample dispensing, **d** mount, **g** plastic insulation material.

MASSMANN achieved detection limits with this system that were typically ten times poorer than those quoted by L'VOV. The reason for this can be found in the forced inert gas flow through the tube which reduces the residence period of the atoms. Also, the heating speed of resistance heating is slower than that of atomization in an arc, so that the atomization time is longer.

For a large number of elements, MASSMANN observed background absorption to varying degrees in his cell [829] [830] which he eliminated by using a dual-channel spectrophotometer.

He also observed that the signal height was dependent on the nature and quantity of the matrix [831] and that this dependence was in part considerable.
WOODRIFF constructed a continuously heated graphite cell into which the sample was sprayed by means of a pneumatic [1351] or an ultrasonic [1353] nebulizer. He found that the graphite tube was attacked by aqueous solutions, so he used absolute methanol solutions exclusively for his investigations. The sensitivities obtained with this procedure were one to two orders of magnitude better than those obtainable in the flame; the detection limits, on the other hand, were not greatly improved because of the relatively high noise level. Later, WOODRIFF introduced the evaporated or collected sample into the heated graphite tube in a small graphite pan and found detection limits in the order of 10^{-10} g for fifteen elements [1122] [1352].
KOIRTYOHANN [684] found that the furnace described by WOODRIFF was relatively free from matrix and background interferences and that it offered a very stable cell for basic studies. Because of its size and complicated construction, however, it is hardly suitable for routine analyses.
WEST proposed a very simplified version of a graphite furnace. This did not use a graphite tube, but for completeness will be described [1311]. A carbon rod, 1–2 mm in diameter and 2 cm long, is clamped between two electrodes in a glass cylinder through which argon flows (figure 32). By slightly altering the glass jacket, the same arrangement can also be used for atomic fluorescence measurements. The sample (e.g. 1 µl) is placed onto the carbon rod by means of a micropipette or syringe. The rod is then brought to glowing by resistance heating (100 A, 5 V). An analysis takes 5–10 seconds and the next sample can be introduced after about two minutes. In a further version, the glass jacket is abandoned and the carbon rod is merely flushed with argon from beneath [1310]. Because of its ease of access, this open version is particularly simple to operate.

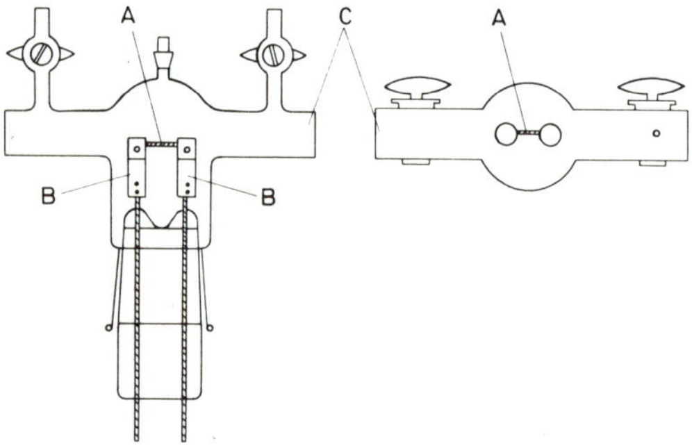

Figure 32. Carbon rod atomizer according to WEST [1311] for flameless atomic absorption. **A** Carbon rod, **B** contacts, **C** glass jacket.

With this simple system, WEST found detection limits of 10^{-10} g for silver and magnesium. Sample introduction is rather critical and requires some experience with this procedure. Furthermore, a large number of interferences have been found [51] [52] which can be traced to the steep temperature gradient between the carbon rod and the environment. This could be confirmed by experiments in which the light beam is focussed directly above the glowing carbon rod [52]. A clear increase in the interferences and a decrease of the sensitivity was observed with increasing distance from the rod.

The commercially available graphite furnaces are all based on the MASSMANN principle, i.e. they are tube furnaces which are heated by resistance heating. They differ from each other in the size of the graphite tube and also in their flexibility and operating convenience [835] [1294]. A detailed description of the individual commercial graphite furnces will not be given here, technical data compiled in the appendix can be referred to. The two most important systems, the Heated Graphite Atomizer HGA-74 and the "Mini-Massmann" CRA-63, are depicted in figures 33 and 34 respectively. A number of typical detection limits are compiled in table 5.

Figure 33. Heated Graphite Atomizer HGA-74 for flameless atomic absorption (Bodenseewerk Perkin-Elmer & Co. GmbH, Überlingen).

KOIRTYOHANN [684] has compared the two systems and found that the "Mini-Massmann" furnace has lower background absorption and higher absolute sensitivity, but it exhibits the strongest matrix effects and is limited to a smaller sample volume. The Heated Graphite Atomizer has the better relative sensitivity and the highest precision, but is subject to greater background absorption problems. MORROW and MCELHANEY [882] found that the small cross-section of the "Mini-Massmann" shadowed the light beam leading to increased noise. Furthermore, they had difficulty in pipetting amounts around 1 µl and found a poorer relative sensitivity. For these reasons, they used a tube of greater diameter with which they did not have these difficulties.

3.2.4. Possibilities of interference

The actual analytical treatment of interferences will first be discussed in chapter 6 "Technique". Only a few typical phenomena which can occur during atomization in a graphite tube will be discussed in this section. A differentiation can be made between interferences caused by the sample itself and those caused by the apparatus or the method.

The following are interferences caused by the sample: Volatilization of the element under study as molecules; hindrance of the atomization through chemical bonding or poor contact with the graphite surface; and non-specific light losses caused by vaporizing matrix material. Interfer-

Table 5. Absolute and relative detection limits in the Heated Graphite Atomizer compared to flame atomic absorption (measured with a PERKIN-ELMER Heated Graphite Atomizer HGA-72 and HGA-74)

Element	Detection limits, graphite tube furnace absolute (g)	relative (μg/l) (sample volume 100 μl)	Detection limits, flame (μg/l)
Ag	1×10^{-13}	0.001	2
Al	2×10^{-12}	0.02	20
As	1×10^{-11}	0.1	150
Au	1×10^{-11}	0.1	10
Ba	5×10^{-11}	0.5	10
Be	3×10^{-12}	0.03	2
Bi	2×10^{-11}	0.2	3
Ca	1×10^{-12}	0.01	1
Cd	1×10^{-13}	0.001	2
Co	5×10^{-12}	0.05	10
Cr	1×10^{-11}	0.1	3
Cu	2×10^{-12}	0.02	1
Fe	3×10^{-12}	0.03	10
Ga	1×10^{-11}	0.1	100
In	1×10^{-11}	0.1	20
Ir	1×10^{-9}	10	1000
K	1×10^{-13}	0.001	1
Li	1×10^{-11}	0.1	0.3
Mg	1×10^{-12}	0.01	0.1
Mn	2×10^{-13}	0.002	2
Mo	3×10^{-12}	0.03	20
Na	5×10^{-13}	0.005	0.2
Ni	1×10^{-11}	0.1	2
P	1×10^{-7}	1000	100000
Pb	2×10^{-12}	0.02	10
Pt	2×10^{-10}	2	100
Rh	1×10^{-10}	1	5
Sb	1×10^{-11}	0.1	40
Se	5×10^{-11}	0.5	200
Si	5×10^{-11}	0.5	20
Sn	1×10^{-11}	0.1	20
Sr	5×10^{-12}	0.05	2
Tc	5×10^{-9}	50	1500
Te	1×10^{-11}	0.1	50
Ti	2×10^{-9}	20	50
Tl	1×10^{-11}	0.1	20
V	1×10^{-10}	1	50
Yb	5×10^{-12}	0.05	5
Zn	5×10^{-14}	0.0005	1

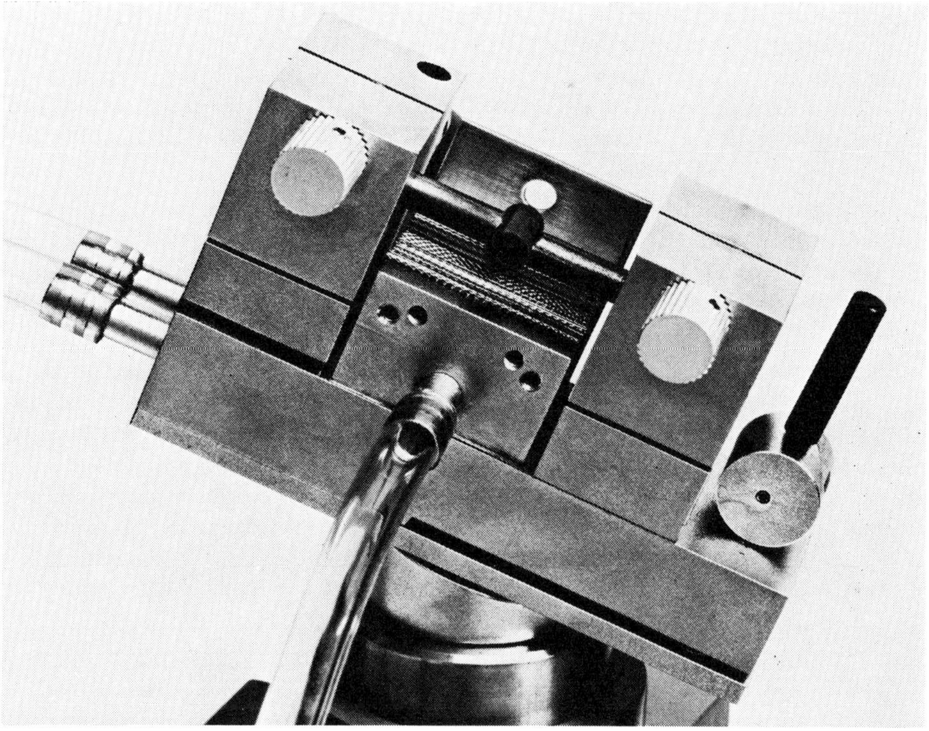

Figure 34. Carbon Rod Atomizer CRA-63 in the 'Mini-Massmann' version for flameless atomic absorption (by kind permission of VARIAN GmbH).

ences which are caused by the apparatus are: Carbide formation on the graphite surface; a reaction of the element under study with the inert gas; and too great direct light emission from the glowing graphite tube.

Since the gas space in a graphite tube furnace compared to atomization in the flame is relatively cool and chemically inert, there is virtually no chance for substances volatilized as molecules to be atomized. They are usually transported from the tube (by diffusion or inert gas flow) and so lost to the determination. The cause of this volatilization can lie either in the physical properties of the compound containing the element under study if this distils or sublimes at relatively low temperatures without decomposing, or in the composition of the matrix.

SEGAR and co-workers in particular [1107] [1109] have drawn attention to the danger of loss of trace elements in sea-water analysis owing to vaporization with sodium chloride and hence to a considerable loss of sensitivity. BINNEWIES and SCHÄFER [128] have recently reported a gaseous metal chloride complex of the type $M_xM'_yCl_2$ which can easily lead to such losses.

The causes for hindrance of the atomization can also lie either in the bonding state of the element under study, or in the matrix if this prevents good contact with the graphite surface.

In both cases the atomization time is increased so much that it is no longer short compared to the residence period, and a smaller signal height is the result. This can be seen most clearly when solids, or solutions with very high solids content, are analysed. Since salts and also finely divided carbon, such as is produced from the thermal destruction of biological compounds, have very poor thermal conductivities, the graphite tube temperature is only slowly transferred to the element under study.

A widely spread cause of interference in atomic absorption is the non-specific light losses caused by molecular absorption and scattering on solid particles. L'vov [8] established that background effects should first occur when the interfering substance is present in concentrations four to five orders of magnitude higher. With the very high sensitivity of the graphite tube technique, however, this is very rapidly reached.

Particular mention is made here of the fact that most of these interferences can largely be eliminated by careful programming of the temperature increase [88]. If losses of the element under study, owing to volatilization as molecules, cannot be prevented by chemical influences (see chapter 6), the pre-treatment temperatures can be so chosen that losses do not occur. On the other hand, careful separation of the interfering partner enables many interferences resulting from non-specific light losses to be avoided, or at least reduced. In this connection, a slow steady increase of the decomposition temperature instead of a stepwise increase (figure 35) for the thermal decoposition of complex matrixes has proved to be of value [1294].

Because of the importance of the programmed temperature increase, a few remarks must be added. After introduction of the sample, the solvent is usually evaporated during the first temperature stage. Temperature and time must be so determined that the solvent is quantitatively evaporated, but not too slowly and without spitting. Other higher boiling components, such as acids, which may be present must be removed by a second temperature stage in a similar manner. For further temperature pre-treatment, the point at which the element under study is just not measurably volatile in the given matrix is particularly important (see 3.1.2). Thermal pre-treatment can, and should, be taken up to this point if the matrix does not vaporize too violently at this temperature. For complicated samples consisting of many components with widely varying boiling, vaporization or decomposition points, the regular increase of the temperature already mentioned can be employed in all cases.

If the matrix can be largely removed by means of thermal pre-treatment, the element can normally be determined without interferences. If a separation cannot be carried out, or is only partially successful because of the high volatility of the element under study, other programme possibilities which reduce the interferences to an acceptable level can be chosen. Buttgereit [200] found that although the signal height is reduced with atomization at relatively low temperature and thereby lower sensitivity is achieved, the vaporization of the matrix is at the same time retarded and so displaced in time from the atomization. Additionally the non-specific light losses occur in a range in which they can easily be corrected by suitable means (see 6.2.5). Lundgren [770] developed a temperature regulated controller with a particularly rapid temperature increase for the graphite tube furnace and was able to atomize cadmium in the presence of 2% sodium chloride at 750 °C. At this temperature, virtually no sodium chloride vaporizes and hence the interference owing to non-specific light losses is negligible.

The *extent* of the interference resulting from non-specific light losses can also be caused in part by the apparatus. With the cell design according to MASSMANN, vaporized matrix products formed during the thermal pre-treatment are conducted over the relatively cool ends of the tube and can condense there. During the atomization stage, the ends of the tube are heated up so much that the condensed matrix material can be vaporized again and so cause non-specific light losses. This interference can be avoided if the tube is evenly heated over its complete length, or if a gas stream flows from the ends of the tube to the middle during the thermal pre-treatment and the decomposition products are transported out through the hottest position, namely the sample introduction hole (figure 36). Nevertheless it should be mentioned that only a part of the interferences can be removed by this means. If the vaporization point of the element under study is so low that the matrix is not noticeably volatile at that temperature, separation by thermal pre-treatment is not possible and other means must then be sought.

The formation of compounds by the element under study with the tube material or with the inert gas is one of the interferences caused by the apparatus. Mention has already been made that a number of elements such as hafnium, niobium, tantalum, thorium, tungsten and zirconium form such stable carbides that their determination in a graphite tube has not been possible up to the present and is also not likely to be possible in the future. A number of further elements that also form carbides can be determined in the graphite tube, but the sensitivity is relatively low and the signals tend towards tailing; that is, heating must be carried out for a prolonged period at high temperature until the element has been quantitatively removed from the tube (figure 37).
One possibility for removing this interference is to avoid contact between the carbide forming element and the graphite tube. RENSHAW [1032] introduced his sample into the graphite tube furnace in a tantalum boat and was thereby able to clearly increase the sensitivity for barium. Later, he found that the sensitivity for this element could be increased by a factor.

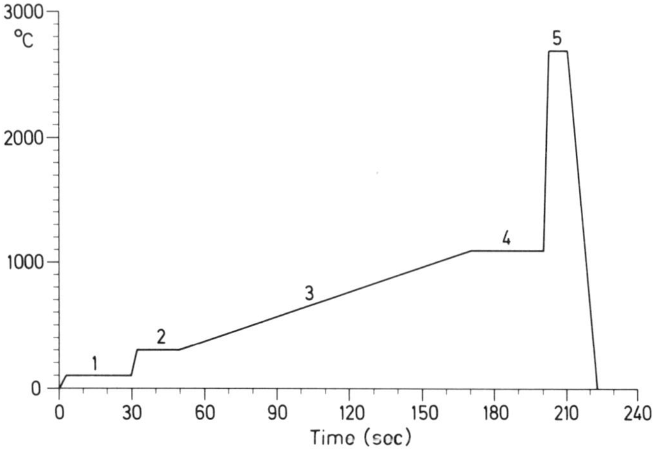

Figure 35. Temperature programming for the pre-treatment of complex matrixes.— After the sample has been dried **1**, the temperature is increased rapidly to point **2** at which the matrix just begins to distill, sublime or decompose. The temperature is then increased slowly at rate **3**, which provides an efficient but not too rapid removal or decomposition of the matrix, to temperature **4** at which the element under study is not yet volatile. After a short isothermal period the temperature is raised rapidly to the atomization temperature **5**.

of twenty when a normal graphite tube was lined with tantalum foil. DAGNALL [249] reported that with tubes saturated with tungsten, the sensitivity for molybdenum is 70% higher and better precision is obtained. KUZOVLEV and co-workers [719] have successfully determined boron in a graphite tube coated with a mixed carbide of tantalum and niobium.

Since an inert gas normally flows around or through the graphite tube, reactions with the gas are seldom and interferences are hardly to be expected. MANNING and FERNANDEZ [802] have reported that the determination of aluminium is three times more sensitive with argon as inert gas than with nitrogen, which would result in the formation of the nitride. CERNIK [216] found better precision for the determination of lead when he used argon as inert gas instead of nitrogen.

AMOS and co-workers [46] found that for the determination of lead with a heated carbon rod, fewer interferences occurred if hydrogen was used as the inert gas. BRATZEL and CHAKRABARTI [157] found, on the other hand, that with a graphite tube furnace the hydrogen took up too much heat from the cloud of atoms, leading to increased interferences.

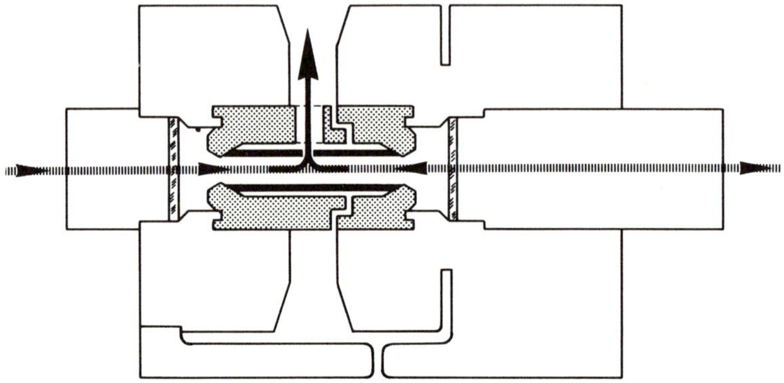

Figure 36. Internal gas flows into the Heated Graphite Atomizer HGA-74 (simplified). The inert gas stream enters at the graphite tube ends and escapes through the sample introduction hole. In this manner the matrix is transported, in the shortest way, from the graphite tube during the thermal decomposition stage and cannot condense on cool places. During the atomization the internal gas flow is shut off or reduced.

As the last possibility of interferences, the intense direct light emission from the graphite tube at high temperatures should be mentioned. Certainly atomic absorption spectrophotometers with their alternating light systems are relatively insensitive to direct light, nevertheless, as will be shown in section 5.2, a high direct light increment can lead to increased noise, or even to saturation of the photomultiplier. According to the extent of the interference, it can lead to a worsening of the precision and the detection limit, or even make a determination impossible. This interference can be avoided by preventing too much direct light from reaching the detector. A pre-requirement for this is the perfect adjustment of the graphite tube in the beam path. MASSMANN [834] has furthermore established that certain optical and electronic requirements in atomic absorption spectrophotometers must be fulfilled for perfect operation with graphite tube furnaces. It is certain that matching the beam path as exactly as possible to the form of the graphite tube, or the reverse, is of advantage. KERBER and co-workers [638] have shown that by means of a suitable diaphragm system and through the choice of a fitting photomultipier, the influence of the emission from the graphite tube can be reduced by orders of magnitude. Accordingly, considerably better precision and detection limits can be obtained with such modified instruments.

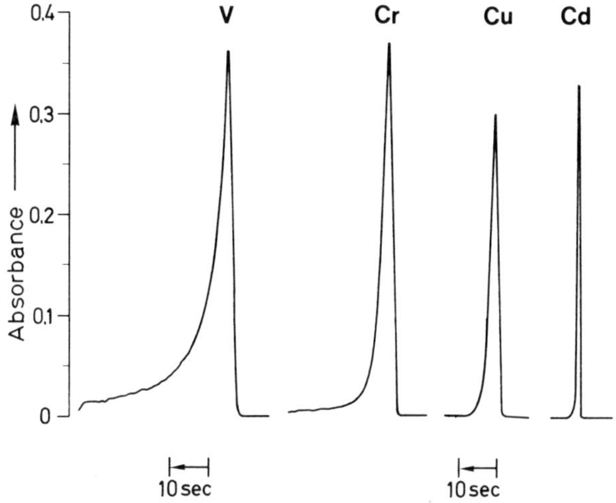

Figure 37. Influence of carbide formation on the atomization signal. While elements such as cadmium and copper exhibit a rapid signal increase and also a rapid signal fall and return to the baseline in less than 10 seconds, carbide formation leads to long tailing which is especially noticeable with vanadium.

3.2.5. Other flameless atomization devices

Although graphite tube furnaces have had the greatest impact in flameless atomization in the last few years because they come closest to the requirements for routine analyses, there are other systems that have had a certain success. A good review over means of flameless atomization has been published by KIRKBRIGHT [653]. In 1960, GATEHOUSE and WALSH [406] reported a demountable vaporization chamber operating on the principle of a hollow cathode lamp in which the sample was vaporized by a glow discharge in a cylinder open at both ends. Later, WALSH reported the determination of phosphorus in copper, and silicon in aluminium and steel by this method [1283]. GOLEB carried out isotope analyses in a similar vaporization chamber with good success [424] [427]. Recently, WALSH described experiments on multielement analyses using this vaporization chamber as the means of atomization [1284]. GOUGH and co-workers examined the cathodic vaporization of metals in a glow discharge and determined a number of elements in standard alloys by atomic fluorescence [433]. This procedure could be used equally well for atomic absorption [1286].

LOFTIN and co-workers conducted air over graphite rods which were brought to glowing in a high frequency field and then determined lead by atomic absorption in an attached heated quartz tube [760]. MORRISON and TALMI used a similar device for the direct analysis of solids [881]. MARINKOVIC and VICKERS investigated the possiblity of atomization in a direct current arc [822] and found that for difficultly atomized elements, such as boron and tungsten, the sensitivity was better than in a flame.

LANGMYHR and THOMASSEN heated a graphite tube by means of a high frequency induction generator and determined rubidium and cesium in standard silicate rocks directly from the solid sample [731]. The temperature of 2200 °C which could be achieved was relatively low and the heating up speed relatively slow, but in this case it appeared to offer certain advantages since the heat transfer time to the sample is small compared to the total heat-up time.

ROUSSELET and co-workers investigated atomization by means of electron bombardment [1062]. In theory, this method offers two advantages; very high temperatures can be achieved (higher than 3400 °C, the melting point of tungsten) and a partial pressure of electrons advantageous for the production of atoms in the ground state can be obtained. The sample in the range 10–100 nl is placed on a tungsten target mounted under the beam path and is heated under high vacuum by bombardment with accelerated electrons produced in a strong electric field (2 kV/cm). The analytical advantage is the use of very small sample amounts combined with a very good absolute sensitivity. In 1967, MOSSOTTI and co-workers reported the use of lasers for atomization [886], an idea which was later used by VULFSON [1278]. The main advantage with this procedure is likewise the small sample requirement (10–100 µg) and the excellent absolute detection limits (10^{-11} g for copper and silver, and 10^{-10} g for manganese). VULFSON and co-workers analysed various geological samples for trace elements by this procedure. DONEGA and BURGESS used an electrically heated boat made of tantalum or tungsten foil in a closed chamber under reduced pressure to determine many elements with detection limits down to 10^{-12} g [309]. This system is relatively simple, has a low current consumption and has the advantage that it requires less than 0.1 second to heat up to 2200 °C (so that the maximum atom cloud concentration can be formed).

This tantalum boat principle was later taken up by HWANG and co-workers and introduced in a slightly modified form as a commercially available accessory. HWANG obtained detection limits of 10^{-13} g, but had problems with difficultly atomized elements because the maximum achievable temperature was under 2400 °C. The use of hydrogen as inert gas was said to increase the lifetime of the tantalum boat and at the same time remove a number of interferences. Besides the analysis of simple aqueous solutions and organic extracts, this system has also been used for the direct analysis of lead in diluted blood [551].

However, various authors have reported a strong dependence of the signal height upon the anion present [334] or upon the acid in whose presence the analysis takes place [827]. These interferences, in a similar manner to the carbon rod of WEST, are probably due to the large temperature gradient between the heated boat and the surrounding atmosphere.

NIXON and co-workers [917] combined vaporization of a sample from a tantalum boat with plasma emission by conducting the vaporization products directly into the plasma. The detection limits were in the range 1 µg/l and less.

3.2.6. Flameless determination of mercury

The unique property of mercury—that it is the only metal exhibiting considerable vapour pressure at room temperature—and that it does not react readily with atomospheric oxygen have led many workers to make endeavours to determine this element flamelessly from quite an early stage. The relatively poor sensitivity of mercury in the flame and the ever increasing necessity of determining this element in very small traces are further stimulants to these endeavours.

In 1939, before the actual rediscovery of atomic absorption by WALSH, WOODSON described an apparatus for the determination of mercury in air [1354]. In a modified form, this apparatus has been used by other workers. BRANDENBERGER developed a procedure for the electrolytic deposition of mercury on a copper spiral with subsequent vaporization in a quartz cell (figure 38) by electrically heating the spiral [152] [153]. Later, besides the dynamic method in which the mercury is transported through the cell by a light gas steam, he developed a static procedure which was simpler to operate [154] [155]. In both cells absolute quantities of 2×10^{-10} g mercury can be determined. HINKLE and LEARNED [514] used a similar procedurc except that they deposited mercury by chemical means on a wire gauze which was then subsequently warmed.

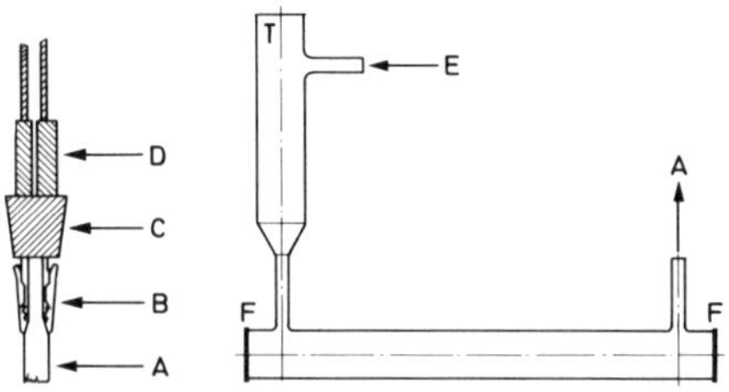

Figure 38. Apparatus for the determination of mercury according to BRANDENBERGER [152].—Left: Holder for the copper spiral, **A** copper spiral, **B** crocodile clip, **C** rubber bung, **D** banana plug. Right: Absorption vessel, **T** funnel to accept the copper spiral, **E** inlet nozzle for air, **A** extraction nozzle for air, **F** quartz windows.

The most successful flameless method for the determination of mercury was discovered by POLUEKTOV and VITKUN [995] [996]. During their investigations on the determination of mercury by atomic absorption, they discovered an unusually large increase in the absorbance, by one to two orders of magnitude, if tin(II) chloride was added to the sample being sprayed. This could be attributed to the reducing action of this reagent which ensured that virtually all the mercury being sprayed passed into the flame. Thereupon, they eliminated the nebulizer and flame, passed air through the sample after tin(II) chloride had been added and then conducted the air through a 30 cm quartz cell mounted in the beam path of an atomic absorption spectrophotometer. The detection limits obtained by this method amounted to 5×10^{-10} g mercury.

POLUEKTOV and co-workers were not the first to describe the reduction of mercury salts to metallic mercury with tin(II) chloride, but they were the first to use this reaction in combination with atomic absorption. HATCH and OTT [481] extended this method and employed it for the analysis of metals in rock and soil samples. Numerous authors have reported the use of this method, occasionally with slight modifications, for the determination of mercury in the most widely varying materials. Details on this will first be given in the practical section of this book.

KAHN [611] described the first commercial system based on the principle of POLUEKTOV and VITKUN (figure 39). Although this system is simple and widely in use, a number of considerations must be observed if false measurements are to be avoided. FRITZE and STUART have drawn attention to the possibility of mercury losses in the drying agent and in the hoses [388], while MANNING has indicated the possibility of interferences resulting from non-specific light losses. Various authors have attempted to avoid these difficulties by amalgamating the mercury formed by reduction with tin(II) chloride on gold foil [1249] [1262] or tin foil [621] and then vaporizing the mercury free of interferences. The simplest method for removing interferences is probably the use of a background corrector (see section 6.2.5).

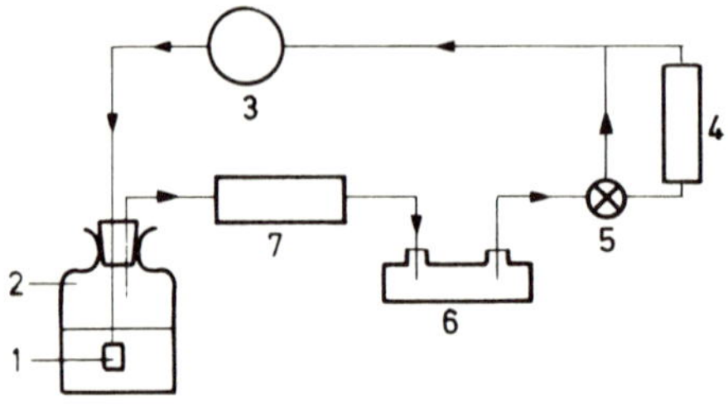

Figure 39. System for flameless mercury determination.—The metallic mercury released in the reaction flask **2** by reduction with tin(II) chloride is conducted through an absorption cell **6** by means of a circulatory pump **3**. **1** Bubbler, **4** absorption agent, **5** three-way valve, **7** drying agent.

3.3. Effectiveness of the Atomization

Despite the enormous advances that have been made in atomic absorption spectroscopy within the last decade, little is still known about the fate of metal atoms and the sample in the flame or graphite tube furnace. Yet it is exactly the processes in the flame, which influence the effectiveness of the atomization, that are of decisive importance for an understanding and the correct estimation of chemical interferences.

ROBINSON also calculated that of the 10^{15} copper ions contained in 1 ml of an 0.1 µg/ml copper solution, only about 10^{-3} are atomized when the solution is sprayed into an air/acetylene flame. This can be attributed to insufficient nebulization, to insufficient thermal dissociation, and to the fact that the flame is not chemically inert, as should be expected from an ideal absorption cell.

These observations are always referred to the total burner unit and are altogether too complex to permit a statement on the actual atomization process to be made. It is therefore certainly of no advantage to include the nebulization of the sample in the considerations. The strongly increasing interest in flameless atomization procedures with their considerably better achievable sensitivities in recent years has re-awakened the question of the effectiveness of the atomization with and without flames.

DE GALAN and WINEFORDNER [285] calculated the portion of free atoms formed in an air/acetylene flame for 22 elements. They found that several elements, such as copper and sodium, were virtually quantitatively atomized. This calculation cannot be extended to all elements however, since it requires a knowledge of the oscillator strength f (see section 1.5). The latter is not known for the majority of elements. WILLIS [1335] verified these findings through independent measurements and mentioned that with the previous calculations, which had led to too low atomization rates, important points had been omitted. KOIRTYOHANN [690] therefore attempted to determine the effectiveness of the atomization experimentally by setting the atomic absorption and the molecular emission in a relationship to each other.

He assumed that when using a mixing chamber burner and a hot flame, the portion of the sample reaching the flame was quantitatively vaporized; i.e. converted into atoms and molecules.
If the portion of free atoms is β and the portion of bound atoms γ, and the above assumption is true, then

$$\beta+\gamma = 1 \tag{3.5}$$

and

$$\frac{A_1}{A_2}\beta+\frac{E_1}{E_2}\gamma = 1\,, \tag{3.6}$$

where A_1 and A_2 are the absorbance values of the atoms of interest and E_1 and E_2 are the emission intensities of the molecules formed in fuel gas weak and fuel gas rich flames respectively.
The relationships are especially easy to obtain when only one molecular species is formed in the flame, the monoxide for example. This is the case for a large number of the rare earth metals in the nitrous oxide/acetylene flame, and KOIRTYOHANN obtained values for these and other elements that were in good agreement with the calculated values. FASSEL and co-workers [350] have especially drawn to attention that the free atoms of metals which form stable monoxides are localized to a high degree in the nitrous oxide/acetylene flame. The total profile of the absorbance observed over the height of the flame alters with the stochiometry of the flame and is in direct connection with the stability of the monoxide. This indicates the importance of this compound for the formation of free atoms. The β values are then only meaningful dimensions for many elements when they are referred to a particular point in the flame and to a determined stochiometry.
RASMUSON and co-workers [1023] found that in a nitrous oxide/acetylene flame there is a determined flow ratio ϱ at which the atomization is complete when the metal forms a monoxide with a dissociation energy of less than 6.5 eV (on condition that the sample is completely vaporized). Elements belonging to this type for example, are aluminium, beryllium, copper, iron, lithium, magnesium and sodium. SMYLY and co-workers, on the other hand, found that in a separated oxygen/hydrogen flame considerably worse values were obtained [1158], which again shows that the chemical environment, and more especially the partial pressure of oxygen, play a decisive role in the atomization processes.
Accordingly it has been shown that for a large number of elements, by choosing a suitable flame and suitable stochiometry, practically quantitative atomization can be achieved. The actual limiting factors are the rate of aspiration and the effectiveness of the pneumatic nebulizer. This can only produce a relatively limited quantity of the finest aerosol per unit time which can be quantitatively vaporized in the flame.
The disadvantage of insufficient nebulization is avoided in the graphite tube furnaces because the sample is introduced directly into the tube. For this reason alone the sensitivity should be around an order of magnitude better. Furthermore, WINEFORDNER has calculated that as a result of the increased residence period of the atoms in the beam path in the graphite

tube for the same rate of atomization as in the flame, the determination in the former must be more sensitive by a factor of at least 600 [1341]. KOIRTYOHANN also found that in a graphite tube a theoretical increase in the sensitivity by a factor of 10000 should be possible [684]. However, according to the investigations carried out to date, the degree of atomization in a graphite tube appears to be at least an order of magnitude poorer than in the flame. This is very probably due to the fact that the atomization takes place exclusively on the graphite surface while the gas space remains largely inert.

Despite the disadvantage of the relatively low degree of atomization, the sensitivity of flameless atomization in a graphite tube furnace is usually two to three orders of magnitude better than in the flame. In the first instance this is due to the considerably longer residence period of the atoms in the beam path.

It would be suitable at this point to briefly mention the advantages and the disadvantages of flame and flameless atomization.

A flame is limited to the use of solutions; generally only diluted solutions can be determined free of interferences. The solution is aspirated at a constant rate and sprayed into the flame; this results in a readout in concentration—a higher signal can only be produced by increasing the concentration (without instrument alterations). To obtain a stable signal, a given minimum sample volume is necessary. It can clearly be seen as a disadvantage of flame atomic absorption that the constancy of the nebulizer and also the composition, pressure and flow of the fuel gas and oxidant can have a considerable effect on the signal. Expensive measures are required to stabilize these parameters. However, the flame is simple to operate, possesses good stability and in many cases offers sufficient sensitivity. With the corresponding care in working, a precision of around 0.2% can be obtained. When the most suitable flame is selected, only few possibilities for interferences exist, so that analyses can be carried out with high reliability.

Flameless atomization cells, on the other hand, offer high absolute sensitivity as a result of the considerably longer period of residence of the atoms in the beam path. They have a largely chemically inert atmosphere and a very low sample requirement. The lower limits are mostly determined by the conventional means of measuring (balance, micropipette) and not by the apparatus. In principle, even solid samples can be directly atomized without previous solution. An advantage of atomization in a graphite tube is that the distribution of the atoms over the tube cross-section is homogenous, i.e. there is no gradient requiring corresponding adjustment as with the flame.

Flameless sources of atomization are not always simple to operate and are almost exclusively designed for trace analyses and small sample amounts. Their precision is rarely better than $\pm 1\%$. Furthermore, nowhere near as much work has been carried out with flameless atomization as with the flame; relatively few comprehensive investigations have been carried out on possible causes of interference, and it will take years before some contradictions can be clarified.

To sum up, it may be said that the two procedures complement each other in an ideal manner. The flame gives the required precision and reliability in the ppm and % ranges, while the flameless procedure serves the absolute trace range where the requirements of precision are not usually as high. The high sensitivity of the flameless method is also of great value when samples in the nl or μg range have to be handled.

4. Optics

The spectral range of interest for atomic absorption spectroscopy begins in the near infrared at 852.1 nm, the wavelength of cesium, and reaches down into the vacuum UV below 200 nm. At the present time, the instrumental limit for a non-flushed instrument and using a flame is 193.7 nm, the wavelength of arsenic. Atomic absorption spectroscopy therefore covers the same wavelength range that is of interest for spectral emission analyses or, alternatively, the range of a UV/VIS spectrophotometer. In principle it should then be possible to employ proven monochromators in atomic absorption. However, it has been shown that the requirements in atomic absorption with respect to resolution and dispersion of the monochromator are different from those of other techniques.

4.1. Spectral Bandwidth

While the quality of an instrument for other analytical methods is often directly dependent upon the spectral bandwidth of the monochromator, for atomic absorption this is not of primary importance. Atomic absorption spectroscopy is selective and free from spectral interferences if a well designed alternating light instrument and an element specific light source are used. The light source emits the spectrum of one (or several) elements and the emission light of the flame (direct light) is eliminated by means of selective amplification. The ability of atomic absorption spectroscopy to differentiate between two elements is solely dependent on the half-widths of the emission lines ($\sim$0.002 nm) emitted from the hollow cathode lamp and of the absorption lines ($\sim$0.004 nm). In other words, values that lie on or beyond the limits of the resolving power of normal monochromators. The monochromator in an atomic absorption spectrophotometer has the sole task of separating the resonance line of the element under study from other emission lines of the lamp. Experience has shown that this can be achieved for practically all elements with a bandwidth of 0.2 nm. In principle, no advantages are gained if smaller slit widths, such as are required to separate other emission lines from the spectral source, are employed. If larger slit widths are employed, atomic absorption does not loose any of its selectivity or specificity, except in the case when multielement lamps are used and the resonance lines of two elements fall simultaneously on the detector. The disadvantages brought by slit widths which are too large in atomic absorption are lower sensitivity and an increasing curvature of the calibration plot. If a further non-absorbable emission line (see page 19) besides the resonance line passes through the exit slit and falls on the detector, this always "sees" the radiation of both lines. With increasing absorption of the light from the resonance line, the intensity of the second line remains unaltered so that the absorption curve does not approach 100% absorption (0% transmission) asymptotically as usual, but approaches a value corresponding to the percentage proportion of the second line. DE GALAN and SAMAEY [286] have shown that the causes for non-linearity of calibration curves are mostly trivial. Non-resolved multiplets (several resonance lines passing through the exit slit) and non-absorbable lines (apart from the resonance line, a further non-absorbable line passes through the exit slit) were recognized as two main causes and this was verified by calculation.

The following figures illustrate this behaviour and show the influence of the spectral bandwidth on the sensitivity, the signal-to-noise ratio and the curvature of the calibration plot. Figure 40 shows the six resonance lines of silicon in the range 250 nm to 253 nm and their sensitivities in atomic absorption spectroscopy. With a sufficiently small slit width every resonance line gives a linear calibration plot within the observed range.

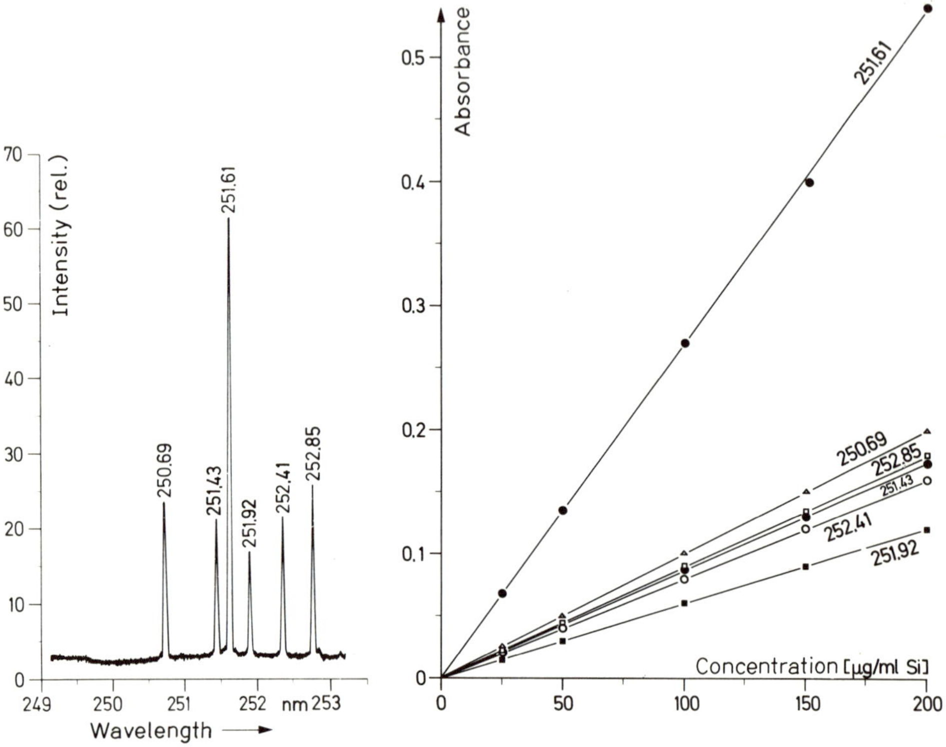

Figure 40. Various silicon resonance lines in the range 250 nm to 253 nm.—a) Emission spectrum of a Si hollow cathode lamp at a spectral bandwidth of 0.07 nm. b) Calibration curves of the six resonance lines measured at a spectral bandwidth of 0.07 nm.

In figure 41, the influence of the spectral bandwidth of the monochromator on the calibration plot and on the signal-to-noise ratio for the 251.6 nm resonance line of silicon is shown. With decreasing spectral bandwidth, the sensitivity clearly increases and the linearity of the calibration plots improves. It the optimum slit width is reached (i.e. the spectral bandwidth at which only the resonance line reaches the detector)—in this case 0.2 nm—a further decrease of the slit width brings no noticeable advantages. As shown in figure 41a, the linearity of the calibration plot and the sensitivity can still be improved somewhat since in practice there is still a small amount of residual scattered radiation present which can be further eliminated by reducing the slit width.

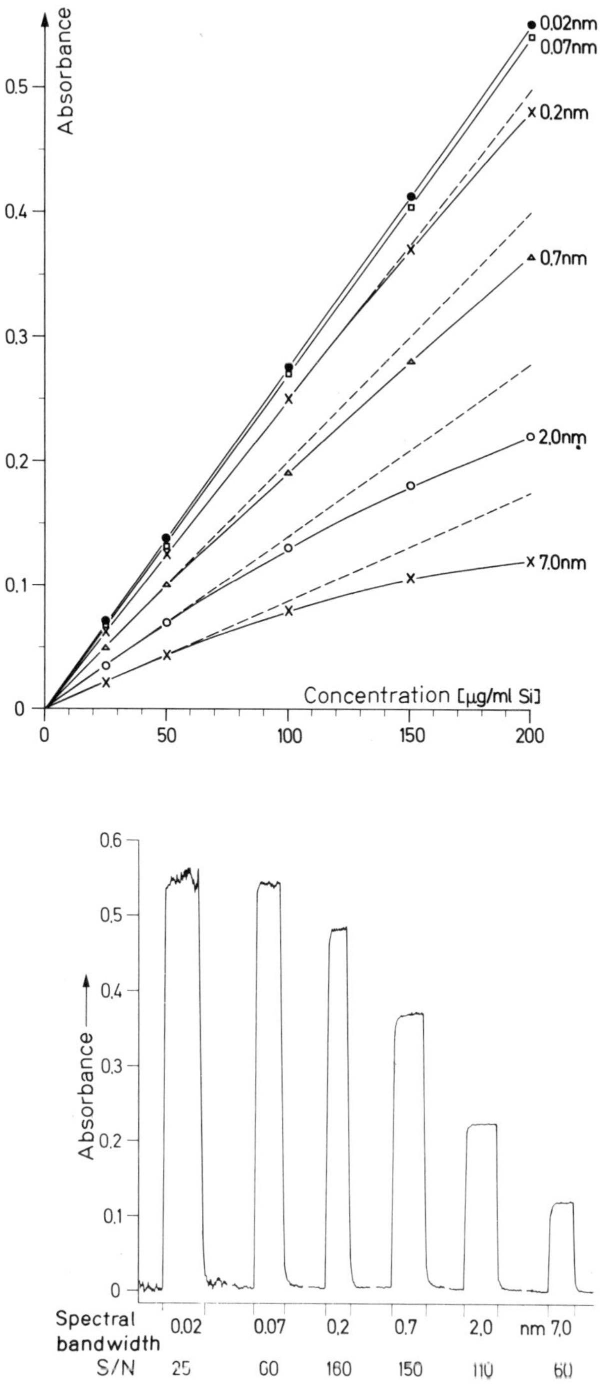

Figure 41. Influence of various spectral bandwidths on the determination of Si at the 251.6 nm resonance line.— a) The sensitivity decreases noticeably at bandwidths greater than 0.2 nm and the curvature of the calibration plots increases. b) The signal-to-noise ration (S/N) is optimum at a spectral bandwidth of 0.2 nm and clearly decreases on both sides. (Standard solution 200 µg/ml Si; nitrous oxide/acetylene flame.)

Nevertheless, this slight improvement in the linearity is only obtained at the expense of an increase in the noise and a noticeably poorer signal-to-noise ratio as clearly shown in figure 41b. This effect will be further discussed in more detail.

In the case of silicon, the effect is caused by the influence of a non-resolved multiplet when spectral bandwidths greater than 0.2 nm are used. The non-linearity of the calibration plot

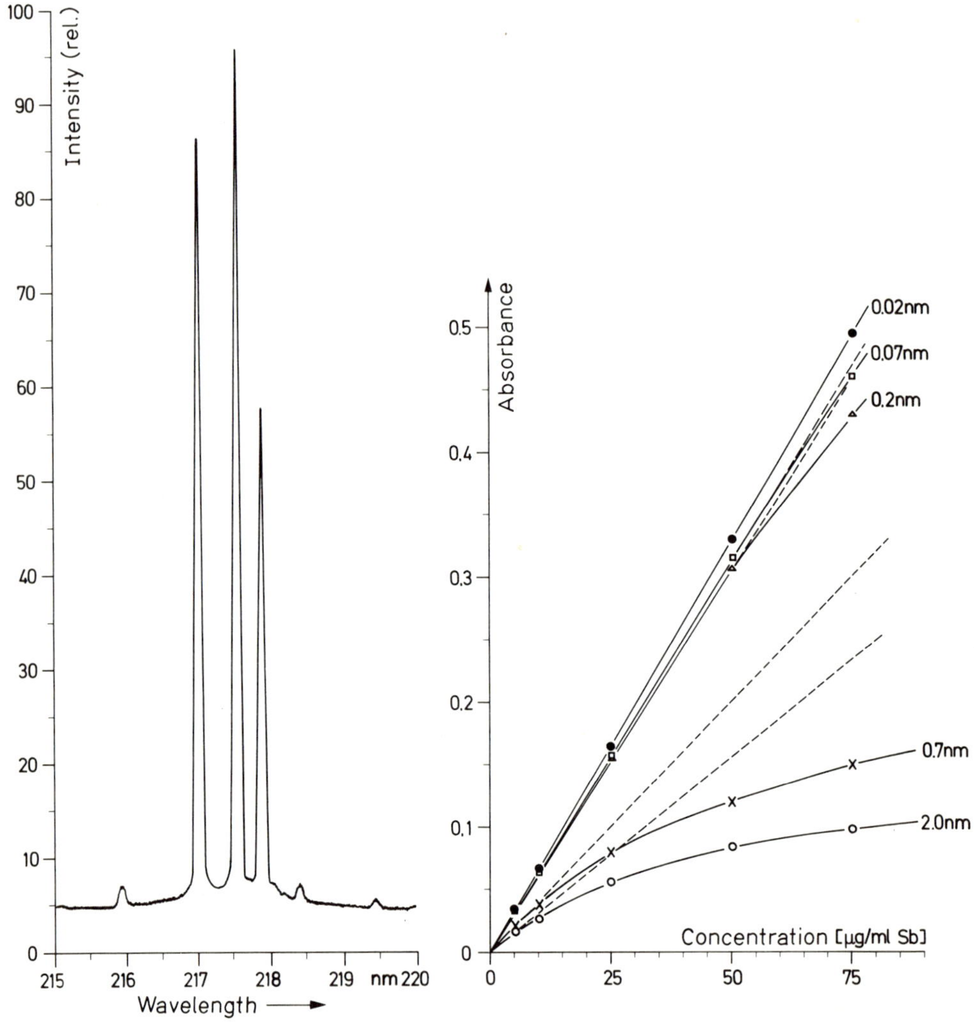

Figure 42. Influence of the spectral bandwidth on the determination of antimony.—a) Emission spectrum of an Sb hollow cathode lamp at a spectral bandwidth of 0.07 nm. Next to the resonance line at 217.6 nm there is an arcing line at 217.9 nm and a sparking line at 217.0 nm. b) At increasing slit widths the sensitivity decreases erratically as soon as non-absorbable lines pass through the exit slit.

is due to the various absorption coefficients $\varkappa_\nu$ of the individual resonance lines (see equation (1.18) page 12). This means that the absorption coefficient in equation (1.15) (page 10) is not constant over the integrated range if several resonance lines pass through the exit slit and fall on the detector.

To illustrate this further, the effect of the spectral bandwidth on the calibration curve of antimony is shown in figure 42. Besides the resonance line of interest at 217.6 nm, an antimony hollow cathode lamp also emits an arcing line at 217.9 nm and a sparking line at 217.0 nm. Neither of these two lines are measureably absorbed by higher antimony concentrations, so that the light from them can be regarded as "non-absorbable".

It can be seen from figure 42b that with smaller bandwidths of up to 0.2 nm, antimony behaves in a similar manner to silicon, i.e. decreasing the slit width brings no further advantage once the resonance line has been separated. With an increase in the spectral bandwidth, there is a sudden decrease in the sensitivity and a strong increase in the curvature as soon as non-absorbable light falls on the detector. This is not surprising because the degree of the curvature is determined by the difference between the individual absorption coefficients. If two lines have exactly the same absorption coefficient (no such case is known), no curvature and no loss of sensitivity should take place. The strongest curvature quite clearly takes place when the second line is completely non-absorbable. In this case, the calibration curve will approach a value corresponding to the percentage transmission of the non-absorbable line asymptotically. DE GALAN and SAMAEY [286] calculated this value to be 55% transmission (absorbance 0.254) for antimony and also found this "limiting absorbance" experimentally for a 1000 µg/ml Sb solution.

While it can be seen that for a number of elements a spectral bandwidth of 0.2 nm is necessary to obtain good sensitivity and linearity of the calibration plot, for other elements the resonance line is more or less isolated so that larger slit widths can be used without disadvantage. It is obvious that in such cases the widest slit width at which the resonance line is just separated from other lines should be used since the signal-to-noise ratio is favourably influenced, as shown in figure 41b.

4.2. Reciprocal Linear Dispersion

The size of the spectral bandwidth should be so selected that only the resonance line of the element under study falls on the detector and all other emission lines from the spectral light source are held back by the slit foils. The spectral bandwidth is thus limited by the resolving power of the monochromator. The resolving power, which gives the smallest spectral bandwidth that can be achieved in a monochromator or, alternatively, the smallest geometric slit width that can be usefully used for reasons of the diffraction, is without great importance; the spectral bandwidths required in atomic absorption lie mostly far above the resolving power of even less complex monochromators.

A further important dimension is the geometric slit width. This gives the effective mechanical widths of the entrance and exit slits in mm or µm for a given spectral bandwidth. The geometric slit width determines the portion of the light emitted from the radiation source that really enters the monochromator.

In an atomic absorption spectrophotometer, the image of the hollow cathode lamp is formed on the entrance slit, i.e. a light beam of several mm diameter falls on the slit. It can be clearly seen from figure 43 that the geometric width of the entrance slit determines the amount of light that falls on the dispersing element and subsequently on the detector. With a wide entrance slit, a relatively large amount of light energy therefore falls on the detector; this means that the noise always present in the signal (the shot noise of the photon current for example) is relatively small compared to the signal. At the same time, lower amplification can be employed so that possible contributions of the electronics to the noise are reduced. For the analyst, low noise means a stable signal and hence good precision and low detection limits.

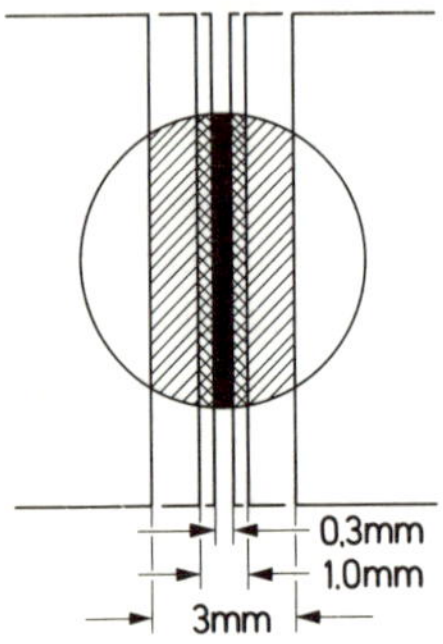

Figure 43. The image of the hollow cathode lamp is formed on the entrance slit. Its geometric width determines the amount of light that falls on the dispersing element.

Two requirements have been placed on the monochromator of an atomic absorption spectrophotometer that are not directly in harmony with each other. Firstly, a spectral bandwidth of 0.2 nm is required, and secondly the entrance slit should have as wide a geometric width as possible. However, the entrance and exit slits of a monochromator must have the same (or at least very similar) mechanical dimensions.

Spectral bandwidth and geometric slit width are linked by the linear reciprocal dispersion of the monochromator, expressed in nm/mm (often also shortened to linear dispersion). A reciprocal linear dispersion of 2 nm/mm means that at a geometric slit width of 1 mm the spectral bandwidth is 2 nm, or a geometric slit width of 0.1 mm is necessary to obtain the desired spectral bandwidth of 0.2 nm.

From this consideration it is thus desirable to have as small a reciprocal linear dispersion as possible, i.e. to use a strongly dispersive element because a relatively large amount of light then passes through the monochromator, even with small spectral bandwidths. On the other hand, because of the significance of the geometric slit width, the largest possible slit width just meeting the requirement for the isolation of the resonance line is always selected.

4.3. Prisms and Gratings

Prisms and gratings serve to disperse the radiation into individual wavelengths. A short examination will be made to determine which best meets the requirements of atomic absorption.

The reciprocal linear dispersion of a prism is fixed by the dispersion of the prism material, i.e. by the wavelength dependence $dn/d\lambda$ of the refractive index. With a prism monochromator therefore, the reciprocal linear dispersion and thus the spectral bandwidth at a fixed geometric slit width are wavelength dependent. In the first approximation, these dimensions become more unfavourable towards the long wavelength end exponentially (figure 44). In practice, this means that a variable slit system must be employed in which the geometric slit width is decreased with increasing wavelength so that a constant spectral bandwidth is guaranteed. At the same time there is a continuous worsening of the light transmittance of the monochromator with increasing wavelength.

The spectral bandwidth and the reciprocal linear dispersion of a grating monochromator are dependent on the grating constants, and these improve as the interval between two grating rulings becomes smaller, or as the number of lines per mm a grating has increases. Next to this, the order of the spectrum naturally plays a part, but in atomic absorption this is not of practical interest since measurements are almost exclusively carried out in the first order.

Both dimensions are virtually independent of the wavelength, i.e. at a fixed geometric slit width, a grating monochromator has practically the same spectral bandwidth and the same reciprocal linear dispersion over the entire wavelength range (figure 44). For atomic absorption spectroscopy, which covers a range approaching 700 nm and for which the accuracy of the analyses as well as the detection limits are dependent upon the linear reciprocal dispersion of the monochromator, the grating monochromator offers considerable advantages over the prism. Furthermore, a good grating nowadays has a ruling density of 2000 to 3000 lines per mm (with a total of around 10^5 rulings), so that with the usual focal lengths employed a reciprocal linear dispersion of 1 nm/mm and less can easily be obtained. As clearly shown

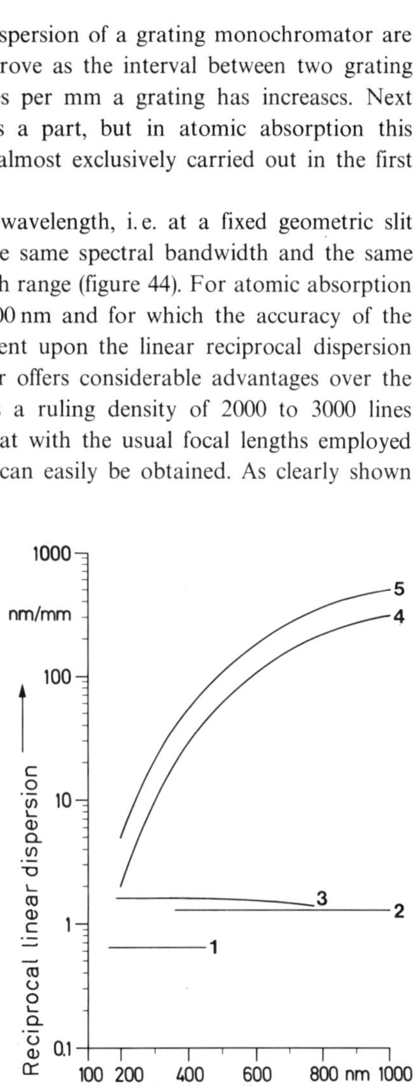

Figure 44. Reciprocal linear dispersion depending upon the wavelength in grating (**1–3**) and prism (**4–5**) monochromators.—**1** PERKIN-ELMER Model 503 in UV range (0.65 nm/mm), **2** PERKIN-ELMER Model 503 in VIS range (1.3 nm/mm), **3** PERKIN-ELMER Model 400 (1.7 nm/mm), **4** and **5** Quartz prism monochromators.

in figure 44, comparable reciprocal linear dispersions are only achieved with normal prism monochromators in the far UV region, while they are already several orders of magnitude poorer in the near UV.

A number of technical solutions are known for the actual construction of the monochromator; the two most important, the simpler LITTROW monochromator and the more complicated EBERT, here in the modified CZERNY-TURNER form, are depicted in figure 45. Both monochromator types are used successfully in atomic absorption.

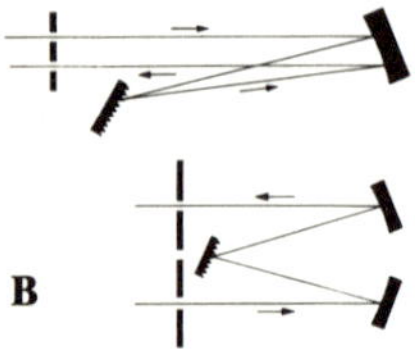

A

B

Figure 45. Two frequently employed monochromator types.—**A** LITTROW monochromator and **B** CZERNY-TURNER monochromator.

In this connection, it should be briefly mentioned that filter monochromators (except perhaps for a few individual elements in the visible range) cannot be employed in atomic absorption since the required resolution cannot be obtained, leading to the difficulties described in section 4.1.

4.4. Resonance Detectors

SULLIVAN and WALSH have reported the use of resonance detectors as "monochromators" in atomic absorption spectrophotometers in various publications [1194] [1195] [1198] [1285].

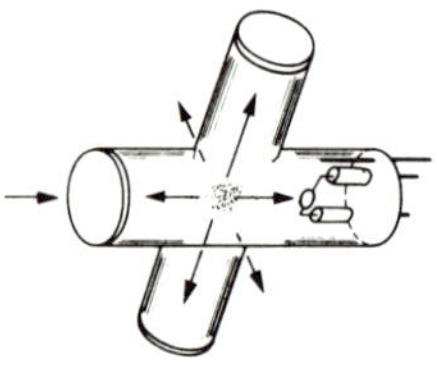

Figure 46. Resonance detector.—After passing through the flame the spectral light emitted from a hollow cathode lamp falls on the atom cloud formed in the resonance detector. These atoms absorb the light of the resonance line and re-emit it in the form of fluorescence radiation. The measurement takes place at right angles to the beam.

The mode of operation of these resonance detectors is depicted schematically in figure 46. After passing through the flame, the spectral light emitted from the hollow cathode lamp falls on the resonance detector. The detector consists of a device for producing a cloud of neutral atoms in the ground state (similar to a hollow cathode lamp). Hollow cathode lamp and resonance detector must then be manufactured from the same metal. If the light which has passed through the flame now falls on a cloud of atoms of the element to be determined, the resonance lines are absorbed while the remaining (non-absorbable) radiation passes through the atom cloud. The absorbed resonance radiation excites the metal atoms to fluorescence which can then be measured by a detector at right angles to the direction of the incident radiation. The fluorescence signal is proportional to the intensity of the resonance radiation.

Since the spectrum of the fluorescence radiation only consists of lines that were absorbed by the cloud of atoms, resonance detectors can take over the function of the monochromator, which in atomic absorption spectroscopy separates the line to be measured.

Because of the absence of non-absorbable background radiation, it was expected that the calibration plots obtained with this type of "resonance monochromator" are linear over a wider range than those obtained when the lamp light falls directly on the monochromator. This expectation was verified experimentally. A disadvantage of the resonance detector is that only a small portion of the fluorescence radiation emitted in all directions falls on the detector.

WALSH suggested the use of resonance detectors for the construction of a multielement atomic absorption spectrophotometer [1198] [1284] in which the simultaneous determination of several elements is possible (figure 47).

The absorption of resonance radiation by a cloud of neutral atoms can also be exploited in another manner to isolate resonance lines through so-called "selective modulation" [149] [193] [758] [1198]. Here, direct light from a hollow cathode lamp is passed through a discharge tube whose open cathode cylinder is made of the same metal as the cathode of the light source. The power supply to this discharge tube is modulated so that a pulsing atom cloud, which absorbs the resonance lines from the hollow cathode lamp at the modulation frequency, is formed. In this way, a beam consisting of modulated resonance lines and non-modulated non-resonance lines is produced. Only the resonance lines are amplified by a selective amplifier after the detector, while ion and filling gas lines are eliminated.

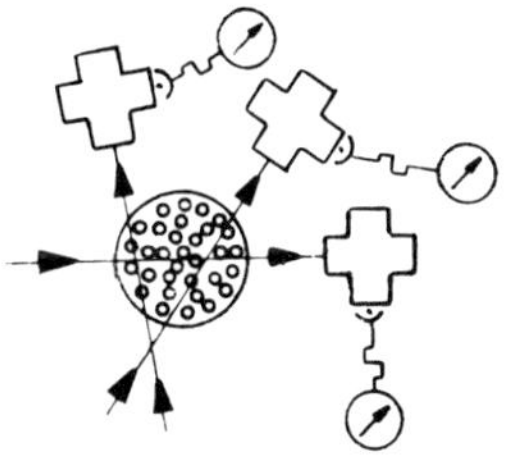

Figure 47. Multielement instrument according to SULLIVAN and WALSH [1198].—By the use of several hollow cathode lamps which irradiate the same flame and whose radiation is captured by resonance detectors, various elements can be simultaneously analysed in one sample.

Nevertheless, for this procedure a monochromator must be used since too much direct light would otherwise fall on the detector, leading to noise. A simple monochromator without special requirements for the resolving power is sufficient; frequently filters can be employed. With this procedure of selective modulation, calibration plots which are linear over a wide range are likewise obtained.

SEBESTYEN [1105] has shown that it is possible to obtain a similar effect in a normal atomic absorption spectrophotometer and with a normal hollow cathode lamp if short pulses of very high current strength are periodically superimposed on the operating direct current of the hollow cathode lamp. Thereby an intense atomic metal cloud is periodically formed in front of the cathode which briefly absorbs the resonance radiation and thus modulates it. With suitable demodulation electronics tuned to this modulation procedure, calibration plots, which were linear into high absorbance ranges, could be obtained for iron and nickel, for example.

5. Electronics and Readout

The demands made on the electronics and the readout in atomic absorption spectroscopy often differ according to the nature of the application. If possible, an atomic absorption spectrophotometer should be able to meet all situations.

Three requirements are frequently demanded nowadays: Speed, high precision, and the lowest detection limits. The last two points can be combined under the requirement that the electronics must be capable of recognizing and displaying the smallest differences in the absorption. This, however, is only then possible and meaningful if the signals are steady and stable; the stability of the signal is primarily dependent upon the stability of the flame and the performance of the optics. This means that all components of an atomic absorption spectrophotometer must be well designed and matched to each other to obtain optimum performance. The light source must burn stably and supply maximum radiation energy. The flame must not interfere in the production of free atoms through unnecessary turbulence or irregularities. The monochromator must have a small reciprocal linear dispersion so that a high radiation flux density reaches the detector. As previously explained, the relative noise of the signal will then be small. Additionally, lower voltages for the photomultiplier (i.e. less amplification) can then be used. This stable electrical signal is a pre-requirement for speed, precision and good detection limits.

A stable signal can often be considerably expanded—a number of atomic absorption spectrophotometers allow an expansion of more than a hundred times—so that the readout accuracy can be improved by a multiple. In modern instruments it is possible to measure differences of 0.0001 absorbance units. If, in addition to high expansion of the lowest absorption ranges (i.e. lowest concentrations) to obtain the best detection limits, any desired partial range of the scale can be expanded in an atomic absorption spectrophotometer (through baseline suppression), analyses of the highest precision can be carried out on samples with high metal concentrations. Naturally with measurements at very high expansion factors, it is necessary to dampen the noise (i.e. the fluctuation of the signal) by an amount corresponding to the expansion and the stability of the signal. The lowest detection limits and the highest precision can only be achieved at the cost of speed.

The possibility of working with very short time constants is equally as important as the ability to expand and dampen signals to clarify the smallest differences in absorption to obtain good detection limits and high precision. A disadvantage of high expansion and damping is the high substance requirement. This can be of crucial importance in conventional atomic absorption with nebulizer and flame when only small sample amounts are available. Lower electronic damping in this case means lower sample consumption. With fast responding instruments, it is possible to carry out precise measurements with less than 0.1 ml sample solution for example, as shown in figure 48.

Under given conditions, even smaller volumes can be used, but then only for special problems [226].

Flameless atomization methods with graphite tube furnaces and the boat technique, which have gained considerable significance in recent years, can only be carried out in instruments with very short response times. The signals produced by these methods generally reach a maximum in about a second and then return to the baseline within a few seconds. These

techniques therefore require instruments with very rapid response times having minimum time constants of 0.2 second at the most.
In the following sections a number of important components of the electronics and the readout will be described with especial mention of their significance to the analyst.

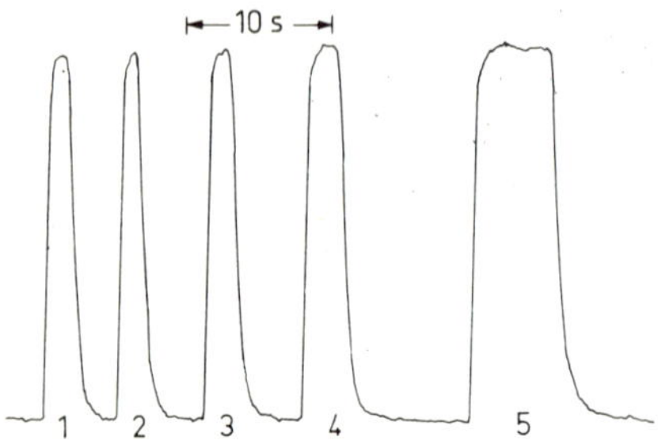

Figure 48. Aspiration times of only 1 second give sufficiently accurate signals with instruments having fast response times. The sample requirement is still below 0.1 ml per measurement. Aspiration times: Signals **1–3**, 1 second each; signal **4**, 2 seconds and signal **5**, 5 seconds.

5.1. Detector

Photomultipliers are mainly used in atomic absorption spectroscopy to convert the light signal into an electrical signal. They consist of a vacuum photocell with an anode, a light sensitive electrode (photocathode), and a number of emission cathodes (dynodes) which have increasing positive potential with respect to the photocathode. Photomultipliers with a total of 10 electrodes are frequently used; in special cases the number of electrodes can be increased to 13.
A photoelectron released from the photocathode is attracted to the first dynode and falls upon it with a kinetic energy that is proportional to the voltage gradient. It then releases a number of secondary electrons which are accelerated and for their part release an even greater number of electrons, and so on, so that the effect is still further amplified. The amplification of a photomultiplier is dependent upon the applied high voltage. This can have values in the order of 1000 V to 1500 V.
The spectral range within which a photomultiplier can be used is dependent upon the light sensitive layer on the cathode and the window material of the tube. It is not easy to find a photomultiplier that has sufficient sensitivity over the entire spectral range of interest for atomic absorption. With good photomultipliers, luminous fluxes of 10^{-6} to 10^{-11} lm can be measured. The sensitivity is around 10–100 A/lm and the maximum current about 10 μA. If, at a given amplification, more light falls on the photomultiplier so that the 10 μA current is exceeded, the signal decreases rapidly. Reversible and irreversible alterations to the dynodes can then take place.
The dark current, which is the current flowing through the tube under the influence of the high voltage when no light falls upon the photocathode, is an important criterium for the quality of a photomultiplier. The so-called dark noise, the fluctuation of the dark current, can under circumstances be an important component of the detector noise; it increases with increasing voltage. Therefore, to obtain good separation of the signal from the dark noise,

the highest possible light intensity on the detector and the choice of a suitable photomultiplier are important.
The quantum efficiency of the photocathode (spectral cathode sensitivity) is of special importance for a photomultiplier since in effect it gives how many photons are required to release an electron from the photocathode. A low quantum efficiency leads to considerable energy losses for the conversion of the photon current into an electrical current and thereby to an increase in the (shot) noise, which cannot be eliminated by higher after amplification.

5.2. Emission Noise

An interference which could be frequently observed earlier, but which is only seen in isolated cases in flame atomic absorption nowadays, is the so-called emission noise. Depending on the cause, this interference occurs either when the flame is ignited or when sample solution is aspirated [176]. With flameless atomic absorption employing graphite tube furnaces (or similar devices), this interference is found relatively often, especially when older atomic absorption spectrophotometers, whose optical-electronic systems are not matched to the requirements of flameless atomic absorption, are used [638]. It can be observed that when the sample is atomized, there is very strong noise, or even a clearly marked shift of the baseline (see figure 49). This is due to the high direct light emission from the glowing walls of the graphite tube.

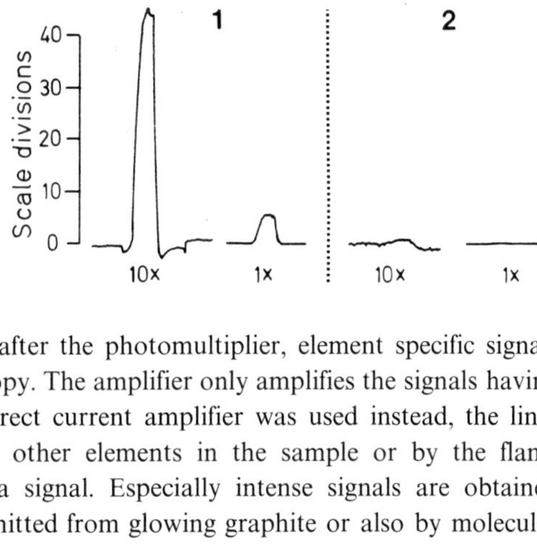

Figure 49. Influence of the direct light emission from the graphite tube on the baseline during atomization.—**1** Atomization at 2200 °C at the wavelength of calcium (422.7 nm) produces a considerable offset of the baseline in an older instrument. **2** In an instrument whose optics and electronics have been matched to the requirements of the flameless technique, the direct light emission plays no further part (422.7 nm, 2500 °C) [638].

By means of a selective amplifier in circuit after the photomultiplier, element specific signals are obtained in atomic absorption spectroscopy. The amplifier only amplifies the signals having a determined modulation frequency. If a direct current amplifier was used instead, the lines emitted within the same spectral range by other elements in the sample or by the flame would also be amplified and give rise to a signal. Especially intense signals are obtained from broad emission bands such as those emitted from glowing graphite or also by molecular bands in the flame.
The selective amplifier eliminates this continuous emission, it "overlooks" the non-modulated components. Nevertheless, this radiation falls on the detector just the same as the modulated radiation and produces a photocurrent. The current leaving the photomultiplier therefore consists of a modulated component, which can be assigned to the hollow cathode lamp, and a non-modulated component, which comes from the emission of either the flame or the graphite tube

furnace. Since the high voltage applied to the photomultiplier is set according to the intensity of the modulated component (i.e. the lamp radiation), overloading of the photomultiplier can occur if the sum of both currents exceeds the maximum current, especially when the lamp intensity is somewhat weak. This overloading causes a loss in sensitivity of the photomultiplier which is observed as a decrease in the signal. However, even when the detector is not overloaded, a high direct light increment causes fluctuations of the signal, leading to poorer detection limits and lower precision of the results.

In principle, various possibilities exist for the elimination or reduction of this emission noise. These are based either on a reduction of the direct light increment or on an increase in the alternating light increment. The increase in the alternating light increment can be most reliably achieved by using a more intense radiation source; occasionally an increase in the lamp current, or more seldom a decrease, can be of help. The reduction of the direct light increment, as far as it emanates from the flame, can be most reliably achieved by choosing another flame when this is possible; occasionally an alteration of the stochiometry or the use of a smaller slit width can bring an improvement.

In the case of graphite tube furnaces, the direct light emission from the glowing graphite is typically several orders of magnitude more intense than the emission from a flame. For this reason the interference is more frequently noticed. In general, the same principles can be applied to eliminate the interference, namely the use of a more intense radiation source (e.g. a single element lamp instead of a multielement lamp, or an electrodeless discharge lamp instead of a hollow cathode lamp), or a reduction of the direct light increment.

The direct light emission from the graphite tube can be very simply and effectively removed by reducing the atomization temperature. The majority of elements can be quantitatively atomized with high sensitivity at temperatures between 2000 and 2500°C; a higher atomization temperature often brings only a small increase in the sensitivity but at the same time increases the emission noise considerably, so that as a whole, a clear worsening of the signal-to-noise ratio is observed.

In contrast with the flame, the emission source in graphite tube furnaces is clearly optically localized; it is in the form of a ring around the cloud of atoms. Thus the possibility exists for largely masking this emission optically. The improvements which can be obtained with such a modified optical system are shown in figure 49. In atomic absorption spectrophotometers of recent manufacture, the corresponding optical and electronic measures, such as the introduction of stops and diaphragms, reduction of the slit height, use of a suitable photomultiplier, etc. have been taken into consideration.

At temperatures above 2500°C, components vaporizing from the graphite surface (e.g. C_2 radicals), or molecules or radicals formed with the inert gas (e.g. CN) begin to emit with increasing intensity. This direct light emission naturally cannot be masked—a further reason for avoiding atomization temperatures that are too high.

5.3. Integration

Simple measurement of the voltage on a suitably calibrated voltmeter, either after the signal has been passed through a demodulator (single-beam instruments) or the ratio between the sample beam and the reference beam has been electronically formed (double-beam instruments), is the method that has been in use the longest. With this type of signal handling, all fluctuations are displayed on the meter in strengths dependent upon the selected time constant. After jumpwise alterations, such as occur when the sample is introduced, the meter only comes to the "middle" measurement result asymptotically after several time constants have elapsed.

More recently, especially in connection with digital displays, true integration of the signals as a function of time has been increasingly applied. This means that a "clock" in the instrument must exactly define a precise integration time since the result is critically dependent upon the length of the integration period. In practice, the reproducibility of the integration time is very much more important than the absolute length. This problem is insignificant compared to the enormous advantage that the measured result is displayed steadily to within a known accuracy at the end of the selected integration period. It is neither necessary to observe the display to see whether a sufficiently accurate asymptotic limiting value has been reached, nor to estimate the mean value of a fluctuating display. The operator is largely relieved of the decision as to when and what should be read.

The precision of the results is dependent upon the integration time in that it is inversely proportional to the square root of the time. Usually, the integrated signal at the end of the measurement is divided by the time unit so that the signal height is independent of the measurement period, and longer integration periods are expressed solely in a lower coefficient of variation. The integrated signal can also be displayed directly, so that longer integration times lead to correspondingly higher signals—nevertheless, the signal-to-noise ratio is not further improved.

While in flame atomic absorption integration of the measured results brings, in the first instance, a saving of time, various authors have reported that integration of the peak area of the signals produced by graphite tube furnaces leads to an improvement in the precision and the linearity of the calibration plots [751] [982].

5.4. Display

Since the decrease in the radiation energy is measured instrumentally in an atomic absorption spectrophotometer, or, expressed more exactly, the radiation flux density $J_{\nu,t}$ after the sample has been introduced is compared to the radiation flux density $J_{\nu,0}$ without sample, the usual readout is in % absorption or % transmission. According to the BEER-LAMBERT law, the concentration is proportional to the negative logarithm of the transmission, i.e. the absorbance (see equation (1.18), page 12). Since linear relationships are easier to handle in practice than logarithmic, the signal in % absorption must be converted into absorbance. By plotting the absorbance values versus the concentration, a linear calibration curve which can be exactly established with a minimum of measurement points is obtained.

The conversion of % absorption (% *Absn*) into absorbance (A) is from the formula

$$A = \log \frac{100}{100 - \% Absn} \quad \text{or}$$

$$A = 2 - \log(100 - \% Absn)\,, \tag{5.1}$$

where 100 − % $Absn \equiv$ % transmission. The relationship between % absorption and absorbance is shown graphically in figure 50.

In modern spectrophotometers a logarithmization of the signal takes place within the instrument, so that the readout on the display is already in absorbance and no conversion is required.

Meters with calibrated scales or digital voltmeters with electronic control are presently employed as display instruments; for documenting the values a strip chart recorder or a printer can be used.

Meters have the advantage that they are usually reasonable in price. Also, fluctuations or alterations in the display can be followed easily. The readout accuracy is however not usually very great; therefore the use of meters for precision and trace determinations is restricted.

Digital voltmeters with displays consisting of light emitting diodes offer an optimum in operating convenience and accuracy. A choice usually exists between selecting a semi-continuous display and readout of individual values. The "continuous" display, in which the individually occurring values are given in rapid sequence (usually less than 1 s), enables fluctuations in the values to be easily observed. This is of value, for example, for selecting the most suitable time constant or integration time, and also for the adjustment of the lamp, nebulizer, burner, etc. For routine operation, requesting the individual values is more convenient since subjective reading errors are thereby excluded. Nevertheless, for this mode of measurement, it must be taken into consideration that every single result has a given error which can only be statistically determined by noting a number of results. Measurement of the blank value should also be carried out from time to time since alterations in the baseline will not otherwise be recognized. In figure 51 a chart recording of a very low concentration measured in "integrated recording mode" is shown in which the blank values were measured between the samples in exactly the same way as for digital readout.

Meters are divided into quality grades to denote the accuracy which can be achieved. With instruments of grade 0.5, which are frequently used in atomic absorption spectrophotometers,

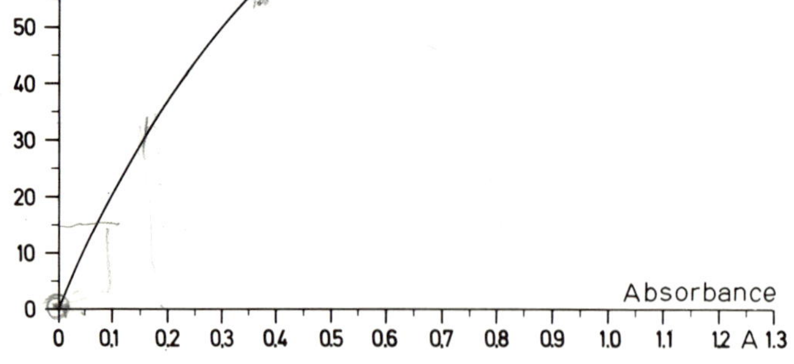

Figure 50. Relationship between % absorption and absorbance.—With increasing absorbance the curve approaches 100% absorption asymtotically.

the achievable accuracy is 0.5% of full scale deflection. Digital voltmeters always have a discrepency of one figure in the last place; the readout accuracy is therefore directly dependent on the number of illuminated digits. A digital voltmeter with three places corresponds to an instrument of quality grade 1, four places then give a quality grade of 0.1.

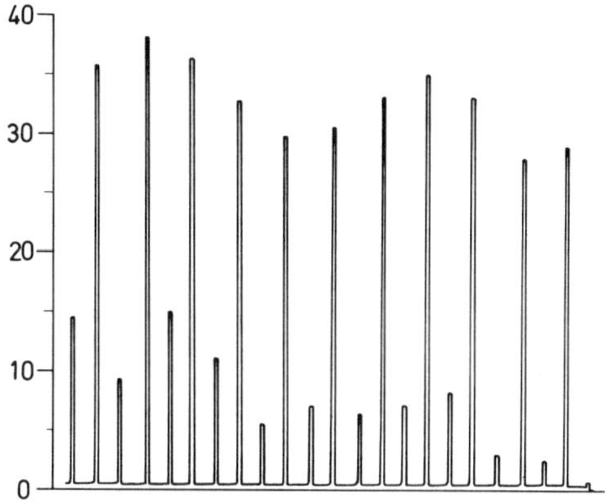

Figure 51. Measurement value formation through integration.—To achieve high precision or for determinations at the detection limit, a series of measurements must be carried out with a measurement of the blank value between each sample. Determination of 0.025 µg/ml barium with 200 times scale expansion (Perkin-Elmer Series 400 spectrophotometer).

The accuracy that can be achieved from the readout, referred to the concentration of an element in a sample, is furthermore dependent upon the mode of the readout. It differs for readout in absorbance and readout in % absorption. The relationships are shown graphically in figure 52 for display instruments of various quality grades. It can be seen that for instruments of the same quality grade, the relative readout error with calibration linear in absorbance decreases with increasing absorbance values and finally at full scale deflection the accuracy corresponding to the quality grade is achieved. With readout linear in % absorption, on the other hand, the relative error, referred to the concentration results, passes through a minimum which lies at approximately 65% absorption (absorbance 0.45). Here, the relative error is about three times the quality grade. The relative error increases with lower and higher absorption values, but the increase with lower values is not as rapid as with a scale divided linearly in absorbance. This means that smaller absorption signals (less than 55% absorption or 0.35 absorbance) can be read with greater accuracy on a scale linear in % absorption.

These considerations are only valid when scale expansion is not used. For the determination of lower concentrations, all atomic absorption spectrophotometers offer the possibility of electronically expanding the signal and thus achieving a better readout accuracy. This expansion must take place in exactly defined steps for displays in % absorption since the expansion factor must be known for the conversion into absorbance. For displays in absorbance, on the other hand, it is advantageous if the expansion is stepless, i.e. continuous, since by selecting a suitable scale, a readout directly in concentration units can be achieved, leading naturally to a considerable saving of time.

A direct display in concentration is then given when every absorbance value within reasonable limits can be digitally assigned to each possible concentration. This is certainly the case when the absorbance display can be steplessly multiplied by any factor between 0.1 and 10, which

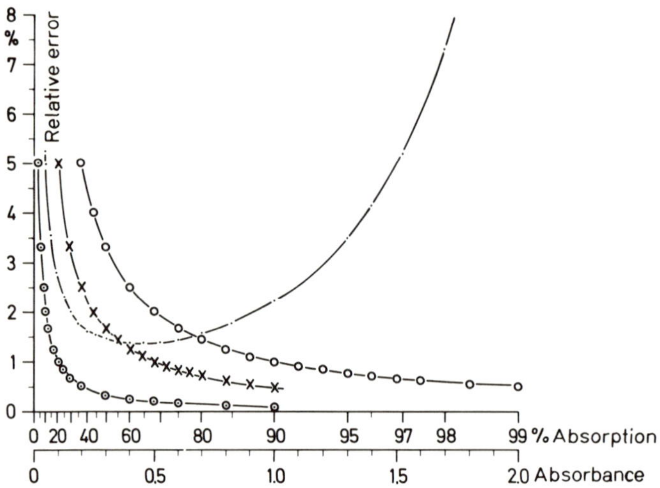

Figure 52. Relative error referred to the concentration results for various desplay instruments. –·– Readout in % absorption; instrument of grade 0.5. × Readout in absorbance; instrument of grade 0.5. ○ Readout in absorbance; digital voltmeter with 3 digits. ⊙ Readout in absorbance; digital voltmeter with 4 digits.

corresponds to a maximum compression of 10 : 1 and a maximum expansion of 1 : 10. Naturally, even higher expansion factors are often reasonable and therefore desirable. Modern atomic absorption spectrophotometers offer considerable convenience for readout in concentration to a pre-selected display value by push button, etc.

Furthermore, with a number of instruments it is possible to compensate for deviations from the linear relationship between absorbance and concentration as given by the BEER-LAMBERT law. This type of correction is particularly important for the direct digital display of concentration, since with non-linear calibration plots the displayed values cannot be used directly.

A useful addition to the display on a digital voltmeter is a printer. The connection of a sequential printer, which prints out all the digits appearing on the display, is the simplest possibility of providing convenient documentation of the analytical results. In this way readout errors are eliminated, which is of considerable importance when large numbers of samples have to be handled. Together with an automatic sampling and sample identification system, largely automated analytical instruments can be obtained.

While a printer, which gives a digital value for each analysis, meets the requirements of modern routine analyses, a strip chart recorder provides the most valuable analytical documentation. The chart recording is practically the only possibility of clearly retaining the conditions occurring during an analysis. It gives information on the aspiration time, the signal height, possible drift, and fluctuations or other alterations to the baseline and the signal. Further to this, it gives information on the signal-to-noise ratio, on the time constant used, and also on possible carry-over or memory effects. In all, the chart recording gives a relatively clear picture of the course of an analysis, even years afterwards; in other words, over things that can only be seen on a meter during the actual moment of readout, and on a digital display under circumstances not at all.

The use of a chart recorder is of particular importance for all techniques that give rapid time-dependent signals, such as flameless atomic absorption, the boat technique and the hydride system. Here, the recording of the actual atomization signal is not of prime importance.

Newer instruments have maximum value storers which store and display the peak height digitally. For working out a new analytical method, however, a chart recorder is indispensable, since the signals occurring during the thermal pre-treatment of the sample and the shape of the atomization signals are also of interest.
The response time of the recorder is of not inconsiderable importance for the linearity of the calibration plot and the signal height of the ballistic signals produced by flameless atomization techniques [982]. The recorder should have a pen traverse time of typically less than 0.5 s for full scale deflection.
The principles already mentioned for a meter scale apply to the accuracy of the readout on a recorder; for example, the readout accuracy is dependent upon the width of the recorder chart paper, and the pen stroke and blacklash must also be taken into consideration. Compared to readout on a meter, readout from a recorder offers considerably more convenience and reliability, and is also time-saving.
With noisy or strongly fluctuating signals, readout from a meter is often very difficult, time-consuming and inaccurate, while the same signals can usually be rapidly and simply evaluated from a recorder tracing. Similarly, when high time constants are being used, it can be ascertained more rapidly from a recorder whether the final value on the display has been reached or not. Analyses at the detection limit are practically only feasible when a recorder is used, since only from the tracing is it possible to distinguish clearly between the signal, which appears very slowly when high time constants are used and is usually small, and any possible drift.
Nevertheless, this is only the case when the recorder tracing is compared to the display on a scale. With modern instruments having integration and four-place digital displays, even measurements at the detection limit can be rapidly and very easily evaluated digitally. The achievable precision is at least as good as with a recorder, and is often better by a factor of two to three.

5.5. Automation

Since atomic absorption spectroscopy is a very rapid analytical method, it is hardly surprising that in 1966, the year in which it received its greatest impetus and general recognition, the first automatic analyser was introduced. GAUMER [408] constructed an instrument for the analysis of trace elements consisting of a commercial atomic absorption spectrophotometer to which an automatic sample changer, a digital evaluator [630] and a printer [1173] were connected. Soon afterwards, the first commercial instrument combination for automatic atomic absorption analyses was introduced [1138]. This consisted of an automatic dilutor, a sample changer for 200 vessels, a double-beam atomic absorption spectrophotometer with a digital evaluator and a printer. All stages of an atomic absorption analysis were carried out automatically and the results were printed out directly in concentration units. Modern atomic absorption spectrophotometers are already fitted with connections for teletypewriters, so that complete analytical reports can be written on punched tape for subsequent computer handling [610].

A decided disadvantage of atomic absorption spectroscopy, which especially impedes genuine automation, is the relatively high time requirement for changing from one element to the next. Multielement lamps offer certain advantages if the corresponding combination of elements is to be determined. Changing a lamp is the most rapid procedure, however. Setting the new wavelength and other instrument parameters, such as slit, amplification, etc., and carrying out new standardization are the most time-consuming procedures. They are also the most difficult to mechanize and automate. In the following, automation always means the automatic execution of all stages necessary for the analysis of one element in a large number of samples. Automation in atomic absorption spectroscopy is only suitable for a large number of similar samples. If many elements have to be determined in a few samples, automation with commercially available instruments is not possible.

Wide recognition of this disadvantage has led to increased endeavours to construct "multielement" instruments in which as many elements as possible in a sample can be determined either simultaneously or in rapid sequence. MAVRODINEANU has described a multichannel instrument which operates with several hollow cathode lamps and detectors [844]. WALSH and co-workers in particular mentioned the use of resonance detectors for multielement instruments [1198] and they constructed operable instruments on this principle [1196] [1197] [1284]. ALDOUS, JACKSON and MITCHELL used a Vidicon detector for the construction of a multichannel atomic absorption spectrophotometer and determined ten or more elements simultaneously [28] [868]. SILVESTER and co-workers [1125] focussed the light from two multielement lamps through the flame or the graphite tube furnace onto a three-way beam splitter. Each of the three beams was then passed through a monochromator with two exit slits onto two detectors so that six elements could be determined simultaneously. RAWSON [1027] also determined six elements simultaneously by combining the light from four hollow cathode lamps through fibre optics and a quartz rod light integrator and conducting it to a monochromator with six exit slits. The element specific light was then conducted from the exit slits by fibre optics to six detectors. The greatest disadvantages of this system are the high light losses due to the fibre optics. BUSCH and MORRISON [188] have published a good review on multielement flame spectroscopy.

6. Technique

Despite the multiplicity of elements and materials that are analysable by atomic absorption spectroscopy, the technique is simple and the possibilities of interference are few. Nevertheless, atomic absorption spectroscopy should not be regarded as being completely free of problems. In this chapter, the various measurement techniques, together with their advantages and disadvantages, and the different possibilities of interference will be discussed in detail. Special reference will be made to the causes of interference and the general methods for eliminating them.

6.1. Reciprocal Sensitivity and Detection Limit

The term "sensitivity" is not used uniformly in atomic absorption spectroscopy. DE GALAN [284] in particular has drawn attention to this confusion and has attempted to introduce a term used uniformly in spectroscopy for sensitivity.

The relation between the measured signal x and the concentration (or amount) C of the element to be determined is given by the *calibration plot*

$$x = f(C)\,. \tag{6.1}$$

On every point of this curve the *sensitivity* S is defined as

$$S = \frac{\mathrm{d}x}{\mathrm{d}C}\,. \tag{6.2}$$

The sensitivity is thus generally dependent upon the concentration (or amount) of the element being analysed. When the calibration plot obeys the BEER-LAMBERT law (equation (1.3), page 1) and gives a straight line, the sensitivity is independent of the concentration and is identical to the gradient.

In contrast to this, there is the *reciprocal sensitivity* S^*

$$S^* = \frac{\mathrm{d}C}{\mathrm{d}x}\,, \tag{6.3}$$

which is the concentration (in µg/ml) or the amount (in g) of an element that gives a signal corresponding to an absorbance of 0.0044 (1% absorption) (compare [821]). Dimensions µg/ml 1% or g/1%.

The principal cause of confusion is that reciprocal sensitivity is often written simply as sensitivity. In the interests of a standardization of definitions, both terms should be carefully separated. In this context, attention is also drawn to the use of such mixed terms as "sensitivity limit" etc., which are nowhere defined and should, therefore, be avoided.

Nowadays, the detection limit is generally defined as the smallest concentration (in µg/ml) or amount (in g) of the element to be determined that can be detected with 95% probability. This is equivalent to the concentration or amount that gives a signal twice the value of the standard deviation σ. To determine the standard deviation, at least 10 measurements must be made on a concentration or amount near the detection limit with a blank value measurement between each determination.

$$\sigma = \sqrt{\frac{\Sigma(\bar{x}-x)^2}{n-1}} \tag{6.4}$$

An example for the determination of a detection limit is shown in figure 51 (page 85). In this case, the standard deviation is 0.002 µg/ml, so that the detection limit (2σ) is 0.004 µg/ml. The reciprocal sensitivities and detection limits that can be achieved in a good atomic absorption spectrophotometer are compiled in table 6.

The two definitions for the reciprocal sensitivity and the detection limit, namely in units of concentration (µg/ml) and units of quantity (g), are derived from the various methods of atomic absorption spectroscopy. Solutions are used almost exclusively in classical atomic absorption with nebulizer and burner with a constant rate of aspiration. The resulting signals are thus directly proportional to the concentration of the element being analysed. From this arises the definition of the relative reciprocal sensitivity and the relative detection limit in units of concentration.

Flameless methods of atomization, such as the boat technique or graphite tube furnaces, are not restricted to solutions and are in principle largely independent of the sample volume. The resulting signals are therefore proportional to the quantity of the element being analysed. The absolute reciprocal sensitivity and the absolute detection limit are obtained in units of weight.

From the definitions for sensitivity and detection limit, it is clear that two dimensions emanating from different pre-requirements are involved. In this chapter, the factors upon which these dimensions depend and how they can be influenced will be discussed in some detail.

The detection limit is naturally dependent upon the sensitivity since this indicates the achievable signal height for a given concentration or quantity. As a second variable, the detection limit contains the fluctuation of the background, hereafter termed noise. While the sensitivity, as will be shown, largely represents a "natural constant", the noise is almost wholly caused by instruments. Although many causes for noise come into question, careful scrutiny will reveal clear relationships which have been discussed in full in earlier chapters and which will only be briefly repeated here.

A frequent cause for increased noise is that the energy is too low. This can signify that the light source emits too weakly at the desired wavelength; this was particularly the case with many earlier hollow cathode lamps, as shown in figure 53. A monochromator of low light gathering power can be a further cause of too low energy; an insufficient reciprocal linear dispersion (high values in nm/mm) denotes a narrow entrance slit, so that only a small fraction of the radiation emitted from the light source reaches the monochromator (see page 73). This and other causes of increased noise have previously been discussed in full.

If the possible contribution of the electronics to the noise in the signal is disregarded, the most important source of undesired noise remaining is the flame. While the contribution of a well designed laminar burner with mixing chamber to the noise is relatively low (except at the start of the vacuum UV range), as demonstrated by BOLING [140] and SLAVIN [1145],

Table 6. Reciprocal sensitivities and detection limits in atomic absorption spectroscopy (measured with a PERKIN-ELMER Model 503 atomic absorption spectrophotometer [1135])

Element	Wavelength nm	Reciprocal sensitivity µg/ml 1%	Detection limit µg/ml	Flame
Ag	328.1	0.06	0.002	Air/C_2H_2
Al	309.3	1.0	0.02	N_2O/C_2H_2
As*)	193.7	0.15	0.02	Argon/H_2 (diff.)
		0.8	0.15	Air/C_2H_2
Au	242.8	0.25	0.01	Air/C_2H_2
B	249.7	40	2.0	N_2O/C_2H_2
Ba	553.6	0.4	0.01	N_2O/C_2H_2
Be	234.8	0.03	0.002	N_2O/C_2H_2
Bi	223.1	0.4	0.003	Air/C_2H_2
Ca	422.7	0.05	0.001	Air/C_2H_2
Cd	228.8	0.03	0.002	Air/C_2H_2
Co	240.7	0.2	0.01	Air/C_2H_2
Cr	357.9	0.1	0.003	Air/C_2H_2
Cs	852.1	0.5	0.05	Air/C_2H_2
Cu	324.7	0.1	0.001	Air/C_2H_2
Dy	421.2	0.7	0.05	N_2O/C_2H_2
Er	400.8	0.9	0.04	N_2O/C_2H_2
Eu	459.4	0.6	0.02	N_2O/C_2H_2
Fe	248.3	0.2	0.01	Air/C_2H_2
Ga	287.4	2.2	0.1	Air/C_2H_2
Gd	407.9	17	1.2	N_2O/C_2H_2
Ge	265.1	2.5	0.2	N_2O/C_2H_2
Hf	286.6	15	2	N_2O/C_2H_2
Hg	253.7	10	0.2	Air/C_2H_2
Ho	410.4	0.7	0.04	N_2O/C_2H_2
In	304.0	1.0	0.02	Air/C_2H_2
Ir	264.0	10	1	Air/C_2H_2
K	766.5	0.05	0.001	Air/C_2H_2
La	550.1	45	2	N_2O/C_2H_2
Li	670.8	0.05	0.0003	Air/C_2H_2
Lu	336.0	7.5	0.7	N_2O/C_2H_2
Mg	285.2	0.01	0.0001	Air/C_2H_2
Mn	280.1	0.1	0.002	Air/C_2H_2
Mo	313.3	0.6	0.02	N_2O/C_2H_2
Na	589.0	0.015	0.0002	Air/C_2H_2
Nb	334.3	40	2.0	N_2O/C_2H_2
Nd	463.4	10	1.0	N_2O/C_2H_2
Ni	232.0	0.15	0.002	Air/C_2H_2
Os	290.9	1	0.1	N_2O/C_2H_2
P	213.6	300	100	N_2O/C_2H_2
Pb	283.3	0.7	0.01	Air/C_2H_2

Table 6 continued

Element	Wavelength nm	Reciprocal sensitivity µg/ml 1%	Detection limit µg/ml	Flame
Pd	247.6	0.3	0.02	Air/C_2H_2
Pr	495.1	25	5	N_2O/C_2H_2
Pt	265.9	2.0	0.1	Air/C_2H_2
Rb	780.0	0.1	0.002	Air/C_2H_2
Re	346.0	15	1	N_2O/C_2H_2
Rh	343.5	0.5	0.005	Air/C_2H_2
Ru	349.9	0.5	0.1	Air/C_2H_2
Sb	217.6	0.5	0.04	Air/C_2H_2
Sc	391.2	0.4	0.02	N_2O/C_2H_2
Se*)	196.1	0.25	0.1	Argon/H_2 (diff.)
		0.5	0.2	Air/C_2H_2
Si	251.6	1.5	0.02	N_2O/C_2H_2
Sm	429.7	10	2	N_2O/C_2H_2
Sn	224.6	1.1	0.02	Air/H_2
		2.5	0.5	Air/C_2H_2
Sr	460.7	0.15	0.002	N_2O/C_2H_2
Ta	271.4	20	1	N_2O/C_2H_2
Tb	432.6	7.5	0.6	N_2O/C_2H_2
Tc	261.5	3		Air/C_2H_2
Te	214.3	0.7	0.05	Air/C_2H_2
Ti	364.3	2	0.05	N_2O/C_2H_2
Tl	276.8	0.5	0.02	Air/C_2H_2
Tm	371.8	0.35	0.01	N_2O/C_2H_2
U	351.5	50	30	N_2O/C_2H_2
V	318.4	2.0	0.05	N_2O/C_2H_2
W	400.9	15	1	N_2O/C_2H_2
Y	410.2	1.8	0.1	N_2O/C_2H_2
Yb	398.8	0.1	0.005	N_2O/C_2H_2
Zn	213.8	0.02	0.001	Air/C_2H_2
Zr	360.1	10	1	N_2O/C_2H_2

* Measured with an electrodeless discharge lamp.

turbulent flames contribute maximum optical noise which is directly transferred to the signal, as the excellent investigations of LEBEDEV [737] have shown (see also section 3.1.3, page 30).

Figure 53. Influence of the radiation intensity on the signal-to-noise ratio; left: Old, low energy hollow cathode lamp; right: Intensitron® hollow cathode lamp.

ROOS attempted to determine the various causes of noise in atomic absorption quantitatively by means of individual functions [1046] [1047] [1050]. He found that for the majority of elements the flame was the biggest single contribution to the noise. This noise can consist of alterations of the atom concentration with time, alterations of the flame absorption (far UV), and emission noise (see page 81).

At this stage, the procedure recommended by a number of authors for increasing the sensitivity by passing the light emitted from the hollow cathode lamp through the flame several times will be briefly mentioned. This procedure is used almost exclusively with turbulent burners since the zone structure in laminar burners reduces the effect considerably. This procedure has little value, since not only the sensitivity but also the noise is correspondingly increased (the latter particularly strongly since the fluctuations in the upper flame zones are substantially greater), when the light is repeatedly passed through the flame. In general—and exceptions often prove the rule—a gain in the sensitivity is only obtained at the cost of a worsened signal-to-noise ratio and poorer detection limits. Furthermore, a large number of mirrors near the flame is always a potential source of increased light losses owing to insufficient reflection.

In a well designed atomic absorption spectrophotometer, instrumental causes for the occurrence of increased noise have been taken into consideration, i.e. the instrument is so constructed that the noise is as low as possible. Small signals can then often be electrically expanded (up to 100 times and more) and thus more easily read. These signals must be electronically averaged, for example by the introduction of RC units (resistor-capacitor combinations) or through intergration (see page 82), because the noise is expanded by the same amount with signal expansion.

Strong damping of a signal always brings an increased measurement period and principally influences the high frequency noise. Long term noise or drift is hardly influenced; for this reason, the use of high time constants is frequently limited by the instability of the instrument. These long term fluctuations are mainly caused by alterations in the emission of the light source or in the sensitivity of the detector; it is here that the double-beam system in atomic absorption spectroscopy shows its greatest advantage.

The second dimension influencing the detection limit of an element is the signal height or, in principle the same, the sensitivity. The sensitivity is dependent upon the length d of the irradiated layer (in cm) and the absorption coefficient κ. The degree of absorption and thereby the sensitivity are given by equation (1.14) (page 10), which simplified, takes the form:

$$\kappa_\nu = \text{const.} \cdot N_\nu \cdot f_\nu \,. \tag{6.5}$$

The only variable in this equation is N_ν, the number of atoms available for absorption of light of frequency ν. The factors governing the dimension N_ν will next be considered and the possibility of incomplete atomization will initially be disregarded.

Since the definition of the reciprocal sensitivity is based on weight, either absolute or weight per unit volume, and the total number of atoms N per unit weight is inversely proportional to the atomic weight, the reciprocal sensitivity must also be inversely proportional to the atomic weight. This is an aspect worth considering since the atomic weight of the lightest (lithium, atomic weight = 6.9) to the heaviest (uranium, atomic weight = 238) metals determinable by atomic absorption spectroscopy alters by a factor of approximately 35. It is then to be expected that the relative reciprocal sensitivity within a single group of the periodic table will decrease with increasing atomic weight. In table 7 an attempt has been made to take this proportionality into consideration and the reciprocal sensitivities divided by the atomic weights have been entered into the periodic table.

Since the absorption coefficient κ_ν of the transition for the line of frequency ν is not dependent upon the total number of atoms N per unit volume, but is dependent upon the number of atoms N_ν that are available for absorption of light of frequency ν, the magnitude of N_ν compared to N should next be considered.

It had already been mentioned in section 1.3 (page 5) that the general statement "free atoms in the ground state are able to absorb light of exactly defined frequency" requires further enlargement for atoms having complicated structures since the "ground state" is not always clearly defined. Rather, such atoms frequently exhibit a fairly large number of low-lying excited states which are more strongly populated at the existing flame temperatures than the actual ground state.

Capacho-Delgado and Manning [208] have published an interesting study which makes these conditions clear in the case of tin. With investigations using various flames, they found a stronger relative increase in the sensitivity for individual resonance lines when using an air/hydrogen flame (2050 °C) compared to an air/acetylene flame (2300 °C). The results are

Table 7. Reciprocal sensitivities of the elements determinable by atomic absorption spectroscopy in μMole/l 1% absorption

Li	Be											B	(C)	(N)	(O)	(F)
7.2	3.3											3700				
Na	Mg											Al	Si	P	(S)	(Cl)
0.65	0.4											37	53	10000		
K	Ca	Sc	Ti	V	Cr	Mn	Fe	Co	Ni	Cu	Zn	Ga	Ge	As	Se	(Br)
1.3	1.2	9	42	39	1.9	1.8	3.6	3.4	2.5	1.6	0.3	32	34	2.0	3.2	
Rb	Sr	Y	Zr	Nb	Mo	Tc	Ru	Rh	Pd	Ag	Cd	In	Sn	Sb	Te	(I)
1.2	1.7	20	110	430	6.3	30	5	4.8	2.8	0.6	0.3	8.8	9	4.1	5.5	
Cs	Ba	La	Hf	Ta	W	Re	Os	Ir	Pt	Au	Hg	Tl	Pb	Bi	(Po)	(At)
3.7	2.9	320	84	110	80	80	5.3	50	10	1.3	50	2.5	3.4	1.9		

Ce	Pr	Nd	Pm	Sm	Eu	Gd	Tb	Dy	Ho	Er	Tm	Yb	Lu	
	180	70		65	4.0	108	47	4.3	4.2	5.4	2.1	0.6	40	
		U												
		210												

compiled in table 8. Lines emanating from the ground state undergo an increase by a factor of about 2.6 in the cooler air/hydrogen flame, while lines emanating from metastable states (1692 *K* or 3428 *K*) experience a lower increase or none at all. This means simply that an alteration of the flame temperature by 250°C brings a distinct shift in the populations of the ground state and metastable, excited states.

The more complicated the structure of the electron shell of an element is, the more numerous are the metastable states. Especially when the hot nitrous oxide/acetylene flame is employed, the probability that all these states are occupied is greater. This necessarily results in a decrease of the ratio N_ν/N, which means that the sensitivity at frequency ν is lower and that, particularly noticeable for the lanthanides (table 24), there are a large number of resonance lines of comparable sensitivity. This is shown in table 9 with titanium as the example.

The ratio N_ν/N is especially unfavourable for elements that must be determined at lines emanating from little populated metastable states or corresponding to transitions of low probability. The resonance lines of mercury (184.9 nm) and phosphorus (177.5 nm) are in the vacuum UV range and thus inaccessible to flame atomic absorption spectroscopy. Mercury must therefore be measured at the less sensitive (by a factor of about 100) intercombination line at 253.7 nm

Table 8. Absorbances measured at various Sn resonance lines in the air/hydrogen and air/acetylene flames [208]

Wavelength nm	Energy level (K)	Absorbance H_2	Absorbance C_2H_2	$\frac{H_2}{C_2H_2}$
224.605	0–44 509	0.821	0.337	2.6
286.333	0–34 914	0.492	0.193	2.7
254.655	0–39 257	0.178	0.074	2.5
207.308	0–48 222	0.013	0.005	2.6
			Mean	2.6
235.484	1692–44 145	0.398	0.239	1.7
270.651	1692–38 629	0.201	0.194	1.7
303.412	1692–34 641	0.148	0.102	1.5
300.914	1692–34 914	0.087	0.052	1.8
219.934	1692–47 146	0.042	0.028	1.5
233.480	1692–44 509	0.041	0.023	1.8
266.124	1692–39 257	0.028	0.013	2.2
			Mean	1.7
283.999	3428–38 629	0.108	0.119	0.9
242.949	3428–44 576	0.105	0.109	1.0
226.891	3428–47 488	0.065	0.075	0.9
317.505	3428–34 914	0.053	0.055	1.0
220.965	3428–48 670	0.023	0.021	1.0
248.339	3428–43 683	0.014	0.009	(1.7)
228.664	3428–47 146	0.006	0.005	1.2
			Mean	1.0

Table 9. Titanium resonance lines [214]

Wavelength nm	Lower energy level (K)	Observed intensity	Reciprocal sensitivity (μg/ml 1% absorption)
260.515	170	4	8.6
261.128	387	7	6.0
264.110	0	4	4.7
264.426	170	5	4.4
264.664	387	8	3.7
294.200	0	16	5.2
294.826	170	17	4.4
295.613	387	21	3.4
318.651	0	117	3.0
319.191	170	165	2.6
319.990	387	185	2.0
334.188 I	0 }	230	3.7
334.188 II	4629 }		
335.463	170	170	2.9
337.145	387	220	2.0
337.748	387 }	93	7.0
337.758	170 }		
363.520	387 }	390	5.6
363.546	0 }		
364.286	170	430	1.8
365.350	387	465	1.6
372.981	0	195	5.8
374.106	170	250	2.6
375.285	387	356	2.5
375.364	170	47	—
394.778	170	175	19.0
394.867	0	273	10.5
395.639	170	286	9.5
395.821	387	426	5.7
398.176	0	500 }	6.2
398.248	0	100 }	
398.976	170	490	4.7
399.864	387	640	4.3

and phosphorus at the 213.6 nm line [809] emanating from a weakly populated metastable state (11361 *K*). As expected, the sensitivities of these elements are poor and fall out of line from the corresponding groups in the periodic table, as shown in table 7.

Apart from these influences on the sensitivity, which are "given by nature", a further important cause for an alteration in the total number *N* of atoms formed is the chemistry of the environment. This influence will be discussed in detail in the following section on chemical interferences. Ever since the beginning of atomic absorption spectroscopy, the most widely varying attempts have been made to find ways of improving the sensitivity. In principle, all these efforts have been directed either to increasing the total number *N* of atoms formed, or to prolonging the residence period of the atoms in the beam path. Details of the various procedures have already been given in earlier chapters and a further discussion is unnecessary. Once again, attention must be drawn to the fact that procedures increasing the sensitivity without improving the signal-to-noise ratio are valueless since the same effect can be achieved through simple electrical expansion of the signal.

It is also worth mentioning that the sensitivity is independent of the radiation intensity. Increasing the emission from the hollow cathode lamp brings no improvement in the sensitivity, but it has an advantageous effect on the detection limit since the signal-to-noise ratio is improved.

6.2. Interferences

Since atomic absorption spectroscopy is a relative method, i.e. quantitative determinations can only be carried out by comparison to reference substances, any behaviour of the sample which is different from the reference substance implies an interference. Depending upon their cause, such interferences are referred to as chemical, physical, background, or spectral.

While background and spectral interferences are based on the measurement of non-specific signals and occur seldomly in atomic absorption, chemical and physical interferences have an influence (positive or negative) on the total number of atoms *N* formed per unit volume.

In the following sections, the various types of interference, the causes of their occurrence, and the general possibilities for their elimination will be discussed. Relevant details will be found under the discussion on the element concerned and under the examples of application for specific problems.

6.2.1. Chemical interferences

Strictly defined, a chemical interference is any formation of a compound that prevents quantitative atomization of the element. This absolute definition, used especially when different methods of atomization are compared, can be replaced by a relative definition. For comparing different samples when the same means of atomization is being used—for example a flame—a chemical interference is understood as an alteration of the total number of free atoms formed per unit volume owing to the formation of chemical compounds. With this definition, it is necessary to state the reference standard against which the alteration is measured; usually it is an aqueous, neutral or weak acid solution of the pure metal or of a pure metal salt (chloride or nitrate). With this definition of chemical interference, it is possible that the effect can be positive or negative, i.e. relative to the standard, more or fewer atoms per unit volume can be formed in a sample. With the absolute definition of chemical interference, the effect can only be negative, i.e. every interference is recognized as a decrease in the total theoretical number of atoms that can be formed.

In principle, there are two causes for the occurrence of chemical interferences: Either the conversion of the sample into atoms is not quantitative, possibly because a difficultly melted or vaporized salt is formed or the molecules formed are not completely dissociated, or the free atoms react spontaneously with other atoms or radicals from the environment so that they are not long enough available for absorption.

The latter cause, spontaneous reaction of the atoms formed with other atoms or radicals, is typical for atomization in the flame. The magnitude of this interference is dependent upon the chemistry of the flame being used and upon the affinity of the atoms under observation to other possible reaction partners present in the flame. It alters according to the stochiometry of the flame and is often quite different in nature and extent for various flame types. For a given flame, this interference is virtually independent of the sample, so that its effect on the sensitivity which can be achieved for a determined element in this flame is not relative but only absolute. In other words, this interference only affects the determination of an element indirectly in that a constant fraction of the atoms formed are withdrawn from the absorption measurement, thus reducing the sensitivity. It does not directly interfere with a determination by positively or negatively influencing the total number of atoms formed compared to a reference substance.

This type of interference is principally caused by the formation of oxides (or oxide radicals) and hydroxides (or hydroxide radicals), or occasionally carbides or nitrides, from the free metal atoms with combustion products of the flame. Since these combustion products are always present in a large excess, the concentration of the metal atoms under study is more or less immaterial and the reaction equilibrium is independent of the concentration.

As a result of this interference, there are about 30 metallic elements that cannot be determined in an air/acetylene flame because they spontaneously form stable metal oxides. Among these elements are all the lanthanides, aluminium, silicon and boron. This interference is also responsible for the extremely low sensitivities of such elements as boron and uranium, even in the relatively hot and reducing nitrous oxide/acetylene flame. These elements are notorious because they form not only stable oxides, but also difficultly melted and dissociated carbides and even nitrides. Similarly, the elements in subgroups IV and V and also tungsten and several further elements can only be determined with relatively low sensitivities (see table 7).

A knowledge of these possible reactions enables a prediction to be made for every element as to which means of atomization gives the highest relative sensitivity and which chemical environment is suitable. It also explains why turbulent burners—regardless of the flame composition—are unsuitable for the determination of elements with a high affinity for oxygen. In turbulent burners, in contrast to laminar burners, thorough mixing with air already takes place in the lowest zone so that a further reaction of the atoms formed to give oxides is strongly favoured and the sensitivity is correspondingly influenced.

Apart from interferences caused by the flame—or the method of atomization in general—which only affect the achievable sensitivity in a given flame and which are independent of the sample matrix, chemical interferences that are dependent upon the sample matrix also exist. These are caused by further ions—the matrix—present in a solution that can influence the atomization of the metal ions under study relative to a reference solution. In this context, only the influence of the formation of chemical compounds on the atomization will be discussed.

The reference solution, which serves as an "interference free" standard, is generally a neutral or weakly acid aqueous solution of the pure metal or a metal salt, preferentially the chloride or the nitrate. These salts melt and dissociate very easily, so it can be assumed that they afford the optimum concentration of free atoms in a given flame. A chemical interference then exists if in a sample molecules are formed which are dissociated to a lesser extent than the standard, so that the concentration of the metal under study is misleadingly low. Occasionally a situation is observed where the molecules formed in the sample are more easily dissociated than the standard; here the signal is increased relative to the reference substance and a positive interference is spoken of. Since such cases, especially in aqueous solution, are relatively seldom, they will only be briefly mentioned here. A more detailed discussion will be found under the individual elements (chapter 8).

A carrier distillation was suggested by ROOS as the mechanism for this type of signal increase in the presence of certain matrix elements, i.e. the interfering matrix is more easily converted into the molecular state in the presence of other substances which act as carriers [1049]. Ammonium chloride or alkaline sulphates can be effective in this direction. WEST and co-workers have reported a second mechanism for the increase in the sensitivity through accompanying substances. This is the reduction of the lateral diffusion [1306]. As soon as an element is present in the gaseous state (as molecules or atoms) it can diffuse freely in all directions, therefore also to the sides. Thus the number of atoms in the centre of the flame (irradiated by the light beam) is reduced. Through additives which delay the atomization, solid particles are retained longer so that the time available for the atoms to diffuse to the edges of the flame is reduced. The atoms under study are therefore concentrated in the centre of the flame and diluted at the edges.

Various causes exist for the occurrence of chemical interferences based upon the formation of difficultly dissociated molecules of the metal under study, which result in a relatively lower number of free atoms per unit volume. When using flames as the means of atomization—and these will be spoken of principally here—it must always be borne in mind that the dissociation of molecules into atoms is a temperature dependent equilibrium reaction and that this reaction can be positively as well as negatively influenced by subsequent and parallel reactions. Next to this, it should not be forgotten that atomization in a flame is a dynamic atomization, i.e. time is a decisive factor. As explained earlier (see page 29), the processes from the drying of the solution droplets to the release of the atoms from the molecular bonding must take place within a few milliseconds, since this is the time it takes to traverse the space from the burner slot to the beam path of the light source. A cause for the occurrence of chemical interferences—or at least for a stronger occurrence—can be that the equilibrium between atoms and molecules has not been reached, or that solid particles, produced by evaporation of the solvent, have not been quantitatively melted and vaporized. Such observations could be made by quantitatively measuring an interference at various heights above the burner slot (compare [31]). While chemical interferences can occasionally be avoided or reduced by measuring in higher flame zones, the interferences caused by formation of compounds with flame components (e.g. oxygen) can be removed by measuring nearer to the burner slot [1048]. The greater the affinity of the element under study for oxygen is, the earlier and more rapidly the atom concentration falls with increasing height of observation. Two factors are then generally of great importance for the occurrence of chemical interferences: The droplet size of the sample aerosol and the burning velocity of the flame. All flames using pure oxygen as oxidant for example, are unsuitable for atomic absorption spectroscopy because their burning velocities are too high. Likewise, direct nebulizing burners are unsuitable because the volumes of the aerosol droplets vary too much. Many existing examples show that chemical interferences unknown to mixing chamber burners occur with direct nebulizing burners [14], and that interferences observed with mixing chamber burners are clearly more intense in direct nebulizing burners (see page 30). In mixing chamber burners the interference is sometimes merely caused by a displacement of the zone of optimum atom concentration, corresponding to a somewhat slower dissociation of the molecules, so that it is possible to eliminate minor interferences by measuring the absorption in a higher flame zone.

Besides these more trivial reasons, two very important factors must be mentioned, these are frequent causes for the occurrence of chemical interferences. A too low flame temperature or an unsuitable chemical environment often displaces the equilibrium between molecules and atoms to the left. It can be generally stated that with increasing temperature the number of interferences decreases. This is particularly noticeable when using the nitrous oxide/acetylene flame, since apart from the relatively high temperature (approx. 2750°C) it also offers the ideal reducing environment for the dissociation of a large number of compounds. In contrast to this, the even hotter oxygen/acetylene flame (approx. 3050°C) shows clear disadvantages and many interferences owing to its strong oxidizing properties.

From these observations on the causes of chemical interference and on its elimination, the view could be advanced that chemical interferences are due to false operation, i.e. to the use of an unsuitable means of atomization.

This view is justified, especially when reference is made to the large number of publications concerned with the most widely varying interferences which occur with elements in the alkaline-earth group (see [1012]). These interferences can virtually be eliminated without exception by the use of a premixed nitrous oxide/acetylene flame, as will be documented later.

Nevertheless, because the choice of means of atomization is not large, especially with flames, a number of interferences cannot be completely eliminated. On occasion, a situation arises in which an existing chemical interference can be eliminated, either in general or for a specific analysis, by choosing an alternative means, but this alternative brings other disadvantages. If this interference can be avoided by choosing another method, it may be the more suitable because it offers the lesser drawback.

Many chemical interferences can be eliminated by raising the flame temperature or by altering the chemical environment. If these methods are not practical or are, for any reason, not desired, chemical interferences can be eliminated chemically. In principle, three methods are available: 1) Addition of the interfering ions to the standard solution, or more generally, assimilation of the matrix in the sample and standard solutions. 2) Binding the interfering anion with another cation which is added in excess to the sample solution. 3) Shielding the cation, to be determined, by complex formation.

Frequently, only one of these possibilities is practical; nevertheless they should all be examined for their relative advantages and disadvantages. The first possibility, assimilation of the matrix, can only be carried out if the interference is not strongly marked and the matrix is simple. This method is frequently used for matching acid contents. This procedure finds further application when a number of minor interferences are to be expected from a complex matrix, but which is, however, the same in many samples. Occasionally, many minor and very minor interferences, which alone are of no importance, come together in a complex matrix so that the cumulative effect can be very noticeable. If a large number of such samples (e.g. stainless steel, blood serum, or industrial products in general) must be analysed on a routine basis, the analyses are frequently carried out against a standard substance. This is an average sample which has been exactly analysed by other methods for its content of the element under study. These standards are prepared for analysis in exactly the same way as the samples and usually permit very precise determinations, even in the presence of certain interferences.

The second method for eliminating chemical interferences is based on the realization that the binding can take place during the evaporation of the solvent, during the first precipitation from the supersaturated solution, or during rearrangement reactions within the solid particles. If a cation which gives a compound of lower solubility product with the interfering anion is added in excess to the sample solution, according to the law of mass action, the interfering ion will be bound almost quantitatively to the added cation. The element under study can then be atomized normally and determined without interferences. This procedure is also suitable for very strongly marked interferences, although it does have the disadvantage obtained through mixing in large excess, that contamination cannot always be avoided. The best known example of this type of interference elimination is the addition of lanthanum to calcium when it is being determined in the presence of sulphate, phosphate, aluminate, silicate, etc. in an air/acetylene flame.

The third procedure for eliminating interferences chemically is based on the same principle as the second method, but is effective in a different way. The cation under study is bound to a strong, chemically resistant and little dissociated complex so that salt formation with the interfering anion cannot take place. The metal passes into the flame within the "protective shell" of the complexing agent (frequently EDTA) which is first destroyed under the influence of the temperature and the metal then released. This procedure can also be widely applied, even though it is seldom described. Its sole disadvantage is that strong acid solutions cannot be used.

Although much has been written here about chemical interferences, and further interferences will be described in the following sections, it must be emphasized that the occurrence of genuine disturbances in atomic absorption spectroscopy is very low, especially when the "correct" means of atomization is used.

A number of the chemical interferences blamed on the flame already occur in the graduated flask or in storage vessels for sample and standard solutions. Neutral solutions of a large number of elements are unstable and have a tendency to hydrolyse—in contrast for example, silicic acid is rapidly precipitated from acidified silicon solutions. It should be borne in mind, especially with strongly diluted solutions or with a very low content of metal, that the precipitate cannot be recognized as such and frequently adheres to the walls of the vessel. Very diluted solutions are often only stable for short periods and rapid exchange reactions with the walls of the vessel can take place. These facts are often insufficiently observed since their effects only become noticeable with genuine trace analyses. It must be concluded, with the rather obvious statement: "Only such elements that are present in a solution can be determined from that solution."

Many of the interferences mentioned are typical for atomization in flames. With flameless atomization procedures, chemically inert or reducing atmospheres are used so that various interferences caused by the properties of the flame are excluded. A further typical characteristic of flameless atomic absorption, especially with graphite tube furnaces (to which the following is mainly referred), is the fact that although the atomization is a thermal dissociation of molecules into atoms (compare 3.2.1), it is not an equilibrium reaction. This is as a result of the extremely small amount of element under study which is typically atomized because the gas space in the graphite tube is chemically inert and considerably cooler than the tube surface.

An interference virtually unknown in flame atomic absorption, namely the sublimation of the element under study as molecules arises from these differences. Since the atomization of a sample in a graphite tube furnace can only take place from the graphite surface, a clear requirement must be made: The sample must remain in good contact with the hot graphite surface for as long as possible to become as completely atomized as possible. This means that the atomization, and thereby the sensitivity of a determination, can be favourably influenced if the sample is converted into a difficultly melted or vaporized form. In contrast to flame atomic absorption, the formation of sulphates, phosphates or oxides in the graphite tube is often advantageous.

Good contact with the graphite surface is nevertheless only desirable as long as the compound can be dissociated into atoms rapidly enough (0.1 s) at temperatures below 2500 °C. If this is not the case, the formation of compounds can hinder the atomization and result in a reduction of the signal height. It has been reported that this interference can be eliminated by integration of the peak area, but investigations of this are only in an early stage.

The most well known interference of this type is probably carbide formation. A differentiation must be made between two kinds of carbide formation. The formation of compounds with the graphite surface affects all samples in the same way and generally influences the sensitivity for a given element. It may be compared to oxide or hydroxide formation in the flame and is therefore to be regarded as an absolute and not a relative interference. Next to this, there is the formation of carbides with decomposition products from the sample, for example with finely divided carbon produced from ashing biological materials. While the former interference can be eliminated by suitably coating the tube material (compare 3.2.4), up to now it has not been possible to undertake measures against carbide formation with the matrix except to externally destroy the matrix beforehand.

A number of further interferences have been reported; however it is not yet clear which mechanism they obey. In all cases a reduction of the signal in the presence of higher salt contents is involved. While in a number of cases, such as the determination of trace elements in natural waters, waste water and digested biological materials [361] [1174] [1302], a plateau is reached when a determined quantity of salt is present (figure 54), i.e. with a further increase in the salt concentration no further reduction of the signal takes place, this cannot be found in other determinations. Frequently, no relationship can be found between the composition of the matrix and the reduction of the signal when determining trace elements in digested rock samples [1101] and similarly for the analysis of lead in urine (figure 55).

In the first case, similar to the introduction of solids directly into the graphite tube [768] [1295], one can probably speak of a general retardation of the atomization owing to poor contact of the element under study with the graphite surface. In the second case, a genuine chemical interference is almost certainly the cause.

The influence of acids upon determinations in graphite tube furnaces, which has often been reported [768] [984] [1293], can largely be traced to the formation of easily volatile compounds of the element under study and hence to sublimation as molecules. Losses owing to deflagration of the salts at increased temperature play a part, especially when using perchloric acid, and occasionally with nitric acid.

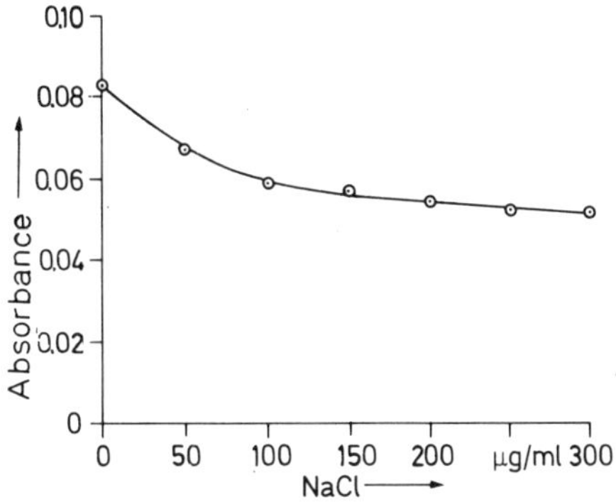

Figure 54. Influence of sodium chloride on the determination of lead in the graphite tube furnace.—After an initial loss in sensitivity of around 25%, the curve reaches a plateau at about 150 µg/ml NaCl. Higher salt concentrations do not further reduce the lead signal [361].

Nothing further will be said about interferences in graphite tube furnaces because the knowledge in this area is still relatively limited. It seems clear, however, that forming a compound of the element under study (through corresponding additions) which withstands the high decomposition temperatures, but which is rapidly and quantitatively dissociated at the atomization temperature, eliminates the majority of interferences. The importance of careful thermal pre-treatment for the separation of the matrix has already been mentioned (see 3.2.4) since this is the easiest way of eliminating interferences.

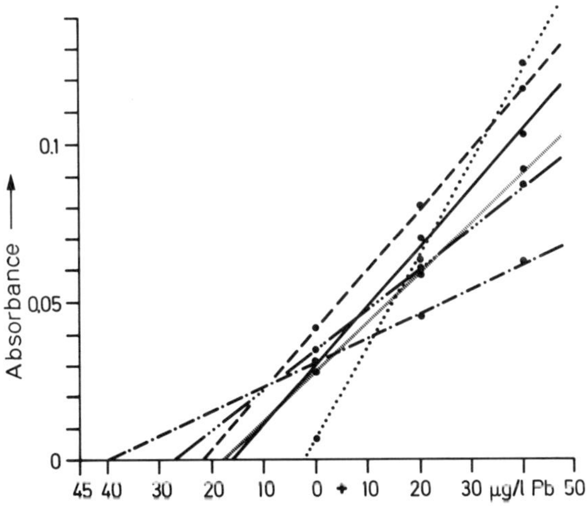

Figure 55. Determination of lead in urine in the graphite tube furnace. Since the influence of the calibration plot on the gradient is different for each urine sample, the method of additions must be used in every case.

6.2.2. Physical interferences

Physical interferences are understood as those caused by differing physical properties of the sample and the reference substance, such as different viscosities, surface tensions, or specific gravities of the solutions or solvents. Organic solvents belong to this category since they frequently exhibit totally different behaviour to aqueous solutions. However, since physical and chemical influences often act together, organic solvents will be treated separately.

The interferences caused by different viscosities or surface tensions, or by different specific gravities of the solutions or solvents are in principle disturbances to the nebulization procedure. If the viscosity (and the specific gravity) of an aqueous solution is increased by adding ever greater quantities of a salt, it can be established that with the same nebulizer settings less and less sample will be aspirated and that at the same time the droplets will become larger. This means that a greater portion of the solution will condense in the mixing chamber, so that less sample will reach the flame. This interference becomes gradually noticeable from about 1% total salt content, and organic macromolecules, such as proteins and sugars, naturally have a stronger effect than pure inorganic salts [683].

HÖHN and UMLAND have drawn attention to the fact that a reduction of the signal in the presence of increasing total salt concentrations must not necessarily be due to physical effects, i.e. to a lower aspiration rate of the nebulizer. Rather, it can be due to increasing embedding of the element in the salt which cannot be further dissociated [520].

In principle, these physical interferences have the same effect as chemical interferences by influencing the total number of atoms formed in the ground state per unit volume. This is also usually a negative influence, i.e. a decrease of the number of atoms formed, if organic solvents are not involved. The difference between chemical and physical interferences is usually simple as shown in figure 56. With a chemical interference, it can be seen that even with

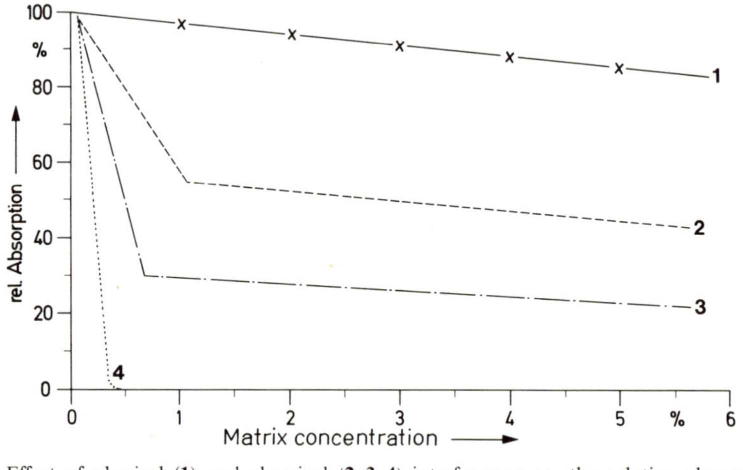

Figure 56. Effect of physical (**1**) and chemical (**2, 3, 4**) interferences on the relative absorption of a sample with increasing matrix concentrations.—While physical interferences usually cause an even decrease in the signal, chemical interferences generally produce a rapid decrease to a determined point where a break in the curve corresponds to a definite compound.

small additions of the interfering anion there is a sharp decrease in the absorbance of the sample, which after reaching a certain minimum value remains virtually constant. The position of the break in the curve is dependent upon the nature and concentration of the metal cations and interfering anions, and mostly corresponds to a defined stochiometric compound which is either not dissociated in the flame or only to a given percentage. Naturally deviations from this "ideal" course of a chemical interference are frequently found. A physical interference, on the other hand, is usually manifested by a slow and regular decrease of the sample absorbance with increasing salt content, or more generally, with increasing viscosity, specific gravity, etc.

In principle, physical interferences can be eliminated by matching the physical properties of the sample and standard solutions. This can be achieved by either diluting the sample solution or, similar to minor chemical interferences, by transferring the matrix of the sample solution to the standard solution. If neither diluting the sample solution nor matching the matrix is possible, the method of additions offers an extremely practical way of solving methodically this problem, as will be explained later.

A frequently recommended method for eliminating physical interferences is the use of an internal standard [353] [1034]. In this method, a known concentration of another element is added to the sample so that the influence can be determined, and this influence is then transferred to the element being analysed. The method can be generally applied to the determination of physical interferences but calls for several pre-requirements; among others, a dual-channel spectrophotometer should be used if possible since otherwise the measurement is too complicated.

Other authors have reported the use of a sample pump with a fixed liquid feed rate instead of a nebulizer to eliminate physical interferences [585] [1044]. The results are astonishingly good, but the system is mostly too slow for routine operation.

The typical physical interferences cannot occur in graphite tube furnaces because their causes lie in the nebulization procedure and samples are pipetted into the graphite tube, not sprayed. A "mechanical" influence was mentioned in the previous chapter and a further one should be mentioned here. Since the graphite tube must accommodate the entire atomized sample, there is a given sample limitation. Typically, not more than 100 mg sample can be atomized without the capacity of the tube being exhausted. If more sample is atomized, the excess is forced out of the tube and lost to the determination, which in effect means a loss of sensitivity.

6.2.3. Organic solvents

In principle, the use of organic solvents causes the same problems as high solids concentration, namely an altered viscosity, a different surface tension and an altered specific gravity. The difference is that the interferences here are mostly positive, i.e. there is an increase in the signal so that the disturbance is regarded as an advantage and not as an interference. For this reason, and because the effect is only partially of a physical nature, this topic will be treated separately.

The use of organic solvents, either mixed with water or pure, to achieve higher sample signals had already been recommended [116] and examined [133] [134] by a number of authors in connection with flame photometry. ALLAN [10] carried out the first investigations in atomic absorption spectroscopy and found similar effects for the sensitivity of the determination. This enhancing effect of various mixtures of organic solvents with water and also pure organic solvents in shown in table 10.

Table 10. Increase in the Cu signal through organic solvents in a pre-mixed flame (according to ALLAN [40])

Solvent		Relative sensitivity
0.1 N Hydrochloric acid		1.0
Methanol	40%	1.7
Ethanol	40%	1.7
Acetone	40%	2.0
Acetone	80%	3.5
Acetone + Isobutanol	20% + 20%	2.35
Ethyl amyl ketone		2.8
Methyl isobutyl ketone		3.9
Ethyl acetate		5.1

Since most organic solvents have a lower viscosity and a lower specific gravity than water, they are more easily aspirated. The surface tension, which is often substantially lower, brings about finer nebulization so that considerably more sample reaches the flame per time unit. These are the physical causes that lead to an increase in the signal; they are typical for all solvents. Next to this is a chemical effect, which causes certain differences from element to element [683].

While the dissociation of water (which always takes place in the flame when aqueous solutions are used) is a strong endothermic reaction that noticeably reduces the flame temperature, the combustion of an organic solvent is generally (except for CCl_4, etc.) an exothermic reaction which increases the flame temperature. A higher flame temperature can lead to better atomization of the element, so that an enhancing effect is also to be expected from this cause. Finally, many metal atoms are presumed to be more easily released from organic compounds than from various inorganic ones, since organic molecules are thermally less stable than inorganic. This fact can also lead to an increase in the signal; this effect, just as the increased flame temperature, means that serious chemical interferences are less probable in organic solutions.

For a number of elements it could be shown that a linear relationship exists between the increase in the sensitivity and the logarithm of the product between viscosity and boiling point [740]. The physical nature of the effect is evident for the element under study.

Since no attempt is usually made to eliminate this "interference" with organic solvents, the only requirement for a quantitative analysis is that the sample and reference substance are dissolved in the same solvent. Furthermore, it is often necessary to adjust the fuel gas flow to the new conditions. Since organic solvents usually function as an additional fuel gas, such flames burn much too "fat"; the actual fuel gas flow must be reduced until a flame with the original stochiometry is obtained. Since many organic solvents are miscible with water to various degrees and a small amount of water can considerably alter the combustion characteristics of such a solution, care must be taken to maintain the constant water content of sample, standard and blank solutions to prevent false measurements. Differences can often be optically recognized through the altered combustion characteristic of the flame.

6.2.4. Ionization interferences

Numerous elements are more or less strongly ionized, especially in hot flames. Since the ionized atoms have an altered electron shell and thereby exhibit a completely different spectrum, they are lost to a determination by atomic absorption spectroscopy at the usual resonance line. Therefore, by influencing the total number of atoms in the ground state available per unit volume for absorption, ionization has the same effect as the other interferences which have been considered.

Accordingly, an ionization interference brings a reduction in the sensitivity which is dependent upon the flame temperature and the ionization potential of the element under study. Furthermore, the extent of the ionization for a given element at a given temperature is dependent upon the concentration of that element. The ionization is stronger at lower concentrations than at higher concentrations, as shown in figure 57 on the barium example. This results in the concave calibration curve (from the concentration axis) which can be explained by the increasingly rapid recombination of ions and electrons at higher concentrations.

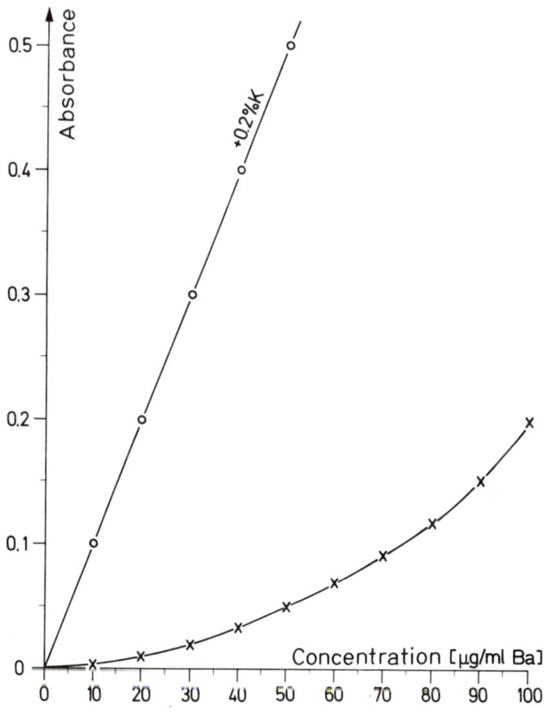

Figure 57. Ionization interferences during the determination of barium in a nitrous oxide/acetylene flame.—The lower curve (×) was measured with aqueous solutions, the upper curve (o) was measured with an addition of 0.2% potassium (as KCl) to suppress the ionization.

For practical analyses a calibration plot which is curved due to ionization interferences has two crucial disadvantages. First, many measurement points are required to clearly determine the curvature. Second, the loss in sensitivity strongly affects the lower concentrations, making trace analyses virtually impossible. For these reasons, elimination of this interference is desirable.

In principle there are two possibilities for eliminating ionization interferences. For example, the element concerned can be atomized in a flame of lower temperature. This procedure is suitable for the alkali metals because although they are markedly ionized in an air/acetylene flame (see table 11), they are virtually no longer ionized in the slightly cooler air/hydrogen flame. Nevertheless, for the majority of elements this method is not suitable because they are either no longer determinable in the cooler flame (lanthanides, etc.), or considerable chemical interferences must be reckoned with (e.g. barium).
The second method of eliminating ionization interferences is to shift the ionization equilibrium

$$Me \leftrightharpoons Me^+ + e^-$$

to the left on the basis of the law of mass action by producing a large excess of electrons in the flame. In practice, this can be carried out quite simply by adding a large excess of

Table 11. Ionization of a number of elements in the air/acetylene and nitrous oxide/acetylene flames

Element	Concentration employed µg/ml	% Ionisation	
		Air/acetylene	N_2O/acetylene
Li	2	0	
Na	2	22	
K	5	30	
Rb	10	47	
Cs	20	85	
Be	2	—	0
Mg	1	0	6
Ca	5	3	43
Sr	5	13	84
Ba	30	—	88
Y	100		25
Lanthanides			35–80
Tm	50		57
Yb	15		20
Lu	1000		48
Ti	50		15
Zr	500		10
Hf	1000		10
V	50		10
U	5000		45
Al	100		10

an easily ionized element (mostly potassium, or also cesium) to the sample and standard solutions. This is largely ionized in the flame and quantitatively supresses the ionization of the element under study (see figure 57).

The fact that the ionization equilibrium can be displaced is another important reason why ionization interferences can become serious and why it is absolutely essential that they be eliminated. If a standard solution containing the element as a pure salt is used, the element will be ionized to a certain degree. If the sample solution contains other metals as well as the element under study, which is usually to be expected, and these are possibly more easily ionized or are present in greater quantities, the ionization equilibrium of the sample will be suppressed, i.e. a relatively higher signal appears. False measurements of not inconsiderable extent are thus possible, and they should under all circumstances be avoided.

Because chemical interferences decrease with increasing flame temperature while ionization interferences increase, a choice often has to be made between various elements as to which interference is more difficult to eliminate. Frequently the choice will fall to the flame of higher temperature because ionization interferences are in general easier to control. However, the decision is also frequently dependent upon the matrix; the choice of the most suitable means of atomization in dependence upon the matrix will be more fully described in the practical section.

KORNBLUM and DE GALAN [698] found that elements with an ionization potential of less than 5.5 eV are almost completely ionized in a nitrous oxide/acetylene flame and thus at least 10 g/l of another easily ionized element (cesium or potassium) must be added to enable measurements free of interferences to be made.

Elements with ionization potentials up to about 6.5 eV require 1–2 g/l and those with ionization potentials over 6.5 eV only require about 0.2 g/l of an easily ionized element to eliminate the interference.

Up to the present, no ionization could be detected in graphite tube furnaces. LUECKE and co-workers [768] even found a clear loss of sensitivity when cesium salts were added to easily ionized elements.

6.2.5. Background absorption

There are two fundamental causes for the occurrence of background absorption effects: Light scattering on solid or liquid particles in the flame and true absorption of the light by molecules or radicals, mainly from the sample matrix. This background absorption is due to a non-specific (not caused by the specific element) light loss which leads to a signal that is too high. In fact, the signal consists of the true absorption of the element under study and the non-specific absorption of the background. Simple separation of both signals is not possible.

The first effect, scattering of light from the hollow cathode lamp on solid or liquid particles in the flame, obeys RAYLEIGH's stray light law in the first approximation, in which the scattering coefficient τ is given by

$$\tau = \frac{J_t}{J_0} = 24\pi^3 \frac{N \cdot v^2}{\lambda^4}. \tag{6.6}$$

According to this, the effect is directly proportional to the number N of scattering particles per unit volume and to the square of the particle volume v, and inversely proportional to the fourth power of the wavelength λ. This scattering effect will thus occur quite strongly with increasing particle size (τ increases by about 64 times for a doubling of the particle radius) and for decreasing wavelength (from 800 nm in the visible range to 200 nm in the UV range, τ increases by 256 times).
From these considerations it can be stated that interferences of this type must be considerably stronger with a turbulent burner. As KERBYSON [639] has established, the maximum droplet volume of a direct nebulizer is about 100 times greater than of a mixing chamber burner. The light scattering must therefore be greater by a factor of about 10,000.
Experience has shown that background effects very seldomly occur with a well-designed mixing chamber burner, especially when the total salt concentration of the solution is under 5% and the wavelength is greater than 300 nm. Certain interferences are to be expected with a larger portion of solids in the solution, particularly at wavelengths below 250 nm.
The second effect, true absorption of the light from the hollow cathode lamp by molecules or radicals in the flame, is nothing other than molecular spectroscopy in the UV (which, nevertheless, is normally carried out in solution). Molecules can be produced in the flame in great number when an easily atomized element is to be determined in the presence of a difficultly atomized matrix in a flame of low temperature, or when an element present in higher concentrations in the matrix forms oxide molecules or hydroxide radicals with the flame gases. A prerequisite for an interference is naturally that the element under study and the molecule absorb in the same spectral range. A typical example is the CaOH band (figure 58) in an air/acetylene flame whose absorption maximum coincides almost exactly with the resonance line of barium.

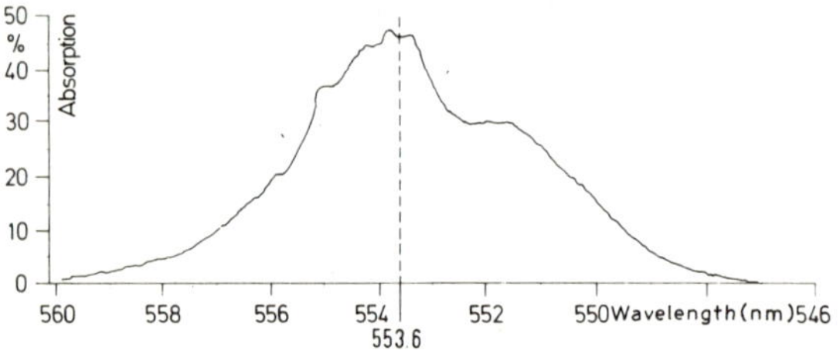

Figure 58. Absorption of the CaOH radical (1% Ca solution) in an air/acetylene flame in the range of the barium resonance line (553.6 nm).

CAPACHO-DELGADO and SPRAGUE [210] found that a 1% calcium solution measured at the barium resonance line (with a Ba hollow cathode lamp) gave an absorption of 50%. This effect could not be observed in a nitrous oxide/acetylene flame.
The best remedy for these interferences caused by molecular absorption would appear to be a flame of higher temperature which prevents the formation of the interfering molecules or which can dissociate stable compounds. However, for reasons already explained under chemical interferences, this method is not always practical.
It is typical for this type of interference, i.e. for all types of non-specific light losses, that the background absorption is a broad band absorption. It does not alter measurably over several nm while atomic absorption is restricted to a very narrow wavelength range (approx.

0.002 nm). All attempts to eliminate the interference are based on this fact. The fact that the constancy of this interfering absorption cannot be guaranteed outside a limited spectral range of a few tenths of a nm is of equal importance. According to RAYLEIGH's stray light law, light scattering at 200 nm can change by about 20% over a distance of 10 nm. With molecular absorption, the alterations are much more rapid, as can be seen in figure 58. A basic requirement for all procedures correcting non-specific light losses is that the correction should take place "as near as possible" to the actual absorption line.
The occasional recommendations for measuring against a "blank solution" which exhibits the same non-specific background absorption as the sample, but which does not contain the element under study, cannot be carried out in practice since such blank solutions are mostly unobtainable. Attempts to prepare these blank solutions synthetically almost always founder on the high purity requirements placed on such substances.
The same is equally applied to the internal standard method, which likewise postulates that the metal serving as reference is not contained in the sample. Thus the "next best" element must be employed to correct the non-specific light losses [1171], i.e. the element whose wavelength lies nearest to that of the element under study. Frequently, the next resonance line is so far removed that only a qualitative, or at the best a semi-quantitative, statement can be made on the light losses at the wavelength of interest.
For this reason, WILLIS [1330] and SLAVIN [1140] recommended that the non-specific light losses should be corrected by measuring the effect at a non-absorbable line and then subtracting the results from the measurement signal. Non-absorbable lines are understood as emission lines ending at a higher excited state, ion or filling gas lines, or in general emission lines which exhibit no measurable absorption for metallic elements in a given flame. While such lines exhibit no measurable specific absorption, they can be attenuated by light scattering, molecular absorption, etc., in other words, exhibit non-specific light losses. The use of non-absorbable lines for the correction of non-specific light losses has the advantage that the absorption measured at such a line can only be non-specific, and further, that many more of these

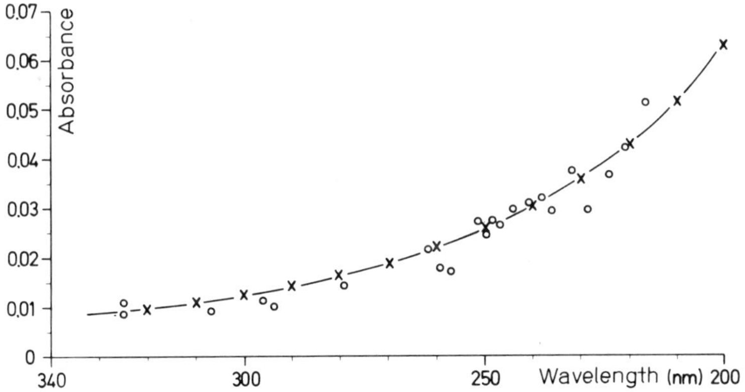

Figure 59. Non-specific light losses caused by light scattering at various wavelengths; 1% Ca solution in an air/propane flame: o values according to BILLINGS [123]— x theoretical curve from the RAYLEIGH stray light law.

lines are available than resonance lines. BILLINGS [123] applied this technique to numerous elements in relatively cool flames and found it quite practical. The stray light absorption determined by BILLINGS at various wavelengths for a 1% Ca solution in an air/propane flame and the calculated curve are shown in figure 59.
However, exact correction is not always possible, because this method has the disadvantage that non-absorbable lines directly neighbouring the resonance line are often unavailable. The precision can be increased by making measurements at several non-absorbable lines on either side of the resonance line; but this procedure is time-consuming and complicated.
For eliminating background effects KOIRTYOHANN [681] and L'VOV [778] used a dual-channel system in which light from the hollow cathode lamp and light from a continuous source was passed alternately through the flame (or the graphite tube). KAHN [608] described the first commercial system of this type which could be used in double-beam atomic absorption spectrophotometers. As depicted in figure 60, light from a hollow cathode lamp and light from a deuterium arc lamp is passed alternately in rapid sequence through the flame by a rotating mirror-chopper. After passing through the monochromator, both beams fall on the same detector. The double-beam electronics then form the ratio of both radiation intensities.

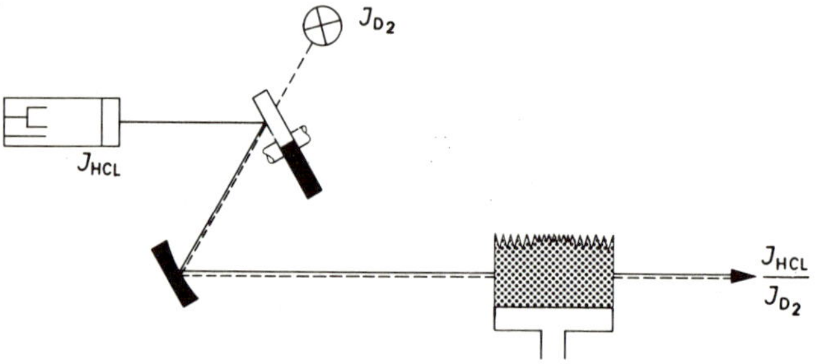

Figure 60. Schematic construction of a deuterium background corrector.—By means of a rotating mirror-chopper, light from a hollow cathode lamp (J_{HCL}) and light from a deuterium lamp (J_{D_2}) is sent alternately through the flame.

The function of this deuterium correction system is shown in figure 61. The exit slit of the monochromator separates the resonance line from the emission spectrum of the hollow cathode lamp, while a band from the continuum of the deuterium lamp, corresponding to the spectral slit width passes through. The half-width of the emission line from the hollow cathode lamp is approximately 0.002 nm; the width of the observed continuum—according to the set bandwidth—is generally 0.2 nm or 0.7 nm. The intensities of the hollow cathode lamp J_{HCL} and the deuterium lamp J_{D_2} are balanced before the measurement so that $J_{D_2}/J_{HCL}=1$ and no reading is obtained on the display.
If normal absorption of the element under study takes place, J_{HCL} will be attenuated by an amount corresponding to the metal concentration. Naturally J_{D_2} is attenuated at the resonance wavelength. However, since the half-width of an absorption line is in the order of 0.003 nm and the observed bandwidth of the continuum is at least 0.2 nm, even with 100% absorption of the resonance line from the hollow cathode lamp, only a maximum of 1.5% of the D_2 radiation will be absorbed. Therefore, for normal measurements the absorption of the deuterium radiation is below 1% and can be ignored. This is especially valid for bandwidths of 0.7 nm and over. While J_{D_2} falls virtually unattenuated on the detector, J_{HCL} is weakened by an amount corresponding to the metal concentration in the sample, so that $J_{D_2}/J_{HCL}>1$ and

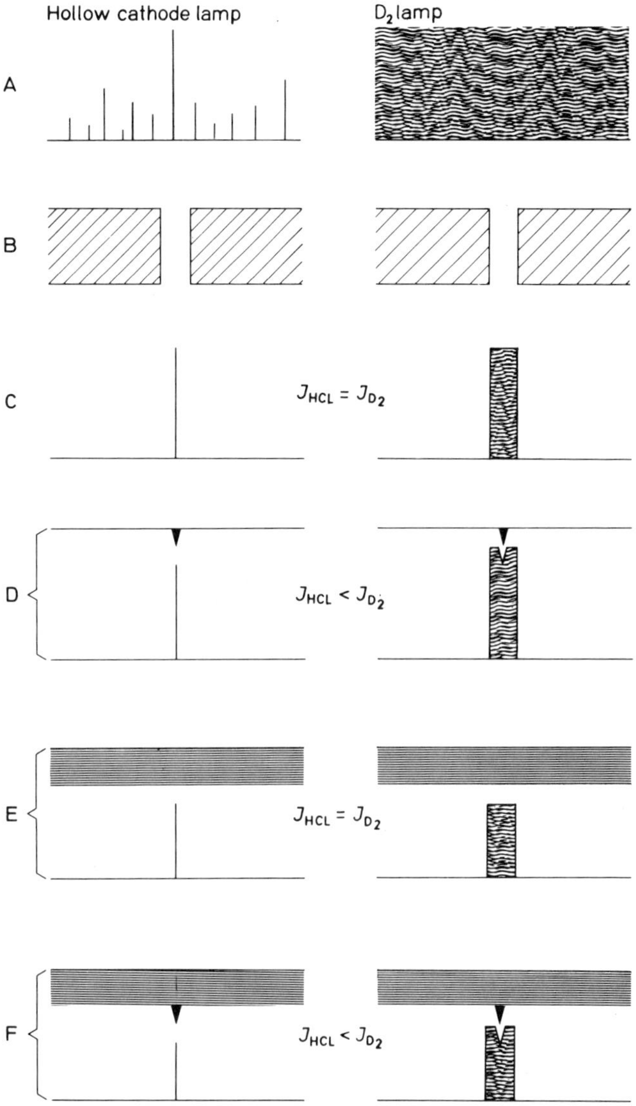

Figure 61. Mode of function of the deuterium background corrector.—**A** The hollow cathode lamp emits a line spectrum; the deuterium arc lamp a continuum. The exit slit **B** of the monochromator isolates the resonance line from the spectrum of the hollow cathode lamp with a half-width of approx. 0.002 nm; it also separates a band from the continuum of the D_2 lamp corresponding to the set slit width (approx. 0.2 nm or 0.7 nm). **C** The intensities of both light sources are equalized within the observed spectral range. **D** In the event of normal line absorption by the element under study, J_{HCL} is attenuated by an amount corresponding to the concentration of this element while J_{D_2} in the first approximation, remains unattenuated. **E** Non-specific, broad band background absorption attenuates the intensities of both lamps to an equal degree. **F** An additional line absorption again attenuates J_{HCL} by an amount corresponding to the concentration of the element under study, while J_{D_2}, for most purposes, is not further attenuated.

a deflection on the display will be observed. In effect, J_{D_2} serves as reference beam in a double-beam system. In the case of non-specific light losses, whether by light scattering or by molecular absorption, both light beams will be absorbed by the same percentage since background absorption is a broad band absorption. Even in extreme cases if the background effect remains constant over the small spectral bandwidth of 0.2 nm (or 0.7 nm), only minute alterations are to be expected and can be disregarded. Since the non-specific light losses weaken J_{D_2} and J_{HCL} by the same degree, $J_{D_2}/J_{HCL}=1$ and no deflection is seen on the display. Thus the background effect is corrected.

If true element specific absorption occurs in addition to the non-specific absorption, J_{HCL} is naturally further attenuated, leading to a readout as for normal absorption.

This correction of non-specific light losses made by comparing the attenuation of the emission line of the hollow cathode lamp to that of a band continuum is decisively advantageous over other procedures because the correction takes place directly on the line instead of near it. Thus, a true correction of the effect takes place with high accuracy and reliability and without additional time consuming operations.

While interferences owing to non-specific light losses occur seldomly in a well-designed mixing chamber burner, i.e. only when using solutions with high total salt contents and in the far UV, they occur quite frequently in flameless atomic absorption. It is hardly surprising therefore, that background correction has also been further developed with the increasing interest in graphite tube furnaces. Thus, single-beam instruments capable of accepting a background corrector have been introduced onto the market [1292], as well as double-beam instruments which operate with background correctors on the double-beam principle in both channels. Furthermore, besides the deuterium lamp, a halogen lamp has been introduced so that correction over the entire spectral range is possible. In this connection it should be briefly mentioned that the use of a hydrogen hollow cathode lamp for correction is not very suitable because it exhibits an intensity which is much too low (figure 62) and leads to strong noise during the measurement.

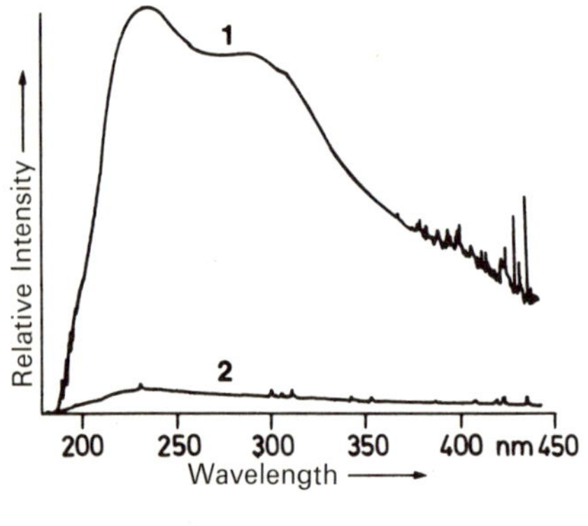

Figure 62. Emission of a deuterium arc lamp **1**, operating current 800 mA, compared to a hydrogen hollow cathode lamp **2**, operating current 40 mA.

Since interferences owing to non-specific light losses usually increase with decreasing temperature, and the flameless determination of mercury is the method which has been examined and employed longest, the majority of investigations on the correction of non-specific light losses have been carried out in this field. JAMES and WEBB used a double-beam instrument with a single light source, two absorption cells and two detectors. The gas stream containing the mercury was split and the mercury was removed from one of the streams with palladium chloride, so that only the non-specific absorption was measured [568]. The difference between the two channels was then the true mercury absorption.

BARRINGER [81] and LING [756] [757] used a correction system in which the centre of the Hg resonance line is completely absorbed from the emission lines by saturated (cold) mercury vapour so that the non-specific absorption is only measured with the flanks of the emission line. This system is very reliable for mercury because the high pressure lamps emit fairly broad lines, while at room temperature mercury is absorbed in very narrow bands. Thus the requirement that the correction should take place as near to the resonance line as possible is fulfilled in an almost ideal manner.

KAHN and MANNING [614] drew attention to the fact that although there are a number of other correction procedures, simultaneous correction with a continuous source (deuterium or halogen lamp) remains the simplest and most flexible method. This is especially true for flameless atomic absorption with its time-dependent signals, since correction through determination of the non-specific signals in a second measurement is extremely dubious.

MASSMANN referred to the fact that correction through comparison with a continuous source is not absolutely reliable in all cases since some molecular absorption bands are so narrow that they can no longer be regarded as continuous. In such cases, the requirements for this correction procedure are not fulfilled, so that it cannot function faultlessly. This information serves principally to show that this system should be used critically and that a check should occasionally be made to see whether all the requirements for faultless correction are given.

For example, it is important to check whether the background absorption (not the absorbance) is wavelength dependent. If a slit width which is smaller than that normally used for the determination improves the correction, the greatest care is required, since the non-specific light losses are certainly not continuous.

On occasions, the non-specific light losses can be so high (e.g. considerably above absorbance 1.0) that correction is no longer possible. This can be checked by measuring a nearby absorption line (at which only the background appears) with the background corrector in operation. If a signal appears, the correction is not complete.

As shown in practice however, these limitations only occur in isolated cases, so that the correction of non-specific light losses through comparison with the absorption of a continuous source is extremely useful, especially in flameless atomic absorption [199–201] (compare figure 22, page 43).

GUÇER and MASSMANN have made detailed investigations on the nature of non-specific light losses in graphite tube furnaces [447]. They recorded the dissociation continua of a large number of molecules that exhibit characteristic broad maxima (figure 63) and whose long

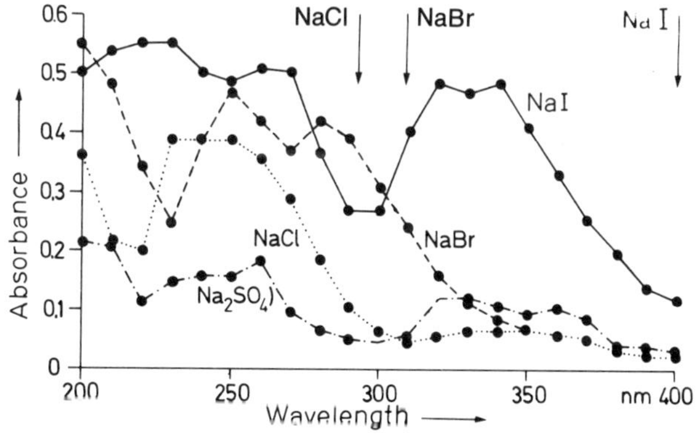

Figure 63. Dissociation continua of NaCl, NaBr and NaI in the graphite tube furnace. Limiting wavelengths for photodissociation are indicated with arrows (according to MASSMANN [447]).

wave limits are determined by the dissociation energy of the molecule. Electron band spectra which can also occur besides this (e.g. CN bands when using nitrogen as inert gas or C_2 bands), can make background measurement with continuous sources unreliable through their sharp band absorption. MASSMANN has drawn especial attention to the NO_2 and SO_2 bands which occur during vaporization of nitrates and sulphates. While molecular absorption exhibits a very reproducible pattern owing to the very rapid (<0.1 s) diffusion of the gaseous particles throughout the tube, light scattering on solid particles is irregular owing to the considerably slower and irregular diffusion. Accordingly, it cannot be reliably determined through a second measurement.

6.2.6. Spectral interferences

Spectral interferences in atomic absorption spectroscopy are virtually impossible as a result of the alternating light system, with which the emission light from the hollow cathode lamp is modulated, and the amplifier which is tuned to the modulation frequency. Nevertheless, spectral interferences can be caused with an alternating light spectrophotometer for two reasons. If multielement lamps which have unsuitable element combinations are employed, i.e. within the observed spectral bandwidth the resonance line of another element as well as that of the element under study falls on the detector, the selectivity and element specific readout of atomic absorption are no longer given. The same can occur if unsuitable cathode support materials are used for easily melted elements. Likewise, interferences can occur with multielement lamps if the slit widths recommended by the manufacturers are not used and resonance lines of several elements fall simultaneously on the detector.
A further potential cause for the occurrence of spectral interferences, which has been drawn to attention by FASSEL [342], is the overlapping of absorption lines. The half-width of an absorption line is reported to be 0.003–0.004 nm; this is by no means the total width of the line however, and frequently the flanks of the line extend over a relatively wide range. For calcium, FASSEL was still able to measure 10% of the absorption which occurs at the centre of the line at a distance of 0.03 nm from the centre. On this basis, the total width of an absorption line can increase to 0.1 nm. If the flank of such an absorption line coincides with the resonance line of the element under study, an effect similar to that described for molecular absorption can occur. Such cases are very rare in practice however, and will not be further discussed here. Furthermore, the few cases of overlapping reported in the literature do not refer to the main resonance lines, but to less sensitive lines which are hardly used in practice. The single exception has been reported by KELLY and MOORE [632], who found overlapping of the iron line at 213.859 nm by the resonance line of zinc, which leads to an interference in the determination of small quantities of zinc in iron.
In general therefore, despite the above limitations, atomic absorption spectroscopy is free from spectral interferences. It is virtually impossible for one element to simulate the presence of another if a well designed alternating light instrument and a faultless hollow cathode lamp are employed. When certain basic rules are observed, atomic absorption spectroscopy is selective and element specific.

6.2.7. Analysis of solid samples

Flame atomic absorption with its pneumatic nebulizer system for introducing the sample into the flame is almost exclusively limited to the use of solutions. While colloidal solutions can usually be determined without interferences [469], suspensions can cause considerable difficulties at times, as shown on the example of iron particles in used lubricating oil [83] [709]. Direct introduction of solid samples into the flame, using a feed screw for example [435] [436], brings a variety of problems and is only practical for a few elements and samples. Relatively high coefficients of variation, which lie around 10–15% [436], are typical for these applications.

Interest in the direct analysis of solids was re-awakened with the introduction of flameless atomic absorption in the graphite tube furnace. BARNETT and KAHN determined copper in fingernails [77]. KAHN and co-workers determined lead in leaves [612], and various authors have reported the direct analysis of easily atomized elements in geological and mineralogical samples [496] [731] [768] [1295]. In all cases, good agreement with other analytical methods or with the examination of dissolved samples could be obtained, i.e. with careful work the accuracy of the determination is guaranteed. In most cases the coefficients of variation were around 10–20%, however.

L'VOV [8] investigated the problem of direct solids analysis very thoroughly and examined the causes of poor reproducibility. He found that the typical error for dispensing liquids was around 0.025 µl, while it was about 0.1 mg for solids. Apart from this, the homogenity of solid samples plays a crucial role.

For example, a grain has a volume of 1.4×10^{-11} cm^3, i.e. a weight of about 4×10^{-11} g (with a specific weight of 3 g/cm^3) with a grain size of 3 µm. With a concentration of 1×10^{-4}% of the element under study, 1 mg of sample then contains 25 grains with the element. The coefficient of variation in the number of grains n in a single sample then corresponds to

$$\frac{\Delta n}{n} = \frac{1}{\sqrt{n}}, \tag{6.7}$$

i.e. when $n = 25$, the coefficient of variation for 1 mg of sample is 20% solely due to the statistical distibution. L'VOV also gives a quantitative relationship between the coefficient of variation V and the mean grain size P of the sample in which

$$V = \frac{1}{\sqrt{P}}. \tag{6.8}$$

The obvious requirement for the finest possible distribution of the sample can be seen from this. Nevertheless, this is not always without problems since every grinding procedure is a potential source of contamination.

If it is further considered that the direct analysis of solids is beset by a number of other difficulties apart from those mentioned above, for example that calibration is not always simple and that multiple determinations involving multiple weighings are required (unless

only semi-quantitative analyses are performed), it is nowhere near as attractive as it would appear at first sight.
It is therefore expedient at this point to state that the best way of homogenizing a sample is always a procedure of solution, and that solutions can be dispensed more reproducibly and with fewer problems than solids. Furthermore, a number of rapid and reliable digestion methods are nowadays available, even for samples that were once considered difficult (see chapter 9), so that dispensing of solutions is to be preferred.
Nevertheless, there are applications for which the direct introduction of solids is completely suitable, such as the semi-quantitative analysis of samples for a determined element in geochemical prospecting. There are also samples in which the element is so homogeneously distributed that the above difficulties do not arise, this is the case with various polymers, which have been directly analysed with very good success [634] [637].

6.3. Analytical Procedure

Mention has already been repeatedly made that atomic absorption spectroscopy is a relative method which gives no absolute values, but which produces quantitative results through comparison with samples having known contents. According to the analytical conditions and the requirements of precision or speed, various procedures for carrying out an analysis are available. The advantages and disadvantages of the individual analytical methods will be briefly examined in the following section. Attention is drawn to the fact that the occurrence of errors can often be traced to the choice of a method which is not entirely suitable [1133].

6.3.1. Routine procedure

The simplest and most rapid method is the direct comparison with standard solutions of known concentration, the so-called routine procedure (figure 64). Three standard solutions of known metal content having absorbance values between approximately 0.2 and 0.8 are prepared (by diluting a stock solution) and a calibration curve is prepared by plotting the absorbances versus the corresponding concentrations. The curve so obtained is mostly linear, i.e. obeys the BEER-LAMBERT law, or at higher absorbance values has a slight curvature towards the concentration axis. The absorbance of the sample, which should be between the standard values, is entered into the plot and the concentration can be read off.
If the atomic absorption spectrophotometer being employed has a readout linear in absorbance and offers the possibility of stepless expansion and compression of the signal, plotting calibration curves can be forgone and the readout can take place directly in concentration. A built-in curvature corrector for the linearization of curved calibration plots considerably extends the range of application of this type of concentration readout.
A prerequisite for this routine procedure is that the sample and standard solutions behave in the same manner in the flame within the tolerances, i.e. no measurable interferences are present, or they have been completely eliminated. This can frequently be assumed when the matrix is known exactly, so that this procedure can be applied in many cases with success.

The obtainable accuracy with the routine procedure lies around 1% relative, which is fully sufficient for a large number of analytical problems.

6.3.2. Precision analysis

The precision of the routine procedure can be further increased if the atomic absorption spectrophotometer being used has sufficient stability and if the corresponding technical equipment is available. MEDDINGS and KAISER [854] found that the precision for routine analyses could be improved to 0.3–0.8% solely through the use of digital readout with automatic mean value formation. KAHN [607] described a procedure which enables higher concentrations to be determined with good precision through the use of baseline suppression and scale expansion.

It is frequently desirable to avoid the higher dilutions which become necessary because of the high sensitivity of atomic absorption and which are a potential source of error. Through the use of a less sensitive resonance line [1] [403] [1078] and by setting the burner head at right angles to the beam path [1217], the majority of samples and the majority of elements can be brought to the desired concentration range with only one dilution. Furthermore, matching the matrix so that potential interferences are eliminated is usually indispensable for a precision analysis. It is therefore recommended, especially with complex matrixes, that the analysis is carried out against a standard substance whose composition has been independently determined and is exactly known [1299].

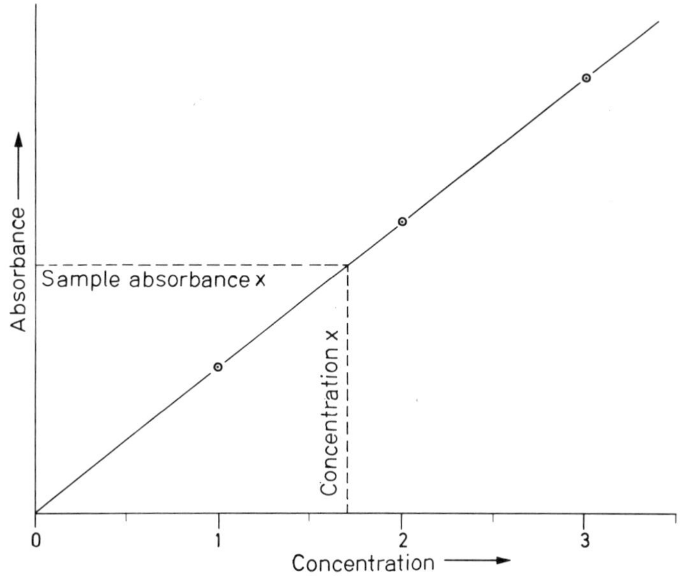

Figure 64. Routine procedure.—The absorbances of various standard solutions of known concentration are plotted against their concentrations. Determination of the unknown concentration x in a sample is by linear interpolation.

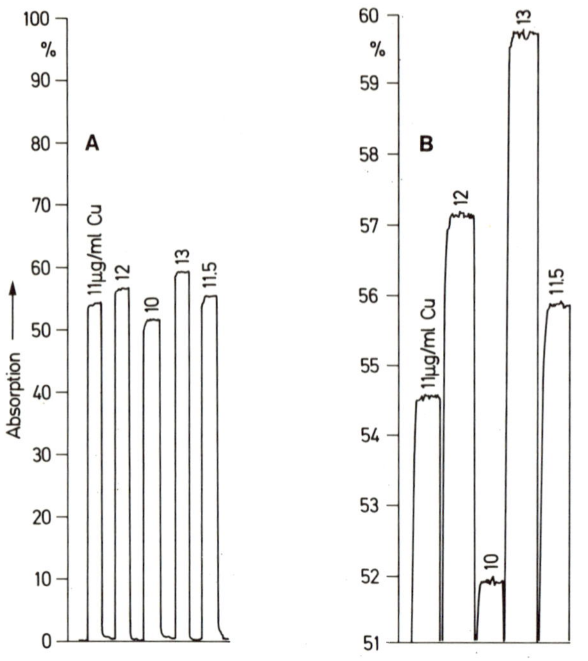

Figure 65. Determination of copper by high precision analysis.—Five concentrations between 10 and 13 µg/ml Cu, **A** without scale expansion, **B** with 10 times scale expansion, increased damping and suppressed baseline. Readout accuracy ±0.005 µg/ml Cu.

Even if these requirements have been fulfilled, the problem still frequently remains for the determination of higher concentrations that only a small range of the calibration plot is of interest and the values here cannot be read with sufficient accuracy. The above mentioned procedure, baseline suppression and scale expansion, best meets the requirements for highest precision within a narrow absorbance range; its function is depicted in figure 65. Once the range of interest is known and the corresponding standard solutions have been prepared, the standard of lowest concentration is used as the zero point, i.e. the lowest expected concentration of an element is set to zero, and then the highest standard (the highest expected concentration) is set to full scale deflection by means of electrical expansion. In this way, the readout accuracy in the range of interest can be increased considerably and it is frequently possible to obtain a relative precision of 0.2–0.3%. A requirement for the successful execution of such a precision analysis is naturally that the instrument has the corresponding facilities for baseline suppression and scale expansion, and that it is stable enough to make this type of determination meaningful. Roos has drawn attention to the fact that for every element there is an optimum concentration range in which the relative error passes through a minimum [1050]. For many elements this range is 90–150 times the reciprocal sensitivity, or in the absorbance range 0.35 to 0.60. For a precision analysis, it is therefore advisable to adjust the concentration of the element under study by dilution and to select the most suitable resonance line etc. so that the analysis can be carried out in this absorbance range. Roos has further pointed out that curvature of the calibration plot always produces a deterioration in the precision (which cannot be eliminated with a curvature corrector), so that as far as possible, precision analyses should be carried out in the linear section of the plot.

6.3.3. Trace analysis

The execution of a trace analysis—in the following this is understood as the analysis of a sample that gives a signal of less than about 0.1 absorbance—hardly differs from that of a routine analysis. The principal difference is that the signal is electrically expanded for better readout. Since the noise is also correspondingly expanded, it is often desirable to suppress it. Various procedures are available: The use of higher time constants, longer integration periods, or automatic mean value formation from as many individual measurements as possible. KERBER and BARNETT [635] developed a computer programme that enables the noise and any possible drift to be controlled; in this way they obtained extremely reliable values for concentrations near to the detection limit.

The detection limit and precision with which small concentrations can still be determined is dependent upon the signal-to-noise ratio; for trace analyses attention must always be paid to an enlargement of the signal and an improvement in the signal-to-noise ratio. Figure 66 shows the smallest concentrations that can be satisfactorily determined with good precision using expansion of the signal and damping of the noise. With stable atomic absorption spectrophotometers, expansions of up to 100 times can be employed without greater problems occurring from drift. This is particularly important because high time constants require a considerably increased analysis time. The requirements placed on the stability of an instrument are particularly high in such cases.

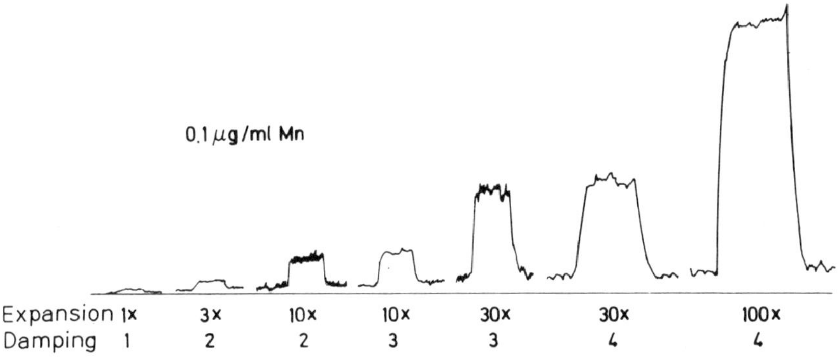

Figure 66. Trace analysis.—When the display is sufficiently stable, even very small signals can be read and evaluated by means of scale expansion with simultaneous suppression of the noise.

6.3.4. Method of additions

Minor chemical and especially physical interferences, such as differing viscosity, surface tension or specific gravity of the sample solution, can be eliminated by matching the matrix of the sample and standard solutions. The preparation of synthetic standards having physical properties corresponding to the sample, or the use of pre-analysed calibration standards, is certainly suitable and useful when large series of similar samples have to be analysed. However, this method is too time consuming for single samples and frequently cannot be employed because the matrix is too complex, too diversified, or simply unknown. In cases where matching the matrix is not possible or not economical, either the internal standard method or the method of additions can be employed.

In the internal standard method, a reference element is added in known concentration to the sample and standard solutions. It is assumed that the reference element behaves in the same way as the element under study, so that the deviation of the signal obtained for the sample solution from the signal obtained for the (mostly pure aqueous) standard solution can be used to correct the element under study [353] [1034].

Two requirements must be fulfilled for the application of the internal standard method: The reference element must not be present in the sample—which must be checked beforehand—and the behaviour of both elements in the flame must be identical. This means that either both elements must be free from chemical interferences in a flame of identical stochiometry or the interferences must be quantitatively identical. This requirement is seldom met in practice, especially in the presence of complicated matrixes—and it is exactly here that the method should function—so that the desired result, the elimination of matrix influences, is rarely or never achieved. Apart from this, this method is complicated and time consuming unless a dual-channel spectrophotometer is available.

The method of additions is, in principle, identical to the internal standard method except that the element under study serves as its own reference. Under these circumstances the above limitations are no longer valid because the reference element behaves identically to the element under study. To carry out an analysis by the method of additions, the sample is divided into three or four aliquots and an identical volume of a (usually aqueous) standard solution is added to each aliquot. The added standard solutions naturally contain different concentrations of the element, and one of them has "zero" concentration, i.e. the sample aliquot is diluted with the corresponding amount of water.

After mixing, all sample aliquots have the same matrix composition so that the influence of the matrix will be the same for all aliquots. The samples only differ in their concentrations of the element under study, and this remains unaltered for one of them. If the absorbance values obtained for the individual aliquots are plotted against the added concentrations, a calibration curve is obtained which does not pass through zero but cuts the absorbance axis at a point greater than zero (figure 67). This absorbance corresponds to the sample without addition, and the concentration of the element under study can be determined by extrapolating to the concentration axis. To determine the concentration of the original sample, only the dilution caused when the addition was made must be taken into consideration. FULLER has described a technique which considerably simplifies the method of additions for instruments having automatic zero setting, automatic calibration and a digital display with sign [394]. The sample is aspirated and the display is set to zero. The sample with the highest addition is aspirated and the concentration is set to the value of the addition (the linearity of the curve can be checked with further standard additions). The blank solution (e.g. water) is then aspirated; the digital display with negative sign corresponds to the concentration of the element under allowance for the dilution factor.

Since the physical and minor chemical interferences mentioned at the beginning are independent of the concentration and only affect the gradient of the calibration plot, they can be reliably and quantitatively eliminated by the method of additions. Concentration-dependent chemical interferences and ionization interferences, on the other hand, cannot be eliminated since both have curved calibration plots which are difficult to control and which make extrapolation to zero very inaccurate. Non-specific light losses due to scattering or molecular absorption also cannot be eliminated, since the two components of the signal, background and true absorption, cannot be separated with the method of additions.

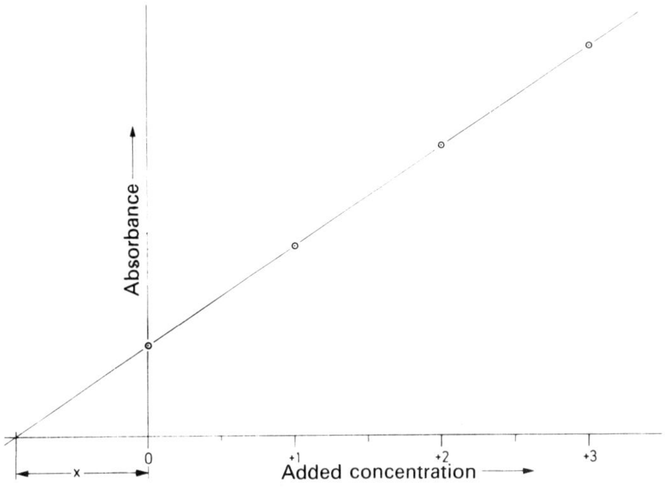

Figure 67. Method of additions.—Standard solutions of various concentrations are added to the unknown sample solution. Through extrapolation of the absorbance values versus zero, the concentration x of the sample solution can be determined. By means of the method of additions principally physical interferences can be eliminated.

6.4. Solvent Extraction

Solvent extraction is a procedure long known to the analyst; for this reason, only the advantages of this technique for atomic absorption spectroscopy will be briefly mentioned. Individual extraction procedures will be especially mentioned in the practical section, when a given determination cannot be carried out directly or only with considerable difficulty. Further details can be found in the relevent literature [880].

Solvent extraction is a process in which two immiscible liquids are brought into such intimate contact that one or more elements are transferred from one liquid phase to the other. Usually, one phase is the aqueous sample solution and the other is an organic solvent which is immiscible with water. On occasions, reversed extraction can also be carried out.

Because of their ionic nature, simple metal salts are normally much more soluble in aqueous media than in organic. To be able to extract a metal with an organic solvent, it must first be converted from the ionic into a non-charged form; generally this is carried out through complex formation with an organic ligand.

Solvent extraction is mostly employed in atomic absorption spectroscopy if the concentration of the element under study in a sample is too low for a determination, or if the matrix is too complex and the resulting interferences cannot be controlled. In other words, solvent extraction serves either for enrichment or for separation and is used when other, simpler techniques fail.

A further application has been described by DELVES and co-workers [294]. By means of selective extraction, eleven elements could be extracted one after the other from 1 ml sample and determined, so that minimum sample volumes could be employed.

Next to this, solvent extraction offers a further, frequently important advantage in atomic absorption. It has already been mentioned that organic solvents bring considerably improved sensitivities and detection limits in the flame. Apart from the enrichment which generally accompanies an extraction procedure, an increase in the signal is also obtained through the use of an organic solvent. These advantages have resulted in the publication of a large number of papers on the use of solvent extraction in atomic absorption spectroscopy, despite the increase in work and time which the extraction process brings.

The combination of ammonium pyrrolidine dithiocarbamate (APDC) as chelating agent, and methyl isobutyl ketone (MIBK) as organic solvent has found wide application, because APDC forms stable chelates with numerous metals over a wide pH range and MIBK exhibits ideal combustion properties in the flame. However, other complexing agents and virtually all organic solvents can be used. Only benzene, which burns with a very sooty flame, and highly chlorinated hydrocarbons, such as chloroform and carbon tetrachloride because of their poor combustion properties and the fact that they frequently form phosgene in the flame, are unsuitable; in principle however, even these solvents can be used. Very volatile solvents are less suitable because losses easily occur making quantitative determinations difficult. Solvents having solidification points only a few degrees below room temperature are also unsuitable because they can freeze on the nebulizer orifice owing to evaporation cooling.

7. Related Analytical Methods

Analytical methods related to atomic absorption spectroscopy will be described in this chapter, special attention will be paid to flame emission and atomic fluorescence spectroscopy since these also employ flames. Before the individual methods are more fully discribed, a preliminary remark would appear to be necessary to remove any false preconceptions. All three methods employ a flame, for all three methods the sensitivity (among other things) is dependent upon the total number of atoms N formed in the ground state, and for all three methods chemical interferences are nothing more than an influence on the total number N of atoms formed. This influence on the free atoms takes place in the flame, and the free atoms are unaware whether we wish to determine them by means of atomic absorption spectroscopy, atomic fluorescence spectroscopy or atomic emission spectroscopy. Therefore, the same chemical interferences are certain to occur to the same degree in all three methods, provided the same flame and the same burner are employed. The frequently expressed view that flame emission spectroscopy is much more strongly subject to chemical interferences than atomic absorption spectroscopy comes from the use of unsuitable flames and burners, which would incidentally also cause the same interferences in atomic absorption.

7.1. Flame Emission Spectroscopy

Much has been written discriminating against flame emission compared to atomic absorption, it has later proved untenable. Thus equation (1.7) (page 7), which represents the dependence of the ratio of excited atoms to non-excited atoms upon the energy of excitation and the absolute temperature, has been used to demonstrate the lower sensitivity of flame emission. It is correct that the energy of excitation is inversely proportional to the wavelength (equation 1.4, page 2) and the proportion of excited atoms, i.e. the ratio N_j/N_0 (equation 1.7) decreases exponentially with increasing energy of excitation. Thus the ratio of excited to non-excited atoms is more unfavourable in the short wave range of the spectrum than in the long wave range (compare figure 4, page 8). However, it is not correct to state that atomic absorption is always more sensitive than flame emission because the number of excited atoms is always smaller than the number of atoms in the ground state (see table 2, page 8). Atomic absorption in only dependent upon the number of atoms N_0 in the ground state; for flame emission, as well as the number of atoms N_j in an excited state, the lifetime of these atoms is important. Experience has shown that atomic absorption is only more sensitive than flame emission when the excitation potential is greater than 3.5 eV; with lower excitation potentials flame emission is usually more sensitive. The detection limits which can be achieved with atomic absorption, flame emission and atomic fluorescence spectroscopy are compiled in table 12 for comparison. It can be seen that atomic absorption and flame emission are, if anything, complementary rather than competitive. Both methods should be considered in this way, especially since the majority of atomic absorption spectrophotometers nowadays enable both methods to be carried out; at least it is no longer necessary to decide which is the method of choice when purchasing an instrument.

Table 12. Comparison of the relative detection limits in atomic absorption, flame emission and atomic fluorescence spectroscopy

Element	Wavelength nm	Relative detection limit (μg/ml) Atomic Absorption [1135]	Flame Emission [345] [363] [689] [976]	Atomic Fluorescence	References
Ag	328.1	0.002	0.02	0.001	[806]
Al	309.3	0.02			
	396.2		0.01	25	[305]
As	193.7	0.15		0.1	[251]
	235.0		50		
Au	242.8	0.01		0.005	[837]
	267.6		0.5		
B	249.8	2	30		
Ba	553.6	0.01	0.001		
Be	234.9	0.002	40		
Bi	223.1	0.003	40		
	302.5			0.05	[258]
Ca	422.7	0.001	0.0001	0.1	[325]
Cd	228.8	0.002		0.0001	[251]
	326.1		2		
Ce	569.7		10		
Co	240.7	0.01		0.005	[377]
	345.4		0.05		
Cr	357.9	0.003		0.005	[251]
	425.4		0.005		
Cs	852.1	0.05	0.008		
Cu	324.8	0.001		0.001	[805]
	327.4		0.01		
Dy	421.2	0.05	0.1		
Er	400.8	0.04	0.3		
Eu	459.4	0.02	0.003		
Fe	248.3	0.01		0.01	[251]
	372.0		0.05		
Ga	287.4	0.1			
	417.2		0.01	0.3	[251]
Gd	368.4	1.2			
	452.0		2		
Ge	265.1	0.2	0.5		
Hf	307.3	2			
	368.2		75		
Hg	253.7	0.2	120	0.08	[171]
Ho	410.4	0.04	0.1		

Table 12 continued

Element	Wavelength nm	Relative detection limit (μg/ml) Atomic Absorption [1135]	Flame Emission [345] [363] [689] [976]	Atomic Fluorescence	References
In	303.9	0.02			
	410.5			0.2	[817]
	451.1		0.005		
Ir	254.4			100	[806]
	264.0	1			
	380.0		100		
K	766.5	0.001	0.003		
La	550.1	2			
	579.1		2		
Li	670.8	0.0003	0.00003		
Lu	331.2	0.7	0.2		
Mg	285.2	0.0001	0.005	0.004	[806]
Mn	279.5	0.002		0.015	[251]
	403.1		0.005		
Mo	313.2	0.02			
	390.3		0.1		
Na	589.0	0.0002	0.0001	100	[305]
Nb	334.4	2			
	405.9		1		
Nd	463.4	1			
	492.4		1		
Ni	232.0	0.002		0.003	[838]
	341.5		0.03		
Os	290.9	0.1			
	442.0		10		
P	213.6	100			
	526.2 (HPO)		0.1		[78]
Pb	283.3	0.01			
	405.8		0.2	0.02	[805]
Pd	247.6	0.02			
	363.5		0.05	0.04	[1212]
Pr	494.0		2		
	495.1	5			
Pt	266.0	0.1	2	50	[806]
Rb	780.0	0.002	0.0003		
Re	346.1	1	0.2		
Rh	343.5	0.005			
	369.2		0.3	3	[806]
Ru	349.9	0.1			
	372.8		0.3	100	[806]

Table 12 continued

Element	Wavelength nm	Relative detection limit (μg/ml) Atomic Absorption [1135]	Flame Emission [345] [363] [689] [976]	Atomic Fluorescence	References
S	394.0 (S_2)		10		[78]
Sb	217.6	0.04		0.05	[257]
	259.8		20		
Sc	391.2	0.02			
	402.0		0.03		
Se	196.0	0.1		0.15	[256]
Si	251.6	0.02	5		
Sm	429.7	2			
	488.4		0.6		
Sn	244.6	0.02			
	284.0		0.3	0.1	[170]
Sr	460.7	0.002	0.0001		
Ta	271.5	1			
	481.3		18		
Tb	432.7	0.6	1		
Te	214.3	0.05		0.05	[256]
	238.3		200		
Th	576.0		150		
Ti	364.3	0.05			
	399.9		0.2	5	[305]
Tl	276.8	0.02			
	377.6			0.04	[806]
	535.1		0.02		
Tm	371.8	0.01			
	410.6		0.2		
U	351.5	30			
	591.5		10		
V	318.4	0.05			
	437.9		0.01		
W	400.9	1	0.5		
Y	407.7	0.1	1		
Yb	398.8	0.005	0.05		
Zn	213.9	0.001	50	0.0001	[818]
Zr	360.1	1	3		
	389.0			10	[305]

PICKETT and KOIRTYOHANN [987] found that atomic absorption and atomic emission spectroscopy together are much better and more universally applicable than the majority of other analytical techniques. Like other authors [351], they found that with the use of hot, premixed flames, such as the nitrous oxide/acetylene flame, emission is just as free of interferences as absorption.

A further censur derived from equation (1.7) is the high temperature dependence of flame emission compared to atomic absorption; small temperature changes in the flame cause a considerable alteration in the number of excited atoms, while the number of atoms in the ground state remains relatively unaltered. It is true that there is an exponential dependence of the number of excited atoms on the temperature, but false conclusions should not be drawn from this. Temperature changes do not occur in the flame unless someone alters the stochiometry. If the stochiometry is altered, the chemical properties are also changed and these can have a strong influence on the production of free atoms; atomic absorption and atomic fluorescence are affected to the same degree as flame emission by such alterations.

It is occasionally considered an advantage of flame emission spectroscopy that no additional light sources (hollow cathode lamp, etc.) are required. For elements that must only be very occasionally determined, this is certainly a financial factor which can be of importance from case to case. Generally however, the specificity of atomic absorption spectroscopy is worth the price of a hollow cathode lamp. The most important difference between the two techniques is still the frequent occurrence of spectral interferences in flame emission which are practically unknown in atomic absorption [176].

Spectral interferences occur in emission when the monochromator is unable to separate the emission lines of two different elements and the determination is thus not element specific. This is essentially a monochromator problem when genuine overlapping of two lines does not take place [176], and many interferences of this type disappear when monochromators having a spectral bandwidth of 0.05 nm, or better, are employed. Naturally such a monochromator must have a correspondingly good reciprocal linear dispersion of at least 2 nm/mm to enable such small slit widths to be satisfactorily used. With such slit widths it is often more meaningful to scan a section of the spectrum with a motorized wavelength drive rather than to carry out the determination at a fixed wavelength since problems with drift, etc. can otherwise occur. BARNETT, KAHN and MANNING [78] have reported the excellent results they obtained for emission measurements using a commercial atomic absorption spectrophotometer. A good review of new developments in atomic absorption and flame emission spectroscopy has been published by KOIRTYOHANN [682]. It can be said that flame emission has profited a great deal from the advances made in atomic absorption spectroscopy within the last few years. Thus the development of new flames and burners has frequently served both methods, and an increasing number of researchers nowadays are not concerned with just one of the methods, but work with all three procedures using a flame. In this way, many more pertinent comparisons are now possible than a few years ago.

The greatest disadvantage of flame emission spectroscopy, apart from the possible occurrence of spectral interferences, is the occurrence of background emission from the flame or from the sample matrix. This effect can influence the detection limit in emission considerably more in the presence of a complex matrix than in absorption spectroscopy. The detection limits quoted in table 12 are valid for pure aqueous solutions and can be noticeably poorer in the presence of complex or concentrated matrixes; the effect is more strongly marked in emission spectroscopy than in absorption spectroscopy. A distinct advantage of flame emission spectroscopy would appear to be that it is more suitable for multielement analyses [189], especially together with plasma "flames" [300] [667].

7.2. Atomic Fluorescence Spectroscopy

Atomic fluorescence spectroscopy is a newer analytical method whose principle—like that of atomic absorption spectroscopy—has been known for a long time but whose significance as an analytical method was first recognized by WINEFORDNER and VICKERS [1347] in 1964; in principle, it is the reverse of atomic absorption spectroscopy.

Atomic fluorescence is produced by exciting atoms with radiation of a suitable frequency, followed by the release of radiation from the deactivation of the excited atoms. The wavelength of the emitted radiation is characteristic for the absorbing atoms, and the intensity is proportional to their concentration and the intensity of the radiation source.

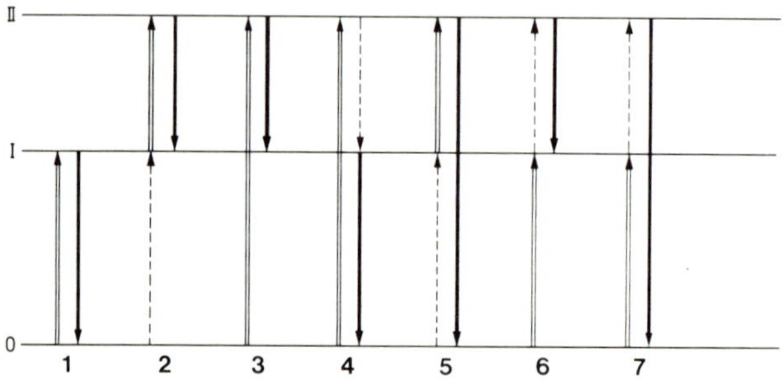

Figure 68. Various types of fluorescence in an atom with two excited levels.—**1** Resonance fluorescence; **2** thermally assisted resonance fluorescence; **3** direct line fluorescence; **4** graduated line fluorescence; **5** thermally assisted direct line (anti-Stokes) fluorescence; **6** thermally assisted direct line fluorescence; **7** thermally assisted graduated line (anti-Stokes) fluorescence.

A distinction is made between various types of fluorescence as shown schematically in figure 68 [867] [1343]. Resonance fluorescence is observed when atoms in the ground state are excited and then return to the ground state by emitting radiation of the same wavelength as used for excitation. The most intense resonance fluorescence is generally observed during the transition from the first excited state to the ground state. This resonance fluorescence will almost certainly find the greatest analytical interest since the intensity is generally much greater than that of other types of fluorescence [1343].

Direct line fluorescence occurs when an atom is converted into a higher excited state and from there returns not to the ground state but to the next lowest excited state. The energy released is emitted in the form of a fluorescence quantum. This always leads to a red shift of the fluorescence line compared to the incident line. Finally, with transitions into higher excited states, an atom can first be deactivated without the emission of radiation and then return from a lower activated state to the ground state with emission of radiation.

OMENETTO [939] [940] and DAGNALL and co-workers [258] have further reported various thermally assisted fluorescende mechanisms in which an atom is brought to a higher energy level by thermal excitation either before the absorption of light energy already exists in an excited, metastable state, or after absorption of light energy.

The intensity of atomic fluorescence, and thereby the achievable sensitivity and detection limit, is dependent upon the number of atoms in the ground state and the intensity of the incident light source. For this reason, research into atomic fluorescence is closely connected with the examination of newer and more intense light sources ([252] [304] [305] [763] [818] [1264] [1342] [1345] [1346]), especially electrodeless discharge lamps ([171] [225] [258] [377]), and also recently lasers [937] [938].

The fluorescence signal, as in atomic absorption, is usually differentiated from an emission signal (thermally produced) by employing alternating light. The light source is modulated at a given frequency and the amplifier is tuned to the same frequency. The observation of the fluorescence takes place at right angles to the incident beam. One difficulty is to differentiate between genuine fluorescence and stray light; if light from the light source is scattered on solid or liquid particles in the flame, this is also modulated and leads to a false signal just like non-specific light losses in atomic absorption. RAINS and co-workers have therefore suggested a correction system for atomic fluorescence similar to that used in atomic absorption [1013]. By alternate irradiation with the continuum of a xenon arc lamp and a line light source, the interferences due to stray light are eliminated. For the same reasons a great deal of work has been carried out on burners and flames in connection with fluorescence ([32] [159] [160] [161] [252] [579] [838] [1156]). The considerable influence of the flame on fluorescence analyses has been emphasized by various authors [578] [837]; attention must be paid to more than the most complete possible atomization and the lowest possible formation of compounds through the chemical composition of the flame. The fluorescence intensity can be strongly reduced in a flame if atoms excited by light energy lose their energy through non-radiative collisions (quenching). These non-radiative collisions are strongly dependent upon the chemistry of the flame. For example, nitrogen exhibits a much stronger quenching effect than argon and cooler flames a lower effect than hotter ones.

An advantage of atomic fluorescence spectroscopy over atomic absorption spectroscopy is that the lowest detection limits can be more easily achieved (at least in theory) by increasing the radiation intensity of the light source. Also a slight instability of the light source should not affect the detection limit to the same degree as in atomic absorption spectroscopy. In atomic fluorescence spectroscopy the detection limit is the technically simple dectection of a very small light signal (similar to atomic emission spectroscopy), while in atomic absorption spectroscopy it is the detection of very small differences between two very intense light signals. With fluorescence spectroscopy a fluctuation in the light source is conveyed proportionally to the signal, while in absorption spectroscopy it appears in its full magnitude in the detection limit and is therefore frequently the limiting factor for a determination.

While the width of the line of excitation from the hollow cathode lamp can enter directly into the sensitivity of a determination in atomic absorption spectroscopy (see page 17) and the presence of non-absorbable radiation from the light source within the observed spectral range can lead to considerable curvature of the calibration plot (see page 69), this is not possible in atomic fluorescence spectroscopy. In atomic absorption spectroscopy, the light from the radiation source (after spectral dispersion) falls directly onto the detector; in atomic fluorescence spectroscopy on the other hand, measurement is at right angles to the incident radiation. Thus it is possible to work with continuous light sources and without extremely small slit widths [243] [1238] since the non-absorbable radiation does not fall on the detector.

In principle, it is even possible to work without a monochromator, as suggested initially by JENKINS [577]. By using an element specific light source and a selective amplifier, the fluorescence of the atoms in the flame was directly measured by a photomultiplier. Nevertheless, it was necessary to use a filter to reduce the strong direct light from the flame which caused higher noise in the detector signal. VICKERS and VAUGHT [1268] used a "solar blind" dectector which was only sensitive between 160 nm and 320 nm. LARKINS and WILLIS used a similar non-dispersive system, with good success, for their measurements [733] [734].

Since the radiation from the light source does not fall directly on the detector in atomic fluorescence spectroscopy, calibration plots which are linear over three to five orders of magnitude are obtained, while in atomic absorption spectroscopy they are frequently only linear over two orders of magnitude. Nevertheless, the course of calibration plots in atomic fluorescence is not always so simple as would appear here. Self-absorption and other effects can occasionally have a considerable influence on the shape of the calibration plots [76] [935] [1375]. Although atomic fluorescence spectroscopy appears to have a number of advantages over atomic absorption and flame emission spectroscopy, nothing final can be said at the present time. Up to now only a few applications employing atomic fluorescence spectroscopy have been published. SYCHRA and MATOUŠEK determined nickel in petrochemicals [1210], PATEL and WINEFORDNER determined silver and gold in jet fuel [962], and SYCHRA and KOLIHOVÁ investigated the determination of zinc and cadmium in blood and serum [1209]. All other applications are based on the use of pure aqueous solutions [430] [693] [839] [1211] [1212] and the study of the fluorescence characteristics of the individual elements. It would appear that this analytical method cannot tolerate such complex matrixes as atomic absorption spectroscopy. Furthermore, hot flames, which can eliminate many interferences and make the determination of many elements first possible, bring certain problems, such as background emission and quenching [76], to atomic fluorescence.

Certainly atomic fluorescence spectroscopy offers advantages over absorption or emission spectroscopy for a number of applications or elements, so that this method can be regarded as complementary to the other two. The detection limits which can be achieved with atomic fluorescence spectroscopy are given together with those for absorption and emission in table 12 (page 126). Further information can be obtained from two excellent articles by WINEFORDNER and MANSFIELD [1343] and WEST [1309]. The conversion of a commercial atomic absorption spectrophotometer for fluorescence measurements has been described by MANNING and HENEAGE [805]. With this instrument they achieved some excellent detection limits [806].

8. The Individual Elements

The properties of greatest importance for an analysis by atomic absorption spectroscopy will be given for a number of elements. The various flames which can be used, possible interferences and their elimination and avoidance, and achievable sensitivities and detection limits will be discussed. "Analytical Methods for Atomic Absorption Spectrophotometry" published by the PERKIN-ELMER CORPORATION [1] serves as the basis. The details have been thoroughly checked and supplemented where necessary.

8.1. Aluminium

Prior to 1966, a number of authors attempted to determine aluminium in oxygen/acetylene flames of varying stochiometry ([223] [310] [791] [1146]); organic solvents were frequently used to increase the senitivity. A completely satisfactory analysis of aluminium was first possible after the introduction of the nitrous oxide/acetylene flame by WILLIS [1333]. Practically no interferences are present in this flame; RAMAKRISHNA [1016] reported that acetic acid increased the absorption of aluminium by about 10% and the presence of titanium is said to increase the absorption by around 25%. According to WEST and co-workers this is due to a reduction of the lateral diffusion and thereby to an increased atom concentration in the middle of the flame [1306]. Silicon on the other hand has a slightly suppressive influence on the absorption of aluminium [238] [365].

The reciprocal sensitivity for aluminium in a nitrous oxide/acetylene flame is approximately 1 µg/ml 1% at the 309.3 nm line, and the detection limit is about 0.02 µg/ml for an aqueous solution. Sensitivities for other resonance lines are compiled in table 13.

Since aluminium is ionized by about 10% in a nitrous oxide/acetylene flame, about 0.1% potassium (as chloride) or another easily ionized metal should be added to sample and standard solutions.

Aluminium can be complexed with oxine at pH 8 and extracted with MIBK. The extraction must be executed within three minutes, otherwise the risk of co-precipitation exists [373].

Table 13. Al resonance lines

Wavelength nm	Energy level (*K*)	Reciprocal sensitivity µg/ml 1%
309.3	112–32437	1.2
396.1	112–25348	2.0
308.2	0–32435	2.5
394.4	0–25348	4.0
257.5	112–38934	16

The determination of aluminium in the graphite tube furnace is largely interference-free since it is heated to around 1400 °C during the thermal pre-treatment and can thus be mostly well separated from the matrix. Nevertheless, aluminium should be converted into the oxide since the chloride is easily sublimed. The detection limit for aluminium in the graphite tube furnace is 2×10^{-12} g. Aluminium is one of the few elements that forms a nitride in the graphite tube. The sensitivity is therefore about three times lower when nitrogen is used as inert gas than with argon [802].

8.2. Antimony

Antimony can be determined in an air/acetylene flame virtually free of interferences. However, little has been published about this element so that possibly not everything is known. MOSTYN and CUNNINGHAM [888] found it of advantage to match the matrix to avoid minor interferences. For the determination of antimony, three resonance lines of similar sensitivity are available as shown in table 14. For the majority of determinations the 217.6 nm line is employed. High copper [888] and lead [1337] concentrations can cause weak spectral interferences at this line. In the presence of these elements and also with higher antimony concentrations it is recommended to use the 231.1 nm line.

Table 14. Sb resonance lines

Wavelength nm	Energy level (K)	Reciprocal sensitivity µg/ml 1%
217.6	0–45945	0.5
206.8	0–48332	1
231.1	0–43249	1.5

The sensitivity of the antimony determination can be considerably increased by using the hydride system [358] [1092]. When $NaBH_4$ is used as reducing agent, the detection limit is 5×10^{-9} g or 0.25 µg/l. KNUDSON and CHRISTIAN were able to improve the detection limit by a factor of five by conducting the hydride into a heated graphite tube instead of into an argon/hydrogen diffusion flame [678].
In the graphite tube furnace approximately 1×10^{-11} g antimony can be determined. It is important to carry out the determination in the presence of sulphuric acid (min. 0.1 N); other acids lead to high scattering and lower sensitivities owing to losses.

8.3. Arsenic

The determination of this element is considerably influenced by the fact that its resonance lines are at 193.7 nm and 197.2 nm at the start of the vacuum UV range and that the available hollow cathode lamps are of rather low performance. Furthermore, since many photomultipliers only exhibit low sensitivity below 200 nm, and light losses owing to increased absorption of lenses and low reflectance of mirrors can occur, arsenic can only be determined satisfactorily in the best atomic absorption spectrophotometers. Correspondingly, little has been published about this element. Only after the development of high performance electrodeless discharge lamps has the determination of arsenic become largely free of problems (figure 69). The reciprocal sensitivity with these lamps in an air/acetylene flame is 0.8 µg/ml 1% and the detection limit is 0.15 µg/ml. With normal hollow cathode lamps, the corresponding values are 1.8 µg/ml 1% and 2.3 µg/ml respectively.

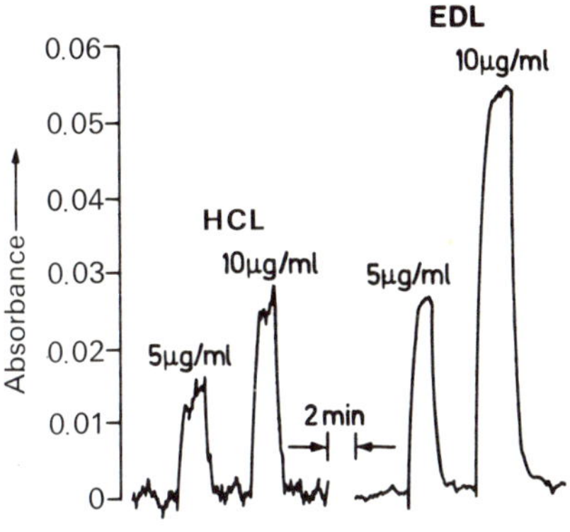

Figure 69. Determination of low arsenic concentrations employing a hollow cathode lamp (HCL) and an electrodeless discharge lamp (EDL) under identical analytical conditions and similar instrumentation in an air/acetylene flame.—A considerably better signal-to-noise ratio is obtained with the electrodeless discharge lamp [872].

Very few chemical interferences can be observed for arsenic in an air/acetylene flame [1154]. The disadvantage of this flame is that it absorbs about 60% of the light energy emitted from the hollow cathode lamp; the air/hydrogen flame exhibits virtually the same absorption (see figure 18, page 34).

The use of an argon/hydrogen diffusion flame brings a considerable improvement in the sensitivity and detection limit for arsenic (table 15). The probability of chemical interferences is much higher in this flame, although nothing has been reported on this up to the present. Background effects are to be expected to a considerable degree for the determination of arsenic, especially with the argon/hydrogen diffusion flame, so that great caution is recommended for the analysis of complex materials.

A further increase in the sensitivity and a simplification of the analysis can be obtained with the hydride system. With $NaBH_4$ as the reducing agent and an electrodeless discharge

Table 15. Comparison: Air/acetylene and argon/hydrogen flames for the As resonance line at 193.7 nm

	Air/Acetylene	Argon/Hydrogen
Absorption of the flame	60%	15%
Reciprocal sensitivity (µg/ml 1%)	0.8	0.15
Detection limit	0.15	0.02

lamp, detection limits of 3×10^{-9} g or 0.15 μg/l respectively can be achieved [358]. A further improvement in the sensitivity and detection limit can be attained by passing the hydride into a heated graphite tube furnace instead of into an argon/hydrogen diffusion flame [678]. DALTON and MALANOSKI [266] have reported that greater quantities of nitric acid and also antimony in amounts above 0.1 mg interfere in the determination of arsenic by the hydride method.
Arsenic can be determined, relatively free of problems, in the graphite tube furnace when it is present in the pentavalent form, since it can then be heated to about 1100°C without losses. The detection limit for arsenic in the graphite tube furnace is 1×10^{-11} g.

8.4. Barium

Barium exhibits numerous chemical interferences in an air/acetylene flame [635] which are stongly reduced or completely eliminated in a nitrous oxide/acetylene flame. Also in a hot flame, the strong background absorption caused by CaOH bands in the presence of a Ca matrix, which BILLINGS [123] observed in an air/acetylene flame, disappears (see figure 58,

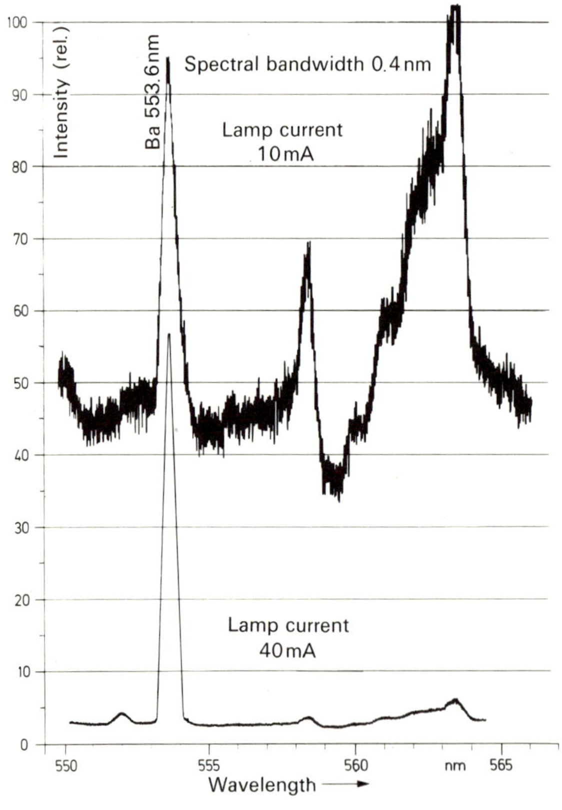

Figure 70. Influence of the lamp current on the determination of barium.—The upper spectrum shows the high portion of emission light is emitted by an air/acetylene flame when a low intensity hollow cathode lamp is employed. The lower spectrum was scanned under identical conditions with an Intensitron® high intensity hollow cathode lamp. For both spectra, zero balance was carried out before the flame was ignited.

page 110). The strong emission of the nitrous oxide/acetylene flame near the Ba line, which brings a considerable contribution to the noise, can occasionally have an interfering influence (see figure 70). Increasing the lamp current and reducing the monochromator slit can be a remedy.

RUBEŠKA found that barium could be determined without difficulties in a nitrous oxide/acetylene flame screened by an inert gas, even in the presence of a large excess of calcium (CaOH emission) [1066].

Barium is ionized to more than 90% in a nitrous oxide/acetylene flame (see figure 57, page 107) so that it is necessary to add 0.2–0.5% potassium (as chloride) or another easily ionized metal to all sample and standard solutions. Barium has only one usable resonance line at 553.6 nm. The reciprocal sensitivity in an air/acetylene flame is about 10 μg/ml 1% and in the nitrous oxide/acetylene flame about 0.4 μg/ml 1%. The detection limits in these flames are 0.7 and 0.01 μg/ml respectively. A ten times better detection limit can be obtained with flame emission spectroscopy [635].

The determination of barium in the graphite tube furnace is inhibited somewhat by carbide formation. In an instrument in which the direct light emission of the graphite tube is effectively screened, 5×10^{-11} g of barium can however be determined. RENSHAW found that the sensitivity for barium could be improved by a factor of 20 by lining the graphite tube with tantalum foil so that carbide formation is prevented [1031].

8.5. Beryllium

Beryllium can only be very poorly determined in an air/acetylene flame, but exhibits an excellent reciprocal sensitivity of approximately 0.03 μg/ml 1% and a detection limit of about 0.002 μg/ml in a nitrous oxide/acetylene flame. The most important resonance line is at 238.4 nm; up to the present, nothing has been reported on the uses of other lines.

The determination of beryllium appears to be largely free of interference in the nitrous oxide/acetylene flame. BOKOWSKI [137] has reported a minor interference in the presence of high silicon and aluminium contents, but found no influences from phosphate or sulphate. RAMAKRISHNA [1016] eliminated the suppressive effect of aluminium by adding about 1.5 g/l fluoride. He found a further increase in the signal of about 20% by adding acetic acid. Beryllium is not noticeably ionized in a nitrous oxide/acetylene flame, so the addition of alkali is not required.

NAKAHARA and co-workers also found the determination of beryllium in the nitrous oxide/acetylene flame to be quite free of interferences. For the analysis of alloys, palladium and silicon occasionally caused interferences. Phosphoric acid increased the signal somewhat, while hydrochloric and nitric acids reduced it.

The detection limit for beryllium in the graphite tube furnace is about 3×10^{-12} g. It is important to neutralize acid beryllium solutions with ammonia in the graphite tube to achieve the full sensitivity and high thermal stability up to 1600°C. In an acid medium, beryllium can only be thermally pre-treated up to 900°C without losses and its sensitivity is reduced to 10–50% depending upon the acid.

8.6. Bismuth

Bismuth can be determined without interferences in an air/acetylene flame. The reciprocal sensitivity at the 223.1 nm line is 0.4 µg/ml 1% and the detection limit is about 0.03 µg/ml. Various other resonance lines are compiled in table 16. An improved signal-to-noise ratio can be obtained in an air/hydrogen flame with a detection limit of about 0.02 µg/ml.

Table 16. Bi resonance lines

Wavelength nm	Energy level (*K*)	Reciprocal sensitivity (µg/ml 1%)	Spectral bandwidth nm
223.1	0–44817	0.7	0.07
222.8	0–44865	1.5	0.07
306.8	0–32588	2	0.2
206.2	0–48489	5.5	0.2
227.7	0–43912	10	0.2
211.0	0–47317	20	0.2
202.1	0–49461	50	0.07

Very low bismuth concentrations can be determined with the hydride system. With $NaBH_4$ as the reducing agent, the detection limits are about 5×10^{-9} g or 0.25 µg/ml respectively [358].

In the graphite tube furnace the detection limit for bismuth is 2×10^{-11} g. Since this element is very volatile, it can only be thermally pre-treated up to about 350°C without losses, and external preparation is often unavoidable. Nitric acid in high concentration causes a signal reduction.

8.7. Boron

Boron can only be determined in a nitrous oxide/acetylene flame, with a reciprocal sensitivity of about 40 µg/ml 1% and a detection limit of 2–5 µg/ml; it is one of the least sensitive elements in atomic absorption spectroscopy. The resonance lines at 249.7/249.8 nm are used almost exclusively.

Various authors [65] [795] have reported the determination of boron in aqueous solution, but various refractory substances interfere in the analysis. For this reason and because of the low sensitivity, various authors have proposed extraction procedures for enrichment and for a separation of the interfering matrix. AGAZZI [21], HOSSNER [536] and WEGNER and co-workers [1288] extracted a 2-ethyl-1,3-hexandiol complex with chloroform. However, this solvent is not very suitable for combustion in a nitrous oxide/acetylene flame. MELTON and co-workers [858] and HOLAK [526] therefore used MIBK as solvent since it proved to be very suitable for this extraction.

MANNING and SLAVIN [808] found that an isotope analysis by means of atomic absorption spectroscopy is only possible for very light and very heavy atoms. MROZOWSKI [891] attempted to separate the ^{10}B and ^{11}B lines by cooling the hollow cathode lamp in liquid air, but he was unsuccessful.

The isotope shift is about 0.001 nm while the line width in a nitrous oxide/acetylene flame is about 0.006 nm, so that separation is impossible. GOLEB [426] was also unable to obtain different absorption values for the two isotopes by using isotope lamps and an absorption tube.

Because of its strong tendency to carbide formation, boron is very difficult to determine in the graphite tube furnace. Only KUZOVLEV and co-workers reported a number of successes with graphite tubes which were coated with a mixed tantalum/niobium carbide [719]. A good coating with pyrolytic graphite, however, seems to give the same result.

8.8. Cadmium

Cadmium can be easily determined without any chemical interferences, worth mentioning, in an air/acetylene flame. At the 228.8 nm line the reciprocal sensitivity is 0.03 µg/ml 1% and a detection limit of 0.002 µg/ml can be easily achieved. In an argon/hydrogen diffusion flame the sensitivity can be improved by a factor of two and a detection limit of 0.001 µg/ml can be achieved. However, the occurrence of chemical interferences and background absorption effects is more probable in this flame.

For the determination of higher cadmium concentrations, the resonance line at 326.1 nm is suitable because the reciprocal sensitivity is about 20 µg/ml 1%; so that excessive dilutions can be avoided.

Since cadmium is very easily atomized, it can also be determined very well be the boat technique and the DELVES system. The detection limits are 1×10^{-10} g and 2×10^{-11} g respectively [218].

Cadmium has a detection limit of 1×10^{-13} g in the graphite tube furnace and is thus one of the most sensitive elements. Because of the high volatility of cadmium, the determination is difficult, especially in the presence of complex matrixes. Since the sample can only be thermally pre-treated up to about 300°C without losses, external chemical preparation or an extraction is often necessary.

8.9. Cesium

The resonance line for cesium is at 852.1 nm, at the start of the infrared, and is therefore outside the normal working range of a number of atomic absorption spectrophotometers. Up to the present no usable hollow cathode lamps have been developed for cesium, so vapour discharge lamps must be used. These have proved to be very suitable. Cesium can be determined in an air/acetylene flame with a reciprocal sensitivity of 0.5 µg/ml 1% and a detection limit of about 0.05 µg/ml without noticeable chemical interferences. Cesium is,

however, strongly ionized in this flame, and large quantities of other alkali salts must be added [569] to suppress the ionization. The ionization is considerably lower in an air/hydrogen flame, so that working with this flame is easier. A number of authors prefer the air/propane flame [296], in which other interferences can occasionally occur.
For higher cesium concentrations the resonance line at 455.6 nm is suitable since the reciprocal sensitivity is about 20 μg/ml 1% [407].
A detection limit of 0.008 μg/ml can be achieved in emission using an air/acetylene or an air/hydrogen flame when 100 μg/ml sodium salt is added [363].

8.10. Calcium

Calcium is one of the elements most frequently determined by atomic absorption spectroscopy, and the number of publications is correspondingly large. The first work by WILLIS [1324] [1328] and DAVID [268] [269] between 1959 and 1961 was concerned with the determination of calcium in serum and urine, and in plants and soil samples respectively.
Calcium can be determined virtually free of interferences in a nitrous oxide/acetylene flame with a reciprocal sensitivity of 0.05 μg/ml 1% and a detection limit of about 0.001 μg/ml. Chemical interferences can only be observed in this flame in the presence of high silicon and aluminium concentrations, and slight ionization can easily be removed by adding small quantities of alkali [800].
In spite of this fact, calcium is almost only determined with an air/acetylene flame and then serves as a prototype for elements disturbed by numerous chemical interferences. RAMAKRISHNA [1014] published a systematic study in which the effect of more than 50 ions on the absorption of calcium was investigated. Only two anions (from 20 investigated) and 15 cations (from 32 investigated) were without any great influence on the absorption of calcium. Nevertheless, it should be emphasized that this work was carried out with a direct nebulizing turbulent burner; with a mixing chamber burner these interferences are much less marked [301]. In a mixing chamber burner the influences of up to 500 μg/ml silicon and 1000 μg/ml aluminium or phosphorus can be well controlled by the addition of 1% lanthanum (as hydrochloric acid solution of the oxide) or 1% EDTA (as di-sodium salt). Calcium can then be determined in an air/acetylene flame at the 422.7 nm resonance line with a reciprocal sensitivity of 0.1 μg/ml 1% and a detection limit of 0.001 μg/ml. Although large spectral bandwidths (about 2.0 nm) are used for the determination of calcium in an air/acetylene flame, this is not recommended for the nitrous oxide/acetylene flame. As shown in figure 71, there is an intense CN emission band between 422 nm and about 410 nm in this flame. If this is not separated in the monochromator, the signal-to-noise ratio is worsened noticeably.
Higher calcium concentrations can be determined at the 239.9 nm resonance line with a reciprocal sensitivity of about 10 μg/ml 1%. The signal-to-noise ratio is not so good at this line however, and it is not recommended for precision analyses.
The determination of calcium in the graphite tube furnace is somewhat impaired by carbide formation. The high direct light emission in the visible spectral range can also lead to considerable interferences with older instruments (see figure 49, page 81). Good detection limits of 1×10^{-12} g Ca can be achieved with an instrument optimized for flameless atomic absorption. Nevertheless,

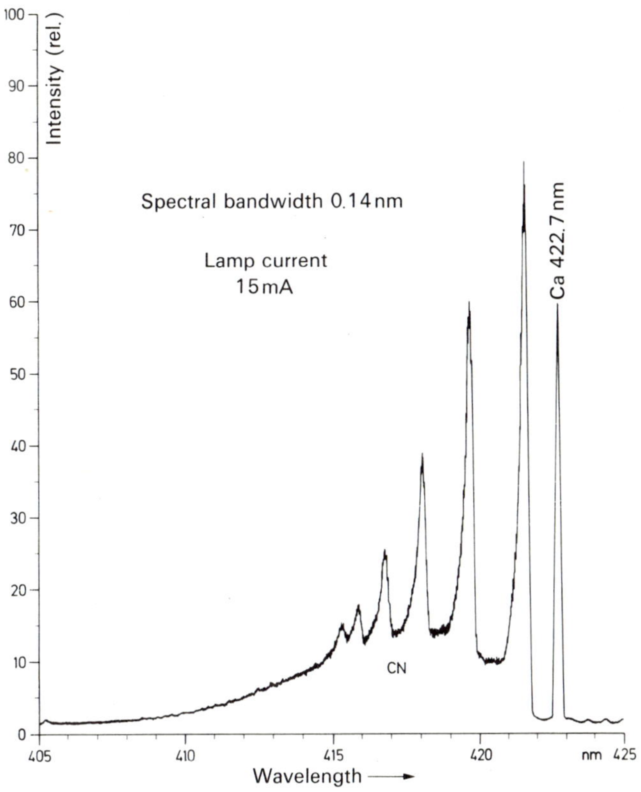

Figure 71. Emission of the nitrous oxide/acetylene flame in the region of the Ca resonance line at 422.7 nm.

it must be pointed out that calcium belongs to those elements strongly subjected to contamination. CHRISTENSEN found no interference by phosphate to the determination of calcium in the graphite tube furnace [231], and no other noticeable interferences appear to exist since it can be thermally pre-treated up to about 1400°C without losses.

8.11. Chromium

Chromium can be determined both in a fuel gas rich air/acetylene flame and in a nitrous oxide/acetylene flame. A comprehensive study on the optimum atom concentration and the absorption profile in fuel gas rich and weak air/acetylene flames has been made by RANN and HAMBLY [1018]; the greatest sensitivity was found with a three-slot burner head [1172] and a faintly luminous flame. Chromium exhibits a large number of resonance lines of similar sensitivity. For the majority of work the 357.9 nm line is used. At this line the reciprocal sensitivity is 0.1 µg/ml 1% and the detection limit is about 0.003 µg/ml. According to GATEHOUSE and WILLIS [407], the resonance line at 425.4 nm with a reciprocal sensitivity of 0.5 µg/ml 1% can be used for higher chromium concentrations.

While chromium can be determined virtually interference-free [989] in numerous samples, in the presence of iron and nickel the absorption is considerably reduced. ROOS [1045] also found that this interference was strongly dependent on the gas mixture; furthermore, he found that chromium(III) exhibited a higher sensitivity than chromium(VI), so that the oxidation state also plays a part. These effects can be largely eliminated by adding 2% ammonium chloride [410], or 2% potassium persulphate [1272], or even 1% ammonium hydrogen fluoride either alone or together with 0.2% sodium sulphate [1008] to the sample and standard solutions. ROOS accounted the elimination of these interferences to a carrier distillation, i.e. iron and other interfering ions distill off much more rapidly in the flame with the added salts [1049]. To eliminate interferences, other authors extracted chromium in MIBK [825] or used ion exchangers to remove the interfering ions [581]. 8-Hydroxyquinoline has also been proposed as a releasing agent to suppress interferences [945].

YANAGISAWA and co-workers separated chromium(III) and chromium(VI) by extraction before analysing the two valency forms by atomic absorption. They used hydroxyquinoline at pH 6 as chelating agent for Cr(III) and diethylthiocarbonate at pH 4 for Cr(VI) [1361].

These chemical interferences, which are often difficult to control, can be most easily eliminated by using a nitrous oxide/acetylene flame [547] [1053]. For precision analyses, even in this flame it is still necessary to match the sample matrix, since acceptable results cannot otherwise be obtained [1199]. The reciprocal sensitivity in the nitrous oxide/acetylene flame at the 357.9 nm line is 0.5 µg/ml 1%.

Chromium can be largely determined without interference in the graphite tube furnace [1101], since chromium can be thermally pretreated up to around 1300°C without losses. The detection limit for chromium is 1×10^{-11} g.

8.12. Cobalt

Cobalt can be determined without difficulties by atomic absorption spectroscopy in an air/acetylene flame. Although it exhibits a large number of resonance lines (the most important are compiled in table 17), only a few are suitable for analytical purposes. Clear spectral interferences can occur with a number of the lines given in table 17 especially when using multielement lamps which contain nickel and with insufficient spectral bandwidths (greater than 0.2 nm).

Table 17. Co resonance lines

Wavelength nm	Energy level (*K*)	Reciprocal sensitivity (µg/ml 1%)
240.7	0–41 529	0.2
242.5	0–41 226	0.3
252.1	0–39 649	0.5
241.1	816–42 269	0.7
352.7	0–28 346	3
345.4	3483–32 431	4

The most suitable line for lower cobalt concentrations is the 240.7 nm line with a reciprocal sensitivity of 0.2 µg/ml 1% and a detection limit of about 0.01 µg/ml. Because of weak self-absorption, which could be observed much more intensely with older lamps [471], and a number of lesser ionization lines which cannot be separated from the resonance line, there is a clear curvature of the calibration plot at higher absorbance values. For higher cobalt concentrations the 352.7 nm line is more suitable; it exhibits good linearity and a very favourable signal-to-noise ratio, thus making good precision possible.
McPHERSON [851] showed that the following elements interfere with the determination of cobalt in an air/acetylene flame: 2000 µg/ml Cr, Ni, W; 1000 µg/ml Cu, Mo; 500 µg/ml Si; 200 µg/ml Mn, V; 100 µg/ml Ti and 50 µg/ml P and S. With high matrix concentrations, matching the matrix can become necessary. With a refractory matrix it can be of advantage to determine cobalt in a nitrous oxide/acetylene flame. The reciprocal sensitivity in this flame is about 0.7 µg/ml 1% at the 240.7 nm line.
SIMMONS complexed cobalt with 2-nitroso-1-naphthol and extracted it with chloroform from plant material digestions [1127]. Because of the poor combustion properties of chloroform, he evaporated the solution and dissolved the residue in MIBK.
Approximately 5×10^{-12} g cobalt can be detected in the graphite tube furnace. Since cobalt can be thermally pre-treated up to about 1100°C without losses, numerous accompanying materials can be removed so that the determination can be carried out largely free of interferences [1101]. Of the various acids, only nitric acid in higher concentrations has a clearly suppressive effect.

8.13. Copper

Copper is one of the elements most frequently and easily determined by atomic absorption. It exhibits no interferences in an air/acetylene flame [646] [989], is virtually independent of the stochiometry of the flame and the lamp current, and is therefore frequently used as a standard to test an instrument or a procedure.
Copper exhibits a number of resonance lines, all of which are analytically suitable (table 18). By selecting the correct resonance line, even higher copper concentrations can be precisely determined without excessive dilution. For lower copper concentrations the 324.7 nm line with a reciprocal sensitivity of 0.1 µg/ml 1% and a detection limit of about 0.001 µg/ml is the most suitable.

Table 18. Cu resonance lines

Wavelength nm	Energy level (*K*)	Reciprocal sensitivity (µg/ml 1%)
324.7	0–30784	0.1
327.4	0–30535	0.3
222.6	0–44916	2
249.2	0–40114	10
244.2	0–40944	55

A somewhat better sensitivity can be achieved by using an air/propane flame, but certain interferences cannot be completely excluded in this flame.
Copper can be complexed with a large number of chelating agents over a wide pH range and extracted with virtually all organic solvents. It is frequently employed as a reference element for extractions to check the completeness of the extraction procedure [39]. Copper can be freed from its matrix—even under extreme conditions—by multiple extraction [1243].

The detection limit for copper in the graphite tube furnace is about 2×10^{-12} g. However, the determination is not so free of interferences as in the flame [1101] since copper can only be thermally pre-treated to about 900 °C without losses, so that it is not always possible to remove the matrix completely enough. Of the various acids, only nitric acid causes a noticeable reduction of the signal.

8.14. Gallium

Gallium can be determined simply in a stochiometric air/acetylene flame; the greatest problem is a good hollow cathode lamp. Since gallium melts at only 30 °C, it is difficult to manufacture a hollow cathode lamp of usual design. MULFORD [892] described a hollow cathode lamp in which the metal was in the molten state, and he also examined various resonance lines for their sensitivities (table 19). At the 287.4 nm line the reciprocal sensitivity is 2.2 μg/ml 1% and the detection limit is about 0.1 μg/ml. With a nitrous oxide/acetylene flame the greatest reciprocal sensitivity is observed at the 294.4 nm line with 1.0 μg/ml 1%.

Table 19. Ga resonance lines

Wavelength nm	Energy level (*K*)	Reciprocal sensitivity (μg/ml 1%)	Spectral bandwidth nm
287.4	0–34782	2.2	0.2
294.4	826–34782	2.4	0.2
417.2	826–24788	3.7	0.2
403.3	0–24788	6.2	0.7
250.0	826–40811	22	0.7
245.0	0–40803	28	2.0

Hitherto, little has been reported on the determination of gallium. GUPTA and co-workers [449] investigated the influence of 45 different ions on the absorption of gallium in an air/acetylene flame. They found that 12 ions reduced the signal slightly when their concentrations exceeded determined values. In general though, gallium appears to be largely determinable without interferences.
Approximately 1×10^{-11} g gallium can be determined in the graphite tube furnace. The addition of an excess of ammonia solution to the sample in the graphite tube eliminates numerous interferences.

8.15. Germanium

Germanium can be determined at 265.1 nm in a nitrous oxide/acetylene flame with a reciprocal sensitivity of 2.5 μg/ml 1% and a detection limit of approximately 0.2 μg/ml [795]. POPHAM and SCHRENK [997] have published a comprehensive study on the absorption of germanium in a nitrous oxide/acetylene flame and in a fuel gas rich oxygen/acetylene flame. They found numerous interferences, especially in the nitrous oxide/acetylene flame, but these were probably due to the fact that they made the measurements only 2 mm above the burner slot, which is certainly insufficient for complete dissociation.
Germanium can be determined with greater sensitivity using the hydride system with $NaBH_4$ as the reducing agent. The detection limits are 2×10^{-7} g or 0.01 μg/ml respectively [358].

8.16. Gold

Gold is very frequently determined in soil samples and galvanic baths by atomic absorption spectroscopy. The most sensitive resonance line is at 242.8 nm with 0.25 μg/ml 1% and a detection limit of about 0.01 μg/ml; frequently the slightly less sensitive 267.6 nm line (0.4 μg/ml 1%) enables better and more accurate analyses to be carried out because of its higher intensity.

Normally, gold is determined in a sharp, fuel gas weak air/acetylene flame in which no interferences worth mentioning occur. Certain differences arise only when cyanide gold solutions are compared with hydrochloric acid solutions; in such cases the sample and standard solutions should be matched. Nevertheless, according to experiences gained in the author's laboratory, minor interferences occur during the determination of gold in complex matrixes—even though no clear chemical interferences can be confirmed—which are strongly dependent upon alterations in the composition of the flame. Thus it is frequently impossible to determine gold in complex matrixes with an accuracy better than ±5%, even with the method of additions. Although a nitrous oxide/acetylene flame reduces the sensitivity of a gold determination, it partially eliminates these difficult-to-control interferences.
The determination of gold in the presence of other noble metals is often relatively difficult. ADRIAENSSENS and VERBEEK found that the majority of interelemental interferences observed in acid solution do not occur in a 2% potassium cyanide solution [17]. SEN GUPTA eliminated the interelemental interferences by adding a copper/cadmium buffer solution [1114]. He has also published a good review on the various methods in noble metal analysis [1116].
Since gold must frequently be determined with even greater accuracy and especially in smaller concentrations, numerous extraction methods are in use [439] [445] [992] [1182] [1231] [1232] which will be described, in part, later.
About 1×10^{-11} g gold can be determined in the graphite tube furnace. Since gold is volatile from about 500–600°C, the matrix frequently cannot be separated and slight interferences can result. The sensitivity of gold is strongly reduced if nitric acid or sulphuric acid are present in higher concentrations.

8.17. Hafnium

Virtually nothing is to be found in the literature on the determination of hafnium; it can be determined in a nitrous oxide/acetylene flame at the 286.6 nm line with a reciprocal sensitivity of 15 µg/ml 1% and a detection limit of 2 µg/ml. A number of further resonance lines of lower sensitivity are compiled in table 20.
The signal for hafnium is clearly increased in the presence of fluoric acid, so that sample and standard solutions should contain at least 0.1% fluoric acid to control this effect and to obtain the best possible sensitivity. BOND [141] favoured the addition of 0.1 M ammonium fluoride solution to increase the signal.

Table 20. Hf resonance lines

Wavelength nm	Energy level (K)	Reciprocal sensitivity (µg/ml 1%)
286.6	0–34877	15
307.3	0–32533	25
289.8	2357–36850	70
296.5	2357–36075	50
295.1	2357–36237	100
294.1	0–33995	150
290.5	2357–36773	150
377.8	0–26464	150

8.18. Indium

Indium may be determined in an oxidizing air/acetylene flame at the 304.0 nm line with a reciprocal sensitivity of 1 µg/ml 1% and a detection limit of about 0.02 µg/ml. Further resonance lines, with their sensitivities, are compiled in table 21. The signal-to-noise ratio, and thereby the achievable detection limit, for indium, like most other easily melted elements, is dependent on whether the construction of the hollow cathode allows the molten metal to be used or not [892].

Table 21. In resonance lines

Wavelength nm	Energy level (K)	Reciprocal sensitivity (µg/ml 1%)	Spectral bandwidth nm
304.0	0–32892	1	0.7
325.6	2213–32916	1	0.2
410.5	0–24373	2.5	0.7
451.1	2213–24373	3	1.4
256.0	0–39048	10	2.0
275.4	0–36302	25	2.0

There is little literature on the determination of indium, but no serious interferences seem to exist. While SATTUR [1079] was unable to detect any chemical interferences, MULFORD [892] found that various elements, especially in higher concentrations, increasingly reduced the indium absorption.
Approximately 1×10^{-11} g indium can be determined in the graphite tube furnace. Higher chloride concentrations and especially hydrochloric acid reduce the signal. Indium can only be thermally pre-treated to about 900°C without losses so that only simple matrixes can be reliably removed.

8.19. Iodine

At 183.0 nm, the resonance line for iodine is already so far in the vacuum UV that it is virtually unreachable with commercial instruments. Nevertheless KIRKBRIGHT and co-workers succeeded in determining iodine by making a number of minor alterations. They flushed the monochromator of a commercial instrument with nitrogen, used a nitrous oxide/acetylene flame screened with nitrogen and an electrodeless discharge lamp for iodine, and reduced the passage of the beam through air to a minimum by using flushed tubes. In this way they achieved a reciprocal sensitivity of 12 µg/ml 1% and a detection limit of about 6 µg/ml [654] [669]. The determination was free of interferences and no differences between the various types of bonding of iodine could be established. Later, the same authors described an improvement of 38 times in the sensitivity by employing the LEIPERT amplification [670]. The iodine is oxidized to iodate which is then treated with an excess of iodide to release six equivalents of iodine. This is extracted in MIBK and aspirated directly into the flame. The reciprocal sensitivity is then 0.32 µg/ml 1%.
L'VOV and KHARTSYZOV determined iodine in a graphite tube furnace [780]. They used the line at 206.2 nm, which is an excited line whose lower energy level is 0.94 eV above the ground state. At an atomization temperature of 2400°C the detection limit for iodine was about 2×10^{-9} g.

8.20. Iridium

For a long period, iridium was considered indeterminable by atomic absorption spectroscopy. The first success was obtained by MULFORD [893] who, on WILLIS's suggestion, examined this element more exactly. Later MANNING and FERNANDEZ [801] made a thorough study and found a reciprocal sensitivity of 10 µg/ml 1% in an air/acetylene flame somewhat rich in fuel gas at the 264.0 nm resonance line. The 208.9 nm line was found to be twice as sensitive, but the signal-to-noise ratio was considerably worse so that the better detection limit of about 1 µg/ml and better precision were obtained from the first line. Several further resonance lines with their sensitivities are compiled in table 22.
FUHRMAN [390] found that the sensitivity of the iridium determination could be doubled by adding 100 µg/ml potassium plus 1000 µg/ml lanthanum; the addition of sodium or potassium alone, on the other hand, reduced the signal considerably. FUHRMAN further reported that

in the presence of potassium and lanthanum, no influences owing to platinum or titanium could be observed on the absorption of iridium.
JANSSEN and UMLAND favoured the addition of 1% sodium and 1% copper to the determination of iridium in the presence of other platinum metals [571]. The sensitivity was said to have been increased by ten times and the interferences caused by other platinum metals eliminated.

Table 22. Ir resonance lines

Wavelength nm	Energy level (*K*)	Reciprocal sensitivity (μg/ml 1%)	+La+K [390]
264.0	0–37872	10	6.5
208.9	0–47858	8	3
266.5	0–37515	15	
285.0	0–35081	18	8
237.3	0–42132	20	
250.3	0–39940	22	
351.4	0–28452	110	

GRIMALDI and SCHNEPPE used the same additions for the determination of iridium in rocks [443]. To eliminate the same interferences, TOFFOLI and PANNETIER preferred the addition of 0.5% lithium with, under certain circumstances, some copper [1234], and SEN GUPTA used 0.5% copper and 0.5% cadmium [1114]. This author also published a good review on the determination of noble metals by atomic absorption [1116].
LUECKE and ZIELKE found that in a nitrous oxide/acetylene flame the numerous interferences present during the determination of iridium disappeared completely [769]. The sensitivity, which was reduced by about 50%, could be brought back practically to the original value by chlorination. In the $(IrCl_6)^{2-}$ complex, iridium exhibits the same sensitivity in the nitrous oxide/acetylene flame as in the air/acetylene flame, and the determination is completely interference free.
In the graphite tube furnace the detection limit for iridium is about 1×10^{-9} g. Since iridium can be thermally pre-treated up to 1200–1300°C without losses, only minor interferences are to be expected. ADRIAENSSENS and KNOOP [16] found that in the presence of other noble metals, only minor interelemental effects occurred during the determination of iridium, and hydrochloric acid had no influence.

8.21. Iron

Iron is one of the elements most frequently determined by atomic absorption spectroscopy; it is determined in a large number of different samples and mostly in lesser concentrations. The determination of iron appears to be largely free of interferences in a stochiometric air/acetylene flame. ALLAN [37] only found an interference caused by silicon, which can be easily eliminated by adding 200 μg/ml calcium [989]. TERASHIMA [1223] found that besides silicon, strontium, aluminium, manganese, citric acid and tartaric acid also reduced the iron signal and that the effect decreased with increasing height above the burner slot. ROOS and PRICE also found a strong influence on the iron signal by citric acid and eliminated the interference by adding phosphoric acid or sodium chloride [1051]. OTTAWAY and co-workers observed a strong reduction of the iron signal by cobalt, copper and nickel [944]. This interference, however, is strongly dependent upon the flame conditions, such as fuel gas/oxidant ratio, height of observation in the flame, and the anion. Addition of 8-hydroxyquinoline has been recommended to eliminate this interference. To eliminate the same interference, MARTIN added lanthanum [826]. For the examination of biological materials, ZETTNER [1380] found no influences from a large number of cations and anions, from the oxidation states of iron and from various chelating agents. FERRIS and co-workers [365] observed a slight reduction of the iron signal by aluminium and silicon.

In strong nitric acid solutions, especially in a slightly reducing flame, a clear reduction of the iron signal, which is only weakly seen in strongly oxidizing flames, can be observed. With a nitrous oxide/acetylene flame the effect disappears completely, although the sensitivity is reduced by about a half, which is acceptable in many cases.

Table 23. Fe resonance lines

Wavelength nm	Energy level (*K*)	Reciprocal sensitivity (μg/ml 1%)
248.3	0–40257	0.2
248.8	416–40594	0.2
252.3	0–39626	0.2
271.9	0–36767	0.3
302.1	0–33096	0.4
	416–33507	
250.1	0–39970	0.4
216.7	0–46137	0.5
372.0	0–26875	0.7
296.7	0–33695	0.8
344.1	0–29056	1.6
293.7	0–34040	2.5
246.3	0–40594	2.5
368.0	0–27167	10
382.4	0–26140	10

For the determination of iron, a large number of resonance lines of different sensitivities are available. The most important are compiled in table 23. The most frequently used line at 248.3 nm gives a reciprocal sensitivity of 0.2 μg/ml 1% and a detection limit of about 0.01 μg/ml at a spectral bandwidth of 0.2 nm.

In the graphite tube furnace 3×10^{-12} g of iron can be determined. Depending on the matrix, thermal pre-treatment can be carried out up to 1000–1200 °C without losses, so that numerous accompanying materials can be removed. The greatest interference to be expected for an iron determination in the graphite tube furnace is contamination, so that analyses must be performed with the greatest of care.

8.22. Lanthanum, Lanthanides

The lanthanides are very similar in their behaviour in atomic absorption spectroscopy even though they are determinable with widely varying sensitivities. All the rare earths require a nitrous oxide/acetylene flame for their determination. They are more or less strongly ionized in this flame so that an addition of approximately 0.1 to 0.3% potassium is always required to suppress this interference. The most important resonance lines of *lanthanum* and the rare earths are compiled in table 24 together with their achievable reciprocal sensitivities and detection limits.

Because the individual rare earth metals do not mutually influence one another in atomic absorption spectroscopy and no chemical interferences of note occur in the hot flame, the ideal means for an absolute specific determination is given for these elements, which are normally so difficult to separate. Emission measurements are more sensitive for a number of rare earths, but because of the copius number of lines an assignment is often made difficult.

Despite various attempts [48] [885], it has not yet been possible to determine *cerium* by atomic absorption spectroscopy. According to a brief reference, cerium is said to exhibit a minimal absorption at 522.4 nm or 569.7 nm; the achievable detection limit is around 100 µg/ml. It is doubtful whether this is genuine atomic absorption of cerium or whether it is an interference, since the measurements are not reproducible. For this reason, JOHNSON and co-workers [582] have recommended an indirect determination via the heteropoly-molybdo-cerium-phosphoric acid which contains six atoms of molybdenum for each atom of cerium. After extraction with isobutyl acetate, the molybdenum is determined, and as a result of the enriching effect a reciprocal sensitivity of 0.1 µg/ml cerium is obtained.

Because it does not occur naturally, *promethium* has not yet been examined by atomic absorption spectroscopy.

A comprehensive work on the determination of *rare earth metals* in the nitrous oxide/acetylene flame has been published by AMOS and WILLIS [48]. MOSSOTTI and FASSEL [885] examined the use of continuous light sources and a turbulent oxygen/acetylene flame for the determination of lanthanides. A great deal of information, especially on the various resonance lines, is given by MANNING [793] [794] and FERNANDEZ [360].

Up to the present, very few investigations on the determination of the lanthanides in the graphite tube furnace have been made. A number of them at least appear to be determinable with good sensitivities. The greatest interferences occur with older instruments owing to the very intense direct light emission in this spectral range, which often makes determinations difficult because of the weak lamps.

Table 24. Resonance lines of La and the lanthanides

Element	Wavelength nm	Reciprocal sensitivity (µg/ml 1%)	Detection limit (µg/ml)
La	550.1	45	2
Ce	—	—	—
Pr	495.1	25	5
	513.3	35	
	473.7	50	
	492.5	50	
	502.7	55	
	505.3	60	
	491.4	85	
	503.3	85	
Nd	463.4	10	1
	489.7	15	
	471.9	20	
Pm	—	—	—
Sm	429.7	10	2
	520.1	25	
	476.0	45	
	472.8	45	
	528.3	50	
	488.4	55	
Eu	459.4	0.6	0.02
	426.7	4	
	466.2	4	
Gd	407.9	17	1.2
	368.4	20	
	405.8	20	
	405.4	20	
Tb	432.6	7.5	0.6
	431.9	9	
	390.1	12	
	406.2	13	
	433.8	15	
	410.5	26	
Dy	421.2	0.7	0.05
	404.6	1.0	
	418.7	1.1	
	419.5	1.4	
Ho	410.4	0.7	0.04
	405.4	2	
	416.3	2.5	
	404.1	7	
	412.7	15	

Table 24 continued

Element	Wavelength nm	Reciprocal sensitivity (µg/ml 1%)	Detection limit (µg/ml)
Er	400.8	0.9	0.04
	386.3	2.5	
	415.1	2.5	
	389.3	4.5	
	397.4	5	
	408.8	6	
	394.4	18	
Tm	317.8	0.35	0.01
	410.6	1.0	
	374.4	1.1	
	409.4	1.2	
	418.8	1.3	
	420.4	2	
	375.2	4	
	436.0	7	
	297.3	9	
	341.0	10	
	304.7	11	
	530.7	14	
	438.6	16	
	473.3	25	
	389.7	25	
	567.6	30	
	391.6	35	
	563.1	55	
	394.9	60	
	341.7	110	
	576.4	170	
Yb	398.8	0.1	0.005
	346.4	0.7	
	246.5	1.5	
	267.2	8	
Lu	336.0	7.5	0.7
	331.2	12.5	
	337.6	15	
	356.8	15	
	311.8	20	
	339.7	55	
	298.9	65	
	451.9	85	
	308.1	160	

8.23. Lead

Lead can be determined in the most widely varying flames without the expectation that chemical interferences would occur. Usually an air/acetylene flame is used, although a number of authors prefer the air/propane flame. DAGNALL and WEST [262] found minor interferences from aluminium, beryllium, zirconium, phosphate and sulphate in this flame which could be largely eliminated by adding EDTA.
The most important resonance lines of lead are compiled in table 25. Although the 217.0 nm line is noticeably more sensitive than the 283.3 nm line, it does not give a better detection limit owing to a poorer signal-to-noise ratio. Furthermore, more background absorption effects occur at the 217.0 nm line, so that the 283.3 nm line is used with greater success.

Table 25. Pb resonance lines

Wavelength nm	Energy level (K)	Reciprocal sensitivity (μg/ml 1%)
217.0	0–46069	0.25
283.3	0–35287	0.7
261.4	7819–46061	15
368.4	7819–34960	30
364.0	7819–35287	80

Since lead must often be determined in very low concentrations, the normal detection limit of 0.01–0.03 μg/ml in the flame is often not sufficient. In such cases, flameless techniques such as the boat technique, the DELVES system and its modifications, and especially the graphite tube furnace are recommended. The relatively high volatility of lead presents certain difficulties; according to the type of bonding and the matrix, lead can only be thermally pre-treated up to 400–700 °C without losses. Chemical preparation and/or a background corrector can then become necessary. The detection limit for lead in the graphite tube furnace is 2×10^{-12} g.

BRIMHALL [163] and KIRCHHOF [647] have reported a possible method for the determination of the three lead isotopes 206, 207 and 208 by atomic absorption spectroscopy, and the construction of isotope hollow cathode lamps.

8.24. Lithium

Lithium can be effectively determined in an air/acetylene flame practically without interferences. At the 670.8 nm line the reciprocal sensitivity is 0.05 μg/ml 1% and the detection limit about 0.0003 μg/ml. High lithium concentrations are more conveniently determined at the 323.3 nm resonance line which has a reciprocal sensitivity of 15 μg/ml 1%.

Although lithium is in the first main group in the periodic table, it belongs to the alkaline-earth elements [689] rather than to the alkali elements because of its behaviour. In contrast to the alkali metals, lithium can only be determined in cooler flames (e.g. air/propane) with considerably reduced sensitivity rather than increased. Lithium also exhibits no ionization in the air/acetylene flame.

PATASSY [961] found no chemical interferences for the determination of lithium in an air/acetylene flame. This was verified by FISHMAN and DOWNS [374] for a number of elements, but they found a minor interference from strontium. This finding was verified by SLAVIN [12] who found an increase of 5% in the lithium absorption in the presence of 100 µg/ml strontium.

The detection limit for lithium in the graphite tube furnace is about 1×10^{-11} g. Since lithium can only be thermally pre-treated to about 600°C without losses, separation of the matrix is often not possible. In the presence of biological materials, lithium forms a carbide which reduces the sensitivity by about a half, but at the same time exhibits a considerably higher thermal stability so that thermal pre-treatment can be carried out up to 1100–1200°C [1301]. Nitric acid and hydrochloric acid in higher concentrations reduce the lithium signal somewhat, sodium chloride reduces it strongly.

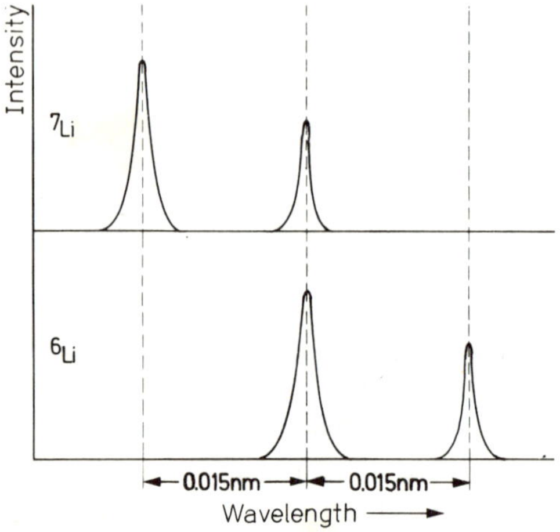

Figure 72. Doublet separation of the lithium isotopes ^{6}Li and ^{7}Li. The weaker doublet line of the more abundent ^{7}Li isotope coincides with the more intense doublet line of the rarer ^{6}Li isotope.

It is possible to determine the isotope ratio ^{6}Li to ^{7}Li by atomic absorption spectroscopy. WALSH [1282] suggested this initially and it was later carried out in practice by ZAIDEL and KORENNOI [1369], MANNING and SLAVIN [808], and GOLEB and YOKOYAMA [427]. ZAIDEL determined the total concentration of lithium by flame emission and then the ^{7}Li concentration using a hollow cathode lamp enriched with this isotope. MANNING used a flame as a light source and sprayed the pure isotopes into it to determine the natural isotope ratio. GOLEB employed cooled hollow cathode lamps and a flameless atomization cell.

The basis of the isotope determination by atomic absorption spectroscopy is the isotope shift in the absorption spectrum which for lithium is 0.015 nm, i.e. considerably greater than the natural width of an absorption line. However, a disadvantage is that the 670.8 nm resonance line is a doublet with a separation of 0.015 nm, so that the weaker doublet line of the more abundant ^{7}Li isotope coincides with the more intense doublet line of the rarer ^{6}Li isotope. This fact, which is shown schematically in figure 72, prevents an independent analysis of both isotopes. However, newer work by WHEAT [1313] [1315] and BUTLER and SCHROEDER [196] have shown that it is easily possible to determine the isotope ratio with a relative accuracy of better than 2% by using a dual-channel atomic absorption spectrophotometer with Li isotope hollow cathode lamps. Therefore, atomic absorption spectroscopy offers, more than just economically, a very attractive alternative to mass spectroscopy for these determinations.

8.25. Magnesium

Magnesium is one of the elements most frequently determined by atomic absorption spectroscopy. This is certainly due in part to the very high sensitivity that can be obtained for this element. At the 285.2 nm resonance line the reciprocal sensitivity in an air/acetylene flame is 0.01 µg/ml 1%, and because of the ideal signal-to-noise ratio, detection limits of 0.0001 µg/ml can be achieved.

This high sensitivity is naturally not always of advantage. Higher magnesium concentrations are more advantageously determined at the 202.5 nm resonance line which gives a reciprocal sensitivity of about 0.5 µg/ml 1% with a three slot burner head and about 5–10 µg/ml 1% [973] with a single slot burner transverse to the beam path. In a laminar air/acetylene flame, magnesium appears to be largely free of chemical interferences. ALLAN [36] found no interferences due to sodium, potassium, calcium or phosphate. WILLIS [1325] [1332] verified this finding, although he found that protein increased the magnesium signal by about 8%. This effect can be eliminated by adding EDTA. DAWSON and HEATON [279] came to the same results. BELCHER and BRAY [98] found that phosphate and silicon up to 100 µg/ml had no influence on magnesium, but that aluminium interfered somewhat. For the analysis of fertilizers McBRIDE [846] found that the absorption of 0.8 µg/ml magnesium is not influenced by even 2000 µg/ml phosphorus (as phosphate). Even with an excess of phosphorus of 10^5 times, SLAVIN [1145] could only find a slight influence due to viscosity, but no chemical interferences.

Work by ZETTNER [1379] showed how dependent these interferences are on the burner design. He modified the burner employed by SLAVIN so that larger droplets reached the flame, and found that with an addition of 10 µg/ml phosphorus the magnesium signal was reduced by 10%. DAVID [269] and STEWART and co-workers [1175] similarly found that interferences due to phosphate and sulphate occurred with other burner designs and that calcium raised the absorption of magnesium. HUMPHREY [541] observed no interferences from numerous cations in considerable excess; only 500 µg/ml silicon reduced the absorption of magnesium by 4%.

These manifold investigations make it clear that through the use of an air/acetylene flame and a well designed burner, only the highest concentrations of silicon and aluminium can interfere with the determination of magnesium somewhat, and that no phosphate interferences, even in the highest concentrations, are to be expected. A very comprehensive study on magnesium spinel ($MgO \cdot Al_2O_3$) inteference has been made by HARRISON and WADLIN [475]. RUBEŠKA and MOLDAN [1069] demonstrated that this mixed oxide really is formed by collecting the particles in an air/acetylene flame and identifying them by direct X-ray analysis. HARRISON and WADLIN found that this interference is very strongly dependent on the height of observation above the burner slot and on the stochiometry of the flame. While 100 µg/ml aluminium reduced the magnesium absorption by 65% directly above the burner slot, the effect was less than 5% at a height of 20–30 mm. Although the sensitivity for magnesium increases with an increasing excess of fuel gas, the interference due to aluminium is more marked in a reducing flame, so that it is preferable to work with somewhat reduced sensitivity in an oxidizing flame in the presence of this element. Titanium and zirconium give similar interferences with magnesium and exhibit similar behaviour to aluminium.

AMOS and WILLIS [48] found that the sensitivity for magnesium was reduced by about 50% in a nitrous oxide/acetylene flame. Furthermore, magnesium is ionized by about 6% in this flame so that the addition of potassium (1000 µg/ml) is recommended. Although this flame is not necessary for the determination of magnesium, it can be of use to eliminate the above interferences [150]. NESBITT [906] verified this for the aluminium interference occurring during the analysis of magnesium in silicate rocks. FLEMING [378] observed that spinel formation was dependent on the stochiometry of the nitrous oxide/acetylene flame and this was verified by HARRISON and WADLIN [475]. According to the fuel gas/oxidant ratio, an increase as well as a decrease in the magnesium signal can take place in the presence of aluminium. HARRISON and WADLIN found that an increase in the signal was dependent upon the height of observation above the burner slot and that at a certain height the influence was practically zero.

While the sensitivity for magnesium is the same in an air/acetylene as in an air/propane flame, a large number of interferences occur in the latter. HORN and LATNER [533] were unable to find interferences due to Na, K, Ca and phosphate in this flame, but JONES and THOMAS [586] reported both increases and decreases due to these elements in the same flame. HALLS and TOWNSHEND [459] found a very marked interference due to phosphate and various other ions in the air/propane flame. WALLACE [1281] found a reduction of 90% in the magnesium signal due to 200 µg/ml aluminium and considerable interferences from Cu, Mn, Ni and Pb. ANDREW and NICHOLS [54] found that even 0.5 µg/ml Al and 0.2 µg/ml Si caused a reduction of more than 50%. FIRMAN [370] and also ELWELL and GIDLEY [7] found influences from a large number of acids and cations. The work of ANDREW and NICHOLS [55] shows how unsuitable cool flames are for the determination of magnesium. These authors examined the influence of two "identical" gases on various interferences and found considerable variations with different gas mixtures. SUNDERMAN and CAROL [1202] examined an air/hydrogen flame for the determination of magnesium and found—in contrast to an air/acetylene flame—a reduction of the signal owing to protein. FLEMING and STEWART [379] found numerous interferences in a turbulent oxygen/acetylene flame, in part with non-uniform courses. RAMAKRISHNA [1014] discovered in a turbulent oxygen/acetylene flame all interferences that also occur during the determination of calcium but to a somewhat reduced extent.

YOZA and OHASHI used atomic absorption as a detector for magnesium in gel chromatography [1364].

About 1×10^{-12} g magnesium can be detected in the graphite tube furnace. However, because of the great risk of contamination the determination is extremely difficult, thus little has been reported up to the present.

8.26. Manganese

Manganese can be determined in an air/acetylene flame without major interferences. The greatest reciprocal sensitivity of 0.05 μg/ml 1% can be achieved at the 279.5 nm resonance line; in practice, the triplet 279.5/279.8/280.1 nm is mostly employed since the use of greater slit widths brings a marked improvement to the signal-to-noise ratio. When using the triplet (approx. 0.7 nm spectral bandwidth), the reciprocal sensitivity is 0.1 μg/ml 1% and the detection limit is about 0.002 μg/ml. Manganese is an exception in that several lines can be used simultaneously without the usual disadvantages. The three triplet lines have almost equal sensitivity so that a greater loss in sensitivity and strong curvature of the calibration plot are avoided. Each individual line naturally gives better linearity. For the determination of higher manganese concentrations, the 403.1 nm resonance line with a reciprocal sensitivity of about 1 μg/ml 1% is more suitable. In this case, a spectral bandwidth of 0.2 nm is necessary to separate the 403.3 nm resonance line, otherwise there is considerable curvature of the calibration plot.

ALLAN [37] found no interferences from Na, K, Ca, Mg, and phosphate for the determination of manganese in an air/acetylene flame. PLATTE and MARCY [989] found a 20% reduction of the manganese signals when small quantities of silicon were present; by adding 200 μg/ml calcium (as chloride), no more interferences due to 1000 μg/ml silicon and numerous other cations could be observed. BELCHER and KINSON [100] found no interference from silicate when only the inner flame cone without the edges was observed. Silicate interference can be eliminated by using a nitrous oxide/acetylene flame in which the reciprocal sensitivity is about 0.3 μg/ml 1%. HUSLER determined manganese in steel in this flame without interferences [547] and BARNETT found virtually no interferences from acids in this flame or in a fuel gas weak air/acetylene flame. Interferences first became noticeable from acid concentrations between 5% and 10% (v/v) [73].

MANSELL [814] has fully described the extraction of manganese and FELDMANN, BOSSHART and CHRISTIAN [354] examined the sensitivity of manganese in four different solvents. MANSELL and EMMEL extracted manganese and other trace elements from a concentrated brine solution [815]. CALKINS [202] separated manganese from a large excess of aluminium by complexing it with 8-hydroxyquinoline and extracting it with chloroform.

OLSEN and SOMMERFELD have drawn attention to the instability of many metal chelates, especially those of manganese. Immediately after extraction, precipitation begins, leading to a loss of sensitivity and blockage of the nebulizer [928]. Directly after the extraction of the diethyl-dithiocarbonate complex in MIBK, these authors evaporated the solution to dryness and then took the residue up in 0.1 N HCl/acetone (1:1). This solution was stable for at least two weeks.

In the graphite tube furnace 2×10^{-13} g manganese can be determined, so that it is one of the most sensitive elements. Since manganese can be thermally pre-treated up to 1000–1100 °C without losses, complete separation of the matrix can frequently be achieved, so that the determination is interference free. Of the usual acids, only nitric acid in higher concentrations leads to a stronger reduction of the signal.

8.27. Mercury

Since mercury must be generally determined in the smallest traces and atomic absorption spectroscopy only gives a reciprocal sensitivity of 10 μg/ml 1% and a detection limit of 0.2 μg/ml at the 253.7 nm line, it is little suited for the analysis of mercury. This low sensitivity is due to the resonance line, corresponding to the transition from the ground state to the first excited state, being at 184.9 nm in the vacuum UV range and thus making the determination impossible with normal instruments. KIRKBRIGHT successfully carried out a determination of mercury at this line by using a modified atomic absorption spectrophotometer with a flushed monochromator and a nitrous oxide/acetylene flame shielded with nitrogen. He achieved a reciprocal sensitivity of 0.05 μg/ml 1% and a detection limit of 0.02 μg/ml mercury [654].

Because of the great significance of mercury in toxicology, and because of the simplicity and specificity of atomic absorption spectroscopy, various procedures have been developed which improve the sensitivity of the determination. Already before the rediscovery of atomic absorption spectroscopy, WOODSON had described a measurement system for mercury [1354] which was later employed by many authors ([70] [71] [755] [847] [874] [948] [949] [1388]). BRANDENBERGER and BADER also developed a flameless method with which 0.2 ng of mercury could be determined [153] [154]. The boat technique described by KAHN and co-workers [616] also improved the detection limit and 20 ng mercury can be determined in a commercial atomic absorption spectrophotometer. WHEAT employed the method described by BRANDENBERGER for the determination of mercury in radioactive samples [1314].

The most successful procedure for the determination of mercury traces, and which is used most frequently nowadays, was proposed by POLUEKTOV and co-workers [996] in 1964 and was later thoroughly investigated by HATCH and OTT [481]. The mercury is reduced to free metal in an acid solution by tin(II) chloride and then conducted to an absorption cell mounted in the beam path of an atomic absorption spectrophotometer. The detection limit, with this procedure, is 0.1 μg/l mercury and is thus approximately three orders of magnitude better than in the flame. A more comprehensive description of the flameless determination of mercury has already been given in section 3.2.6 (page 64), so that further details will not be given here. A review on the various flameless methods for the determination of mercury has been given by MANNING [797], who also draws attention to the risk of occurrence of non-specific light losses with the low temperature procedure [798]. Otherwise, this method appears to be largely free of interferences; it fails solely in the presence of an easily reduced matrix. The applications of this procedure are therefore correspondingly large, especially in medicine and environmental control ([57] [61] [576] [753] [786] [895] [1024] [1251]).

Besides the flameless procedures, many extraction methods to increase the sensitivity of mercury have been described. Frequently ammonium pyrrolidine dithiocarbamate is used as complexing agent and MIBK as extracting solvent [111] [1323].

OSBORN and GUNNING [942] determined the mercury isotope ratio by atomic absorption spectroscopy.

8.28. Molybdenum

Molybdenum can be determined both in a fuel gas rich air/acetylene flame and in a nitrous oxide/acetylene flame. At the 313.3 nm resonance line the reciprocal sensitivity in an air/acetylene flame is about 1 µg/ml 1% and in a nitrous oxide/acetylene flame about 0.5 µg/ml 1%; the detection limits are 0.1 µg/ml and 0.02 µg/ml respectively. Besides its high sensitivity, the nitrous oxide/acetylene flame is also largely free of interferences, thus it is being used more and more nowadays.

The earlier studies of DAVID [271] and MOSTYN and CUNNINGHAM [887] were all carried out with a very fuel gas rich air/acetylene flame. However, as the investigations of RANN and HAMBLY have shown [1018], even under fuel gas rich conditions the flame zone with the highest atom concentration is very limited, so that a large portion of the hollow cathode radiation passes through areas in which the formation of compounds is possible. ROOS [1045] also found that iron caused considerable interferences in the determination of molybdenum; also DAVID [271] and MOSTYN and CUNNINGHAM [887] found interferences in the analysis of steel which they could control by adding 2% ammonium chloride.

The possible chemical interferences in the determination of molybdenum in fuel gas rich air/acetylene flames have been little investigated and matching of the matrix is frequently necessary. A number of interferences occur in a nitrous oxide/acetylene flame which can lead either to a reduction or to an increase in the signal. VAN LOON found that the addition of 0.1 to 0.3% aluminium eliminated all interferences [1254]. WEST and co-workers accounted this effect of aluminium to an inhibition of the lateral diffusion of molybdenum, leading to an increased atom concentration in the middle of the flame [1306]. Especially with organic solvents, which give considerable difficulties in the "fat" air/acetylene flame, the hot nitrous oxide/acetylene flame shows its advantages.

KIRKBRIGHT [655] complexed molybdenum with 8-hydroxyquinoline in the presence of fluoride and EDTA as masking reagents and extracted it into butanol, since this burns well in the nitrous oxide/acetylene flame. URE [1250] treated soil extracts with tri-N-octylamine and 2% n-octanol in petrol and then conducted the extracted molybdenum with ammonia gas back into the aqueous phase in which he determined it in a nitrous oxide/acetylene flame. DELAUGHTER [289] used toluene-3,4-dithiol as the complexing agent for extraction into MIBK, and BUTLER and MATHEWS [195] complexed molybdenum with ammonium pyrrolidine dithiocarbamate and also extracted into MIBK. However, it was found to be necessary to mask iron with citric acid before the extraction.

The detection limit for molybdenum in the graphite tube furnace is about 3×10^{-12} g. Although molybdenum can be thermally pre-treated without losses up to about 1900°C, the determination is not completely free of interferences, and the method of additions must frequently be employed [1101].

Higher concentrations of nitric acid and sulphuric acid lead to a slight signal reduction. The strongest interference, however, is carbide formation which leads to a strong reduction of the signal and which is strengthened in the presence of biological materials. During the thermal destruction of biological and other organic materials, finely divided carbon is often formed, it greatly favours carbide formation.

DAGNALL reported an increase in the sensitivity by about 70% when the graphite tubes were "saturated" with tungsten [249] and KUZOVLEV and co-workers reported that the sensitivity increased by a factor of 2.5 when the graphite tubes were coated with a tantalum/niobium mixed carbide [719]. However, it must be pointed out that the reasons given for carbide formation above cannot be influenced in this way.

8.29. Nickel

Nickel is one of the metals frequently determined by atomic absorption spectroscopy. The best reciprocal sensitivity of 0.15 μg/ml 1% is obtained in an air/acetylene flame at the 232.0 nm resonance line; the detection limit is 0.002 μg/ml. To be able to use the 232.003 nm line meaningfully, a spectral bandwidth of 0.2 nm is necessary since the two strong emission lines at 231.716 nm and 232.138 nm otherwise cause a considerable curvature in the calibration plot and a strong reduction in the sensitivity. Since the 232.0 nm line even with a spectral bandwidth of 0.2 nm gives a markedly curved calibration plot, it is not very suitable for the determination of higher nickel concentrations. For this, the 341.5 nm resonance line with a reciprocal sensitivity of 3 μg/ml 1% is more suitable. This line even permits the use of greater spectral bandwidths (up to 0.7 nm) and then gives a good signal-to-noise ratio. Further resonance lines are compiled in table 26. However, slight spectral interferences can occur at several of these lines, especially with multielement lamps which also contain cobalt; a critical check of the lines and the signals is recommended.

Table 26. Ni resonance lines

Wavelength nm	Energy level (*K*)	Reciprocal sensitivity (μg/ml 1%)
232.0	0–43090	0.15
231.1	0–43259	0.3
234.6	0–42621	1
341.5	205–29481	3
305.1	205–32973	4
300.3	205–33501	5
346.2	205–29084	5

The determination of nickel in an oxidizing (fuel gas weak) air/acetylene flame appears to be largely free of interferences. KINSON and BELCHER [645] found no interferences owing to high concentrations of Co, Cr, Cu, Mn, Mo, Al, V and W. Equally, HCl, HNO_3, H_2SO_4 and H_3PO_4 had no influence. These findings were verified by PLATTE and MARCY [989], although great care must be taken in an exact adjustment of the burner and flame gases

[73] [1199]. SUNDBERG related the most varying effects (enhancing and suppressing) of other metals (Co, Cr, Cu, Fe, Mn, Zn) with non-exact gas settings and height of observation to the dissociation energies of the corresponding monoxides [1199]. In a less sharp flame, interferences from iron and chromium become noticeable. For high precision analyses, WELZ and SEBESTYEN [1299] found it necessary to match the matrix to maintain a relative accuracy of 0.3%. The potential interferences of iron and chromium can be eliminated by the use of a nitrous oxide/acetylene flame; the reciprocal sensitivity at the 232.0 nm resonance line is about 1 μg/ml 1%.

Nickel can be complexed with ammonium pyrrolidine dithiocarbamate and extracted with MIBK [1170]. BURREL [183] used multiple enrichment steps such as co-precipitation with iron(III) hydroxide, which was then later removed [307], to determine the smallest traces of nickel.

About 1×10^{-11} g nickel can be detected in the graphite tube furnace. Since nickel can be thermally pre-treated up to about 1200 °C without losses, the matrix can often be largely separated. However, the determination of nickel is not completely interference free, so that frequently the method of additions must be employed [570]. Higher concentrations of nitric acid also lead to a strong reduction of the signal and to great scattering, while sulphuric acid only exhibits a slight influence.

8.30. Niobium

Niobium can be determined in a nitrous oxide/acetylene flame with a reciprocal sensitivity of about 40 μg/ml 1% and a detection limit of about 2 μg/ml at the 334.3 nm resonance line [795] with a spectral bandwidth of 0.2 nm. Since niobium is noticeably ionized in the hot flame, the addition of about 0.1% potassium (as chloride) to sample and standard solutions is recommended. Up to the present, nothing has been reported on chemical interferences.

8.31. Non-Metals

Iodine, phosphorus and sulphur are treated individually since there are direct atomic absorption procedures for these non-metals. The remaining non-metals cannot be determined directly by atomic absorption spectroscopy since their resonance lines lie in the vacuum UV range. DAGNALL, THOMPSON and WEST used molecular emission for the determination of sulphur at an S_2 band [254] and phosphorus at an H—P—O band [259]; in the presence of indium they determined the halogens by means of InCl, InBr and InI emission bands [261]. Similar work with special application to biological samples has been published by GUTSCHE and HERRMANN [450–453]. Employing continuous light sources measurement of the absorption at these bands is conceivable, but has not been reported up to the present.

For the indirect determination of non-metals by atomic absorption spectroscopy, two different ways have been followed: The utilization of chemical interferences, or the use of known precipitation reactions with metals. CHRISTIAN and FELDMAN [252] used the first method to determine *orthophosphates*, *sulphates*, *sulphides*, *iodates*, *iodides*, *glucose*, *protein*, *8-oxiquinoline* and other complexing agents by examining their influence on the determination of calcium, iron and chromium and placing the extent of this influence in a relationship to the concentration of the interfering ions. KUNISHI and OHNO investigated the indirect determination of sulphate very thoroughly and employed the suppressive effect on the determination of iron and its elimination by lanthanum for the determination of sulphate. With an ion ratio of $La^{3+}:SO_4^{2-}=2:3$ the original sensitivity of iron is again obtained. The authors speak of a sharp "change" which makes very precise determinations possible [717]. BOND and O'DONNEL [142] determined *fluoride* ions in a similar manner by quantitatively evaluating their influence on the determination of magnesium in an air/coal gas flame and on zirconium in a nitrous oxide/acetylene flame.

The second procedure, formation of an insoluble precipitate with a metal cation with subsequent determination of this metal, was employed by EZELL [336] and WEST and ERLUND-HELMERSON [1312] for the determination of *chloride* in various materials. The chloride ion was precipitated with silver nitrate and either the precipitate was dissolved in ammonia solution and analysed for its metal content, or the excess metal was measured in the solution [82] after an exactly measured addition of silver nitrate.

GAMBRELL determined chloride in water through end point determination with silver nitrate by atomic absorption [404]. Two different quantities of silver nitrate were added to the sample solution, the absorbance for silver was measured and extrapolated against zero.

MANAHAN and KUNKEL determined *cyanide* by utilizing the solubility of Cu(II) from basic copper carbonate in alkaline medium. The complexed copper, as $[Cu(CN)_3]^-$, was determined by atomic absorption; the reciprocal sensitivity was 2×10^{-15} Mole CN^- [790]. MANAHAN and JONES also described a specific detector system for *chelating agents* for liquid chromatography [789]. A solution containing a chelating agent is conducted through a short column containing a chelate ion exchanger in the Cu form and then directly to an atomic absorption spectrophotometer set for copper. The quantity of copper is directly proportional to the quantity of chelating agent; the detection limit is 5×10^{-7} mMole.

KUMAMARU and co-workers determined *nitrate* [716] and organic molecules through co-extraction with metal complexes. Nitrate and *phthalic acid* [713] [714] were both extracted as ion pairs from bis-(neocuproin)-copper(I) with MIBK. *Pentachlorophenol* [1359] was also extracted as an ion pair with tris-(1,10-phenanthroline)-iron(II) in nitrobenzene; the concentrations of copper and iron respectively in the extracts are proportional to the concentrations of the ions or molecules under study and can be determined by atomic absorption spectroscopy.

8.32. Osmium

SLAVIN [12] first reported in 1968 that WILLIS had found strong absorption for osmium at various wavelengths. Thereupon FERNANDEZ [356] examined ten resonance lines for their sensitivity and found a reciprocal sensitivity of 1 μg/ml 1% at the 290.9 nm resonance line employing a nitrous oxide/acetylene flame. The detection limit appears to be little better than 0.1 μg/ml. Further resonance lines, with their sensitivities, are compiled in table 27.

By using a fuel gas rich air/acetylene flame a reciprocal sensitivity of about 5 μg/ml 1% can be achieved [356]. MAKAROV and co-workers [788] found a reciprocal sensitivity of 17 μg/ml 1% in an air/propane/butane flame. OSOLINSKI and KNIGHT [943] determined osmium

Table 27. Os resonance lines

Wavelength nm	Energy level (*K*)	Reciprocal sensitivity (μg/ml 1%) N$_2$O/Acetylene [356]	Air/Acetylene [356]	Air/Propane [788]
290.9	0–34365	1.0	5.0	17
305.9	0–32685	1.6	6.4	20
263.7	0–37909	1.8		
301.8	0–33124	3.2		
330.2	0–30280	3.6		
271.5	0–36826	4.2		
280.7	0–35616	4.6		
264.4	0–37809	4.8		
442.0	0–22616	20		
426.1	0–23463	30		

in thiourea complexes and in chloroform extracts. They draw particular attention to the toxicity of OsO_4 and the necessary safety precautions.

8.33. Palladium

Palladium can be determined in a very sharp (fuel gas weak) air/acetylene flame with a reciprocal sensitivity of 0.3 μg/ml 1% and a detection limit of about 0.02 μg/ml. Two resonance lines at 247.6 nm and 244.8 nm give approximately equal results. However, the latter, even with a spectral bandwidth of 0.2 nm, gives a very curved calibration plot; for practical work the line at 247.6 nm is to be preferred. Two further resonance lines at 276.3 nm and 340.4 nm each give a reciprocal sensitivity of about 1 μg/ml 1%. Instead of an air/acetylene flame, LOCKYER and HAMES [759] and STRASHEIM and WESSELS [1181] used an air/propane flame which is also preferred nowadays by a number of authors for the determination of palladium [296]. Although these authors observed no interferences, the determination of palladium, even in the air/acetylene flame, does not appear to be completely free of interferences; especially high concentrations of other noble metals can exert certain influences. The acid concentration should be matched.

JANSSEN and UMLAND eliminated the chemical interferences which occur during the analysis of noble metals by adding 1% sodium and 1% copper [571]. TOFFOLI and PANNETIER preferred 0.5% lithium with the possible addition of copper [1234]. SEN GUPTA, who has published a good review on noble metal analysis [1116], added 0.5% copper and 0.5% cadmium to eliminate interelemental interferences [1114]. ADRIAENSSENS and VERBEEK found that in 2% potassium cyanide solution the majority of interferences which are observed in acid solutions did not occur. Only gold and plantinum influenced palladium somewhat; these effects dissappeared, however, after addition of silver [17].

HARRINGTON analysed colloidal palladium solutions and found that these gave the same absorbance values as true palladium(II) solutions when the burner height was optimized [469].

ERINC and MANGEE [330] described an extraction of palladium from acid solution with hexone after complexing with pyridine thiocyanate.

8.34. Phosphorus

The resonance lines for phosphorus are at 178 nm in the vacuum UV range so that it cannot be determined with normal atomic absorption spectrophotometers. However, KIRKBRIGHT successfully determined phosphorus in a modified atomic absorption instrument with a flushed monochromator and a nitrous oxide/acetylene flame shielded with nitrogen [654]. Of the three resonance lines 177.5 nm, 178.3 nm and 178.8 nm, the 178.3 nm line gave the best reciprocal sensitivity of 5.4 μg/ml 1%. KIRGHBRIGHT determined phosphorus in meat extracts and milk powder by this procedure.

WALSH [1283] determined phosphorus in an atomization cell with a vacuum spectrophotometer. This procedure has found no practical applications up to the present. MANNING and SLAVIN [809] reported the successful determination of phosphorus at the 213.55 nm/213.62 nm and 214.91 nm lines which emanate from metastable states. In a nitrous oxide/acetylene flame and with a phosphorus hollow cathode lamp, they found a reciprocal sensitivity of 290 μg/ml 1% at the 214 nm line and 540 μg/ml 1% at the 215 nm line. Thus phosphorus concentrations of 0.2–5% can be well determined. Finally, the authors point out that phosphorus interferes with numerous cations in atomic absorption spectroscopy and it is thus equally possible that numerous cations can interfere in the determination of phosphorus. KERBER and co-workers [636] compared the determination of phosphorus by atomic absorption and flame emission spectroscopy at the HPO band at 526 nm. They found only slight interferences with atomic absorption spectroscopy. SYTY found that this HPO emission is considerably increased when a cooled plate is mounted behind the flame [1213] and he used this procedure to determine phosphorus in phosphate rocks [1214].

Various indirect procedures for the determination of phosphorus have been proposed [544]. They are based on the formation of ammonium phosphorusmolybdate or the free heteropolyacid with the subsequent determination of molybdenum by atomic absorption spectroscopy. WILSON [1338] dissolved the ammonium phosphorusmolybdate precipitate in ammonia solution and determined the molybdenum directly from this solution. ZAUGG and KNOX [1372] shook acid ammonium molybdate solution and 2-octanol with the phosphorus containing sample and determined molybdenum directly from the organic phase. In a later report [1373] these authors described the application of this extraction procedure to various biological problems. A similar extraction with butyl acetate as solvent has been described by KUMAMARU and co-workers [715].

MANNING and FERNANDEZ observed an absorption for phosphorus in the graphite tube furnace at a lead ion line at 220.3 nm, which they attributed to a PO_x radical [804]. At 1×10^{-7} g 1%, the reciprocal sensitivity is certainly markedly poorer than for metallic elements, but the determination could be meaningfully applied to the analysis of micro samples for biochemistry and agricultural chemistry. The determination is specific for phosphorus, and interferences caused by various types of bonding can be brought under control by adding 100–250 μg/ml calcium (as nitrate).

8.35. Platinum

Platinum has a multitude of resonance lines of which the line at 265.9 nm with a reciprocal sensitivity of about 2 µg/ml 1% is the most sensitive. The detection limit is 0.1 µg/ml in a sharp, fuel gas weak air/acetylene flame. Several further important resonance lines are compiled in table 28.

Table 28. Pt resonance lines

Wavelength nm	Energy level (*K*)	Reciprocal sensitivity (µg/ml 1%)
265.9	0–37591	2
217.4		3
306.5	0–32620	5
264.7	0–37769	11
299.8	776–34122	12
271.9	824–37591	18
304.3	824–33681	35
244.0	0–40970	100

LOCKYER and HAMES [759] found no interferences for the determination of platinum in an air/propane flame. On the other hand, STRASHEIM and WESSELS [1181] reported interferences from numerous cations, which they could largely eliminate by adding 2% copper (as sulphate). According to a number of findings, both strong acids and other noble metals interfere in the determination of platinum; matching the matrix and the use of a nitrous oxide/acetylene flame largely eliminate these interferences.

JANSSEN and UMLAND found numerous interelemental effects for the determination of platinum in the presence of other noble metals which they eliminated by adding 1% sodium and 1% copper [571]. At the same time they found that these additions increased the sensitivity for platinum by around 50%.

TOFFOLI and PANNETIER preferred the addition of 0.5% lithium with possibly some copper [1234], while SEN GUPTA employed 0.5% copper and 0.5% cadmium as buffer [1114].

ADRIAENSSENS and VERBEEK found that in 2% potassium cyanide solution the majority of interelemental interferences observed during the analysis of noble metals do not occur [17]. They used a fuel gas weak, sharp air/acetylene flame and achieved an accuracy of 2%, even in the presence of considerable excesses of other noble metals.

PITTS, VAN LOON and BEAMISH found numerous interferences in an air/acetylene flame which they were able to eliminate by the addition of lanthanum [986]. As an alternative, these authors recommend a nitrous oxide/acetylene flame since no more interferences occur even though the sensitivity is reduced to a fifth. Even high concentrations of other noble metals no longer interfere [987].

Approximately 2×10^{-10} g platinum can be determined in the graphite tube furnace. Since platinum can be thermally pre-treated without losses up to about 1400 °C, only minor interferences are to be expected; the signal can be greatly decreased simply by adding higher concentrations of nitric acid.

Adriaenssens and Knoop found only slight interelemental effects for the determination of platinum in the presence of iridium and rhodium in the graphite tube furnace [16]. Janoušková and co-workers determined platinum on catalysts (aluminium oxide) and found no influences even with a 100 times excess of other ions [570]. Simply adding large amounts of strontium or 0.5 N nitric acid reduced the signal somewhat.

8.36. Potassium

Potassium is frequently determined with sufficient sensitivity and precision in simple flame photometers. However, atomic absorption spectroscopy with the choice of various resonance lines and flames offers greater variability and hence optimization for each analytical problem. Furthermore, in the presence of complex matrixes, emission is subjected more to interferences than absorption.

The detection limits in atomic absorption and flame emission spectroscopy are very similar; while the best values in emission are 0.003 μg/ml, for atomic absorption the reciprocal sensitivity at the 766.5 nm resonance line is 0.05 μg/ml 1% and the detection limit about 0.001 μg/ml. Nevertheless, this detection limit is very dependent upon the instrument, since many photomultipliers only have low energy at the K wavelength and thus exhibit an unfavourable signal-to-noise ratio.

Beside the most sensitive line at 766.5 nm, the second doublet line at 769.9 nm is frequently of advantage; it gives approximately half the sensitivity (0.1 μg/ml 1%) and a calibration plot with better linearity. For higher potassium concentrations the doublet at 404.4 nm and 404.7 nm with a reciprocal sensitivity of about 10 μg/ml 1% is available.

Potassium is frequently determined in an air/acetylene flame. There is a slight ionization interference which can be eliminated by adding 0.1% cesium. The determination of potassium is simpler in a cooler flame in which no ionization can be observed. The air/hydrogen flame has proved very advantageous. It gives a greatly improved signal-to-noise ratio especially at the secondary doublet at 404.4/404.7 nm and thereby better precision.

A frequent cause for a noisy display during the analysis of potassium (and especially sodium) is contamination of the laboratory air through dust or tobacco smoke. If these pass into the flame, considerable interferences can result, especially with trace analyses. This interference can be easily established by measuring the noise signal with and without a flame; with greater differences this is the cause. A simple method for the elimination of this interference—in case the cause cannot be eliminated—is the use of a multi-slot burner head which gives a broader flame. The edges of the flame, in which the interference can be principally observed, are not irradiated by the beam and thus do not affect the analysis.

Manning and co-workers [810] made a detailed examination of the difference between vapour discharge lamps and hollow cathode lamps for the determination of potassium and found that the discharge lamps were superior, especially for the secondary line doublet. After the later developments in the lamp sector, it can however be said that both lamp types are qualitatively equivalent.

JOSEPH and co-workers [595] reported a procedure for determining ^{40}K by atomic absorption spectroscopy.
YOZA and OHASHI used atomic absorption as a detector for potassium in gel chromatography [1364].
Approximately 1×10^{-13} g potassium can be determined in the graphite tube furnace, but the determination is subject to strong contamination risks. The presence of sodium especially causes strong disturbance leading to a considerable loss in sensitivity and with a 10000 times excess of sodium, the potassium signal is almost completely suppressed. Various acids and ammonia also lead to sensitivity losses for the determination of potassium.

8.37. Rhenium

Rhenium can be determined in a nitrous oxide/acetylene flame at the 346.0 nm resonance line with a reciprocal sensitivity of 15 µg/ml 1% and a detection limit of about 1 µg/ml. At the 346.5 nm and 345.2 nm resonance lines the reciprocal sensitivities are 20 and 35 µg/ml 1% respectively. Up to the present, no chemical interferences have been reported in the nitrous oxide/acetylene flame. SHRENK and co-workers [1097] found that in a turbulent oxygen/acetylene flame the rhenium signal was markedly reduced by various cations in concentrations over 200–300 µg/ml as follows: Calcium by about 75%, manganese by 50%, aluminium by 35%, iron by 15% and molybdenum, lead and potassium by about 10%. This can almost certainly be traced to the burner and flame.

8.38. Rhodium

Rhodium can be determined in a nitrous oxide/acetylene flame at the 343.5 nm resonance line with a reciprocal sensitivity of 0.5 µg/ml 1% and a detection limit of about 0.005 µg/ml. A sharp, fuel gas weak air/acetylene flame gives a better sensitivity and detection limit, but considerable chemical interferences are to be expected. A number of other resonance lines, with the reciprocal sensitivities attainable in both flame types, are compiled in table 29.

Table 29. Rh resonance lines

Wavelength nm	Energy level (K)	Reciprocal sensitivity (µg/ml 1%) Air/Acetylene [500]	N$_2$O/Acetylene [63]
343.5	0–29105	0.25	0.5
369.2	0–27075	0.6	1.4
339.7	0–29431	0.7	1.7
350.3	0–28543	1.5	2.5
365.8	1530–28860	2	2.5
370.1	1530–28543	3	5.5
350.7	2598–31102	8	2.5

The recorded details on rhodium are, in part, quite contradictory. While LOCKYER and HAMES [759] claim to have found no interferences in an air/coal gas flame, STRASHEIM and WESSELS [1181] reported a multitude of chemical interferences in an air/propane flame which they could not eliminate simply, as with platinum. DEILY [288] found no difficulties for the determination of rhodium in organic solvents in a sharp air/acetylene flame. ZEEMAN and BRINK [1376] reported numerous interferences in an air/acetylene flame, and GINZBURG and co-workers [417] obtained similar results.

KALLMANN and HOBART found no interferences from phosphoric acid in a sharp, fuel gas weak flame, but found strong increases from alkaline aluminium and zinc sulphates, which they traced to the formation of a rhodium alum [622]. JANSSEN and UMLAND found that the addition of 1% sodium and 1% copper increased the sensitivity for rhodium by 50% and at the same time interelemental effects from other noble metals were eliminated [571]. TOFFOLI and PANNETIER preferred 0.5% lithium and some copper to eliminate the same interferences [1234] and SEN GUPTA used a buffer consisting of 0.5% copper and 0.5% cadmium [1114].

ATWELL and HERBERT [63] carried out systematic investigations in which they examined the influence of 14 cations in concentrations up to 300 μg/ml and four acids up to a concentration of 10% on a rhodium solution containing 30 μg/ml in an air/acetylene flame and in a nitrous oxide/acetylene flame. In the air/acetylene flame only 300 μg/ml lead did not cause any interferences while all remaining elements exerted influences to varying degrees. In the nitrous oxide/acetylene flame, on the other hand, only interferences from iridium and ruthenium could be established, while all the other elements and acids examined had no influence. The authors found that the interferences from iridium and ruthenium could be eliminated by adding 0.5% zinc to the sample solutions. These authors further found that the stochiometry of the flame, the height of the light beam above the burner head and similar parameters were not critical and that the calibration plot is completely linear up to 100 μg/ml.

ADRIAENSSENS and KNOOP determined rhodium in the graphite tube furnace and found only slight interelemental effects and a slight signal reduction owing to nitric acid [16]. The detection limit was 10^{-10} g.

8.39. Rubidium

Very little has been written on the determination of rubidium. At the 780 nm resonance line, a reciprocal sensitivity of 0.1 μg/ml 1% and a detection limit of about 0.002 μg/ml can be achieved in an air/acetylene flame. The sensitivity can be increased by a factor of about 2 in cooler flames. The second doublet line at 794.8 nm ought to exhibit about half the sensitivity, but like the doublet at 420.2/421.6 nm, which should be suitable for higher concentrations, nothing has been reported about this.

The only coherent work on the behaviour of rubidium is by SLAVIN and co-workers [1151]. In this work, the problem of suitable light sources—up to the present no usable rubidium hollow cathode lamps are available—is fully discussed and the use of flames of lower temperature is reported. Like other alkali metals, rubidium is noticeably ionized in an air/acetylene flame, so that another easily ionized metal—preferably cesium—must be added to suppress this

interference. In an air/hydrogen flame rubidium is hardly ionized, so that this flame is possibly to be preferred.
A detection limit of about 0.0003 μg/ml can be achieved by flame emission [363].
JOSEPH and co-workers [595] used atomic absorption spectroscopy for the determination of ^{87}Rb in various natural samples.

8.40. Ruthenium

Ruthenium can be determined in a nitrous oxide/actylene flame at the 349.9 nm resonance line with a reciprocal sensitivity of 0.5 μg/ml 1% and a detection limit of about 0.1 μg/ml. SCHWAB and HEMBREE found that the sensitivity for ruthenium was improved by 100% by the addition of 0.13 Mole lanthanum nitrate and 0.8 Mole hydrochloric acid. All interferences were eliminated by this addition [1100]. ROWSTON and OTTAWAY determined ruthenium in an air/acetylene flame and used a mixture of 0.5% copper sulphate and 0.5% cadmium sulphate to eliminate the numerous interferences [1063].

8.41. Scandium

Scandium can be determined in a nitrous oxide/acetylene flame at the 391.2 nm resonance line with a reciprocal sensitivity of 0.4 μg/ml 1% and a detection limit of about 0.02 μg/ml. AMOS and WILLIS [48] and MANNING [794] examined the sensitivities of numerous other scandium resonance lines, some of which are compiled in table 30. Scandium is noticeably ionized in the nitrous oxide/acetylene flame, so that the interference must be eliminated by adding 0.1 to 0.2% potassium (as chloride) to sample and standard solutions.

Table 30. Sc resonance lines

Wavelength nm	Energy level (K)	Reciprocal sensitivity (μg/ml 1%)
390.8	0–25585	0.6
402.4	168–25014	0.6
402.0	0–24866	0.9
327.0	0–30573	1.4
327.4	168–30707	2

KRIEGE and WELCHER [705] investigated the influence of numerous cations and anions on the absorption of scandium in a nitrous oxide/acetylene flame and found considerable chemical interferences. To eliminate these interferences the authors only recommended matching the matrix.

8.42. Selenium

The most sensitive selenium resonance line is at 196.1 nm at the start of the vacuum UV [221] and thus out of the detection range of normal spectrophotometers. With good atomic absorption instruments, using an air/acetylene flame and an electrodeless discharge lamp, a reciprocal sensitivity of 0.5 μg/ml 1% and a detection limit of 0.2 μg/ml can be achieved at this line. With the simultaneous use of a background corrector, concentrations under 0.1 μg/ml can be determined. This effect is explained by the fact that the air/acetylene flame absorbs about 55% of the radiation from the light source at the selenium resonance line and small fluctuations in the flame absorption enter directly into the result. With simultaneous background correction, the flame absorption and the related noise are eliminated.

ALLAN [42] employed the resonance line at 204.0 nm, which however only gives a reciprocal sensitivity of 5 μg/ml 1%. Determinations are easier in an argon/hydrogen diffusion flame because only about 10% of the light energy is absorbed at the 196.1 nm line. A reciprocal sensitivity of 0.25 μg/ml 1% and a detection limit better than 0.1 μg/ml can be achieved in this flame. While no chemical interferences have been established in an air/acetylene flame up to the present [1017], the cooler diffusion flame must be used with caution. In particular, background effects owing to light scattering can occur markedly in this flame. Furthermore, organic solvents cannot be burnt.

MULFORD [894] extracted selenium with ammonium pyrrolidine dithiocarbamate in MIBK and obtained a detection limit of 0.1 μg/ml. SEVERNE and BROOKS determined selenium in biological materials by co-precipitation with arsenic and hypophosphoric acid. The precipitate is dissolved in nitric acid and analysed in the flame [1118].

The hydride system offers a very rapid, simple and sensitive method for the determination of selenium [1092]. With an electrodeless discharge lamp and $NaBH_4$ as the reducing agent, 3×10^{-9} g or 0.15 μg/ml selenium can be determined [358].

In the graphite tube furnace the detection limit for selenium is about 5×10^{-11} g, but the determination is neither easy nor free of interferences because of the relative high volatility of this metal. IHNAT found that a mixture of sulphuric acid and perchloric acid completely suppressed the signal for the determination of selenium in feeding stuffs, while nitric acid/perchloric acid increased the signal [553]. KOPP found that there was an increase in the signal of 50% due to various interferences [699] for the determination of selenium in river water in the graphite tube furnace.

8.43. Silicon

Since the introduction of the nitrous oxide/acetylene flame, silicon is being increasingly determined by atomic absorption spectroscopy because the analysis can be carried out well and with sufficient sensitivity. At the 251.6 nm line the reciprocal sensitivity is 1.5 μg/ml 1% and the detection limit is about 0.02 μg/ml. Various other resonance lines are compiled in table 31. To achieve the quoted reciprocal sensitivity and a calibration plot of satisfactory linearity, a spectral bandwidth of 0.2 nm is absolutely essential because silicon has a large number of resonance lines lying close together (see figure 40, page 70).

Table 31. Si resonance lines

Wavelength nm	Energy level (K)	Reciprocal sensitivity (µg/ml 1%)
251.6	223–39955	1.5
250.7	77–39955	3
252.9	223–39760	4
251.4	0–39760	4
252.4	77–39683	4
221.7	223–45322	4.5
251.9	77–39760	5.5
221.1	77–45294	8
220.8	0–45276	16

The determination of silicon in a nitrous oxide/acetylene flame is virtually free of interferences. However, the fact that silicon is rapidly precipitated from acid solutions, and thus lost to the determination, quite frequently causes difficulties. MEDLIN and co-workers [855] have recommended a lithium metaborate solubilizer, which keeps both silicon and other cations in solution, for the analysis of silicates. PARALUSZ [950] has examined various organic silicon compounds for their suitability as standards in organic solvents. Nowadays various firms offer water soluble silicone oils which are ideal as standards in both aqueous and organic solutions.

About 5×10^{-11} g silicon can be determined in the graphite tube furnace, but the determination is strongly subject to contamination risks. The use of glass apparatus is to be avoided at all costs. With nitrogen as inert gas the sensitivity for silicon is about 50% higher than with argon [802].

8.44. Silver

Silver can be determined exceedingly well by atomic absorption spectroscopy. In an air/acetylene flame the reciprocal sensitivity at the 328.1 nm resonance line is about 0.06 µg/ml 1% and the detection limit is about 0.002 µg/ml. In flames of lower temperature the sensitivity can be improved by a factor of about two.
According to existing experiences, the determination of silver is free of interferences. WILSON [1339] and BELCHER and co-workers [101] investigated a large number of anions and cations for their possible effect on the absorption of silver and found no noticeable interferences, even in flames of lower temperature. Especially for the determination of silver traces, the absence of chloride ions must be carefully observed since precipitation can otherwise easily take place. RAWLING and co-workers [1026] found that in a 6 N hydrochloric acid solution at least 25 µg/ml silver remains in solution. Naturally such acid concentrations require a chemically resistant nebulizer and burner. GREAVES [439] added diethylene triamine to silver solutions to prevent precipitation.

Since silver must frequently be determined in the smallest traces, the detection limit achievable with atomic absorption is occasionally insufficient. WEST and co-workers [1308] extracted silver as the dithizone complex with ethyl propanate and were thus able to greatly improve the sensitivity. EMMERMANN and LUECKE [329] [766] employed the boat technique for the determination of silver traces in soil samples. With this technique 10^{-10} g silver can be determined.

In the graphite tube furnace about 1×10^{-13} g silver can be determined. Since it can only be thermally pre-treated up to about 450–500 °C without losses, various interferences are to be expected. Thus, the silver signal is strongly reduced by higher concentrations of nitric acid and sulphuric acid, while hydrochloric acid and sodium chloride only cause a slight reduction.

8.45. Sodium

Sodium is routinely determined in simple flame photometers, and the detection limit with flame emission is about ten times better than with atomic absorption spectroscopy. Nevertheless, atomic absorption offers certain advantages even for this element, since sodium must rarely be determined in such small traces as to make the detection limit of this method insufficient; further, atomic absorption spectroscopy appears to be free of interferences in the trace range in the presence of complex matrixes, which can be of great importance in practice; finally, atomic absorption spectroscopy offers a number of less sensitive lines which make the determination of higher sodium concentrations possible without excessive dilutions.

Sodium is usually determined at the 589.0/589.6 nm doublet. The somewhat differing sensitivities of the two lines results in a slightly curved calibration plot and reduced sensitivity, but the signal-to-noise ratio is very favourable. The more sensitive doublet line at 589.0 nm can be resolved with spectral bandwidths of 0.2–0.4 nm when better linearity of the calibration plot is necessary.

Employing greater bandwidths and both doublet lines, the reciprocal sensitivity for sodium in an air/acetylene flame is 0.015 µg/ml 1% and the detection limit is about 0.0002 µg/ml. Higher sodium concentrations can be advantageously determined at the second doublet at 330.2/330.3 nm. The reciprocal sensitivity is about 3 µg/ml 1%.

MANNING [796] found that it is possible to determine sodium with a good zinc hollow cathode lamp and thus reduce the sensitivity even further. Zinc has a doublet at 330.259/330.294 nm whose second line is very close to the second line of the sodium doublet at 330.232/330.299 nm. The distance between both lines of only 0.005 nm causes slight overlapping of the flanks so that the sodium atoms are able to absorb a portion of the zinc radiation. The reciprocal sensitivity is about 140 µg/ml 1%. MANNING further pointed out that there is no risk of spectral interferences since the observed zinc line is not a resonance line and only occurs in emission spectroscopy.

Sodium is somewhat ionized in the air/acetylene flame so that another easily ionized element must be added to sample and standard solutions to suppress this interference. The addition of a cesium salt is the most effective. Nevertheless, various authors prefer the use of a lower temperature flame such as an air/propane or an air/hydrogen flame [296] [810] to suppress the ionization. The latter flame has the advantage, especially at the secondary doublet at 330.2/330.3 nm, of giving a considerably improved signal-to-noise ratio, so that the precision is improved and the analyses are easier.

Sodium is one of the "present everywhere" elements in a laboratory so that the greatest care must always be taken, especially with trace analyses. Sodium is easily dissolved from normal laboratory glassware so that "distilled" water, sample and standard solutions should only come briefly, if at all, in contact with glass. Water should only be distilled in quartz apparatus, or a mixed bed ion exchanger should be employed. Particular care is recommended with the usual washing and cleaning agents since these frequently contain large amounts of sodium. For storing sample and standard solutions, only plastic vessels are suitable. Particular care must be taken with all chemicals that either contain sodium or, frequently the case with acids, are withdrawn from glass bottles (blank values!). The effects of dust and cigarette smoke on flame noise have already been fully discussed under potassium. Everything said there is even more valid for sodium (see page 166); the use of a 3-slot burner head is recommended in every case.

In the graphite tube furnace the reciprocal sensitivity for sodium is about 1×10^{-12} g 1% so that in principle a detection limit of 1×10^{-13} g should be achievable. However, everything already said about the risk of contamination for the flame is valid in a much higher degree for the graphite tube furnace. For this reason, the above detection limit has not been achieved to date. Very useful determinations of sodium can be carried out at the secondary doublet at 330 nm with a reciprocal sensitivity of about 10^{-10} g. The problem of contamination can be controlled to some extent in this case, although naturally the analyses must be carried out with the greatest care and cleanliness.

Various works have been published in which vapour discharge lamps have been compared to hollow cathode lamps for sodium [12] [810] [1240]. Nowadays both lamp types appear to be equivalent, although hollow cathode lamps are always easier to operate.

8.46. Strontium

In its behaviour in a flame, strontium lies between calcium and barium, i.e. it can be determined in both air/acetylene and nitrous oxide/acetylene flames. As expected, there are a large number of chemical interferences in the cooler flame which have been investigated by various authors. In the hot flame, strontium is considerably ionized. The most sensitive resonance line is at 460.7 nm; in the air/acetylene flame the reciprocal sensitivity is 0.15 µg/ml 1% and the detection limit about 0.002 µg/ml; in the nitrous oxide/acetylene flame with suppression of the ionization, the reciprocal sensitivity is 0.1 µg/ml 1% and the detection limit about 0.002 µg/ml. FULTON and BUTLER [397] found a maximum reciprocal sensitivity of 0.075 µg/ml 1% in a mixed air/nitrous oxide/acetylene flame, but this is hardly suitable for routine applications.

LOKEN and co-workers [761] and PARKER [951] found that strontium is ionized by about 10% in an air/acetylene flame. During a detailed investigation, INTONTI and STACCHINI [557] established that an addition of 50–100 µg/ml rubidium was sufficient to reliably eliminate this effect. All remaining alkali metals exhibited irregular effects with increasing concentration.

AMOS and WILLIS [48] found about 80% ionization in the nitrous oxide/acetylene flame, this could be eliminated by the addition of 0.1–0,2% potassium (as chloride).
TRENT and SLAVIN [1242] thoroughly examined the possibilities of interference in an air/acetylene flame and found numerous chemical interferences. The majority could be eliminated by adding 1% lanthanum to the sample and standard solutions. This was verified by BELCHER and BROOKS [99], who however, added only 0.2% lanthanum. The following are some of the most important interfering substances: Si, Al, PO_4, and also Ca, Fe, Mg and HNO_3. Higher concentrations of HCl, $CuCl_2$, $CoCl_2$, NH_4NO_3, $CdNO_3$ also have a marked suppressive influence [1242].
Approximately 5×10^{-12} g strontium can be determined in the graphite tube furnace. Since it can be thermally pre-treated up to about 1500°C, no major interferences are to be expected; only nitric acid, in higher concentrations, reduces the signal somewhat. The high direct light emission of the graphite tube can also interfere in older atomic absorption spectrophotometers.

8.47. Sulphur

The direct determination of sulphur by atomic absorption spectroscopy employing commercial instruments has not yet been possible because the resonance line is at 180.7 nm in the vacuum UV. KIRKBRIGHT succeeded in determining sulphur at this line in a modified atomic absorption spectrophotometer with a flushed monochromator, a nitrous oxide/acetylene flame shielded with nitrogen and an electrodeless discharge lamp [654]. The reciprocal sensitivity was 9 µg/ml 1% and the detection limit about 5 µg/ml. KIRKBRIGHT found no influence from the bonding state of sulphur and he used the procedure for the direct determination of sulphur in oil. A number of indirect methods, which can be widely employed, have also been described. ROE, MILLER and LUTWAK [1042] converted organically bound sulphur into sulphate by means of the SCHÖNINGER or BENEDIKT digestion methods, precipitated this as barium sulphate and determined the barium in the precipitate after solution in EDTA solution. VARLEY and CHIN determined water-soluble sulphate in soil extracts by adding a known quantity of barium chloride and determining the excess barium by atomic absorption [1261]. ROSE and BOLTZ [1057] determined sulphur dioxide by oxidizing it to sulphate, precipitating as lead sulphate, and after centrifugation, determining the excess lead in the supernatant liquid by atomic absorption. DUNK, MOSTYN and HOARE [315] added exactly measured quantities of barium chloride solution to sulphate solutions and determined the excess barium. CHRISTIAN and FELDMAN [232] utilized the occurrence of chemical interferences for the determination of a number of non-metals including sulphur. For example, they found a direct relationship between the absorption of calcium and the sulphate concentration in the solution under analysis. Naturally other ions interfere strongly in this determination and it is not specific.
FUWA and VALLEE [400] have described a procedure for the determination of sulphur by molecular absorption spectroscopy at an SO_2 absorption band. An air/hydrogen flame was directed into a 273 cm long heated Vycor tube and a hydrogen lamp served as the light source. The absorption was measured at 207 nm with a spectral bandwidth of about 0.7 nm. The reciprocal sensitivity achieved was about 10 µg/ml 1%.

8.48. Tantalum

Tantalum can be determined in a nitrous oxide/acetylene flame at the 271.4 nm resonance line with a reciprocal sensitivity of 20 µg/ml 1% and a detection limit of about 1 µg/ml. The absorption of tantalum is markedly increased in the presence of fluoric acid and iron, so that sample and standard solutions should contain at least 0.1% fluoric acid to eliminate this interference and to obtain the best sensitivity. BOND [141] preferred the addition of 0.1 M ammonium fluoride to increase the sensitivity.

8.49. Technetium

HARELAND, EBERSOLE and RAMACHANDRAN have recently described the first determination of technetium. In a fuel gas rich air/acetylene flame a reciprocal sensitivity of 3.0 µg/ml 1% can be achieved at the 261.4/261.6 nm doublet. Interferences from alkaline-earth metals can be eliminated by adding aluminium (about 100 µg/ml per 50 µg/ml interfering ions) [468].

BAUDIN and co-workers determined technetium in the graphite tube furnace and found the best reciprocal sensitivity of 1×10^{-8} g 1% at the 429.7 nm resonance line. The atomization temperature was about 2200 °C [87].

8.50. Tellurium

Tellurium can be easily determined in an air/acetylene flame, which is free of interferences. At the 214.3 nm resonance line the reciprocal sensitivity is 0.7 µg/ml 1% and the detection limit is about 0.05 µg/ml. In an air/hydrogen flame a somewhat improved sensitivity and a detection limit of 0.03 µg/ml can be obtained.
WU, DROLL and LOTT [1358] examined the influence of 25 various ions on the absorption of tellurium and found no chemical interferences worth mentioning. They also studied the extraction of tellurium with various complexing agents and organic solvents very thoroughly. They obtained very good results with ammonium pyrrolidine dithiocarbamate/pentan-2,4-dione.

Higher tellurium concentrations can be determined at the 225.9 nm resonance line with a reciprocal sensitivity of about 5 µg/ml 1%.
The hydride system brings a considerable increase in the sensitivity. With $NaBH_4$ as reducing agent, approximately 5×10^{-9} g or 0.25 µg/ml tellurium can be determined [358]. In the graphite tube furnace the detection limit for tellurium is 1×10^{-11} g.

8.51. Thallium

Thallium can be determined easily and without interferences in an air/acetylene flame. At the 276.8 nm resonance line the reciprocal sensitivity is 0.5 µg/ml 1% and the detection limit is about 0.02 µg/ml.

Higher thallium concentrations can be determined at the 377.6 nm resonance line which gives a reciprocal sensitivity of about 3 μg/ml 1%.
About 1×10^{-11} g thallium can be determined in the graphite tube furnace, but because of the high volatility of this element, interferences are to be expected.

8.52. Tin

Tin can be determined in air/hydrogen, fuel gas rich air/acetylene and nitrous oxide/acetylene flames with comparable sensitivity. The corresponding values for the three most important resonance lines are compiled in table 32 (see also table 8, page 95). The best reciprocal sensitivity of 1.1 μg/ml 1% can be achieved in an air/hydrogen flame at the 224.6 nm resonance line; the detection limit is about 0.02 μg/ml.

Table 32. Sn resonance lines

Wavelength nm	Energy level (*K*)	Reciprocal sensitivity (μg/ml 1%) Air/H_2	Air/Acetylene	N_2O/Acetylene
224.6	0–44509	1.1	2.5	3.0
286.3	0–34914	1.8	5.0	5.4
235.4	1692–44145	2.2	4.0	3.8

ALLAN [42] and GATEHOUSE and WILLIS [407] employed a strongly reducing air/acetylene flame which gave a very favourable signal-to-noise ratio. After the introduction of the air/hydrogen flame by CAPACHO-DELGADO and MANNING [206] [208] and hollow cathode lamps containing molten tin by VOLLMER [1276], this element can be determined very well by atomic absorption spectroscopy. Nevertheless, various chemical interferences appear to influence the determination, so that the use of a nitrous oxide/acetylene flame is recommended for complex matrices. CAPACHO-DELGADO and MANNING [208] found a strong reduction of the signal in the air/hydrogen flame caused by phosphoric and sulphuric acids and a slight influence from nitric acid. Nitric acid also had a slightly negative influence in the air/acetylene flame, while the other two acids increased the tin signal. Interferences from cations appear to be negligible. AGAZZI [20] found interferences from phosphate and pyrophosphate in an oxygen/hydrogen flame with a turbulent burner and also an increase of the signal from high sodium concentrations. On the other hand, AMOS and WILLIS [48] found a reduction of 15% in the tin signal by 5000 μg/ml sodium in an air/hydrogen flame. LEVINE and co-workers [742] found that even small concentrations of potassium increased the signal by 10% in a fuel gas rich air/acetylene flame. BOWMAN [148] discovered a slight increase in the tin signal in a nitrous oxide/acetylene flame owing to ammonium iodide.
The highest reciprocal sensitivity of 0.6 μg/ml 1% can be achieved in an argon/hydrogen diffusion flame, but numerous other elements cause interferences in this flame. NAKAHARA and co-workers found that the majority of these interferences could be effectively eliminated by the addition of 0.1% iron(III) chloride [903]. RUBEŠKA and MIKŠOVOKY found that acids

reduced the tin signal in this flame and that virtually all metallic elements increased the signal, in some cases considerably (up to 100%) [1068]. These authors accounted these effects to an easier heat transfer owing to the additional salts and thus to better atomization of tin.

KAHN and SCHALLIS [617] were unable to observe an absorption signal for 100 µg/ml tin in MIBK when using an air/hydrogen flame. Thereupon HARRISON and JULIANO [472] investigated the influence of organic solvents on the absorption of tin. They found that even small concentrations of alcohols, ketones and organic acids reduce the absorption of tin in an air/hydrogen flame considerably. This effect grows with increasing chain length and branching of the solvents. For example, 10% methanol reduces the absorbance of 50 µg/ml tin by about 25%, 10% ethanol reduces it by about 70% and 10% n-propanol by 80%. In the presence of 10% n-butanol, hardly any further tin signal can be observed. 10% acetone also suppresses the tin signal almost completely. These authors attributed this effect to a reduction of the hydrogen in the flame by organic materials; hydrogen appears to play an important role in the atomization of tin, either indirectly by reducing the monoxide to metal or directly through the intermediate formation of the hydride [1067].

These effects exclude the determination of tin in organic solvents in an air/hydrogen flame. Since the fuel gas rich air/acetylene flame is difficult to operate with organic solvents, only the nitrous oxide/acetylene flame remains. Nevertheless, an ionization interference can be observed in this flame, this must be eliminated by the addition of alkali. MENSIK and SEIDEMANN separated tin from rock and soil samples by sublimation with ammonium iodide ($SnO_2 + 4\,NH_4I \rightarrow SnI_4 + 4\,NH_3 + 2\,H_2O$). After solution in hydrochloric acid, the tin was extracted with trioctylphosphine oxide/MIBK and determined in a nitrous oxide/acetylene flame free of interferences [861].

The smallest tin concentrations can be determined with the hydride system. With $NaBH_4$ as reducing agent, approximately 4×10^{-9} g tin or 0.2 µg/l can be determined [358]. This determination is strongly dependent upon the acid concentration, which must be maintained at a constant level. 0.2 N hydrochloric acid is the most suitable.

Approximately 1×10^{-11} g tin can be determined in the graphite tube furnace. It is necessary, however, to neutralize acid samples in the graphite tube with an ammonia solution. If this is not done, strong acid effects that reduce the sensitivity by about one order of magnitude can be observed. ROSCHNIK found that tin could not be atomized in the graphite tube furnace in the presence of sugar or citric acid. The addition of strontium (for the determination of tin in oil) or phosphoric acid eliminated these interferences [1054].

8.53. Titanium

Titanium can be determined in a nitrous oxide/acetylene flame at the 364.3 nm resonance line with a reciprocal sensitivity of 2 µg/ml 1% and a detection limit of about 0.05 µg/ml. A multitude of other resonance lines were examined by CARTWRIGHT and co-workers [214] (compare table 9, page 96), of which a number are compiled in table 33.

AMOS and WILLIS [48] found that the titanium signal is noticeably increased in the presence of iron or fluoric acid. An increase in the concentration of fluoric acid from 0.2% to 4%

brought a 50% higher signal, and 0.2% iron doubled the absorption. It is therefore recommended to maintain the concentrations of these substances as constant as possible; with higher iron and fluoric acid concentrations the effect appears to be largely controlled. BOND [141] found that 0.1 M ammonium fluoride solution increased the signal quite strongly and so he recommended this reagent.

Table 33. Ti resonance lines

Wavelength nm	Energy level (K)	Reciprocal sensitivity (μg/ml 1%)
364.3	170–27615	2
335.5	170–29971	3
399.9	387–25388	5
399.0	170–25227	5
395.6	170–25439	10
394.8	0–25318	20

RAO determined titanium, free of interferences, in ores after a lithium metaborate digestion in which he maintained concentrations of 2% lithium metaborate, 0.8% ammonium fluoride and 0.05–0.2% SiO_2 (corresponding to 12.5–50% SiO_2 in the original sample). Under these conditions the maximum sensitivity for titanium was achieved [1022].
Approximately 2×10^{-9} g titanium can be determined in the graphite tube furnace.

8.54. Tungsten

Tungsten can be determined in a nitrous oxide/acetylene flame at the 400.9 nm resonance line with a reciprocal sensitivity of 15 μg/ml 1% and a detection limit of about 1 μg/ml. Although AMOS and WILLIS [48] and MANNING [794] mention various resonance lines with better sensitivity, the 400.9 nm line is to be preferred because of its better signal-to-noise ratio.

8.55. Uranium

Uranium can be determined in a nitrous oxide/acetylene flame at the 351.5 nm resonance line with a reciprocal sensitivity of 50 μg/ml 1% and a detection limit of about 30 μg/ml. The resonance lines at 358.5 nm and 356.7 nm exhibit better sensitivity, but a more unfavourable signal-to-noise ratio. MANNING [794] has examined further uranium lines and determined their sensitivities.
Since uranium is markedly ionized in a nitrous oxide/acetylene flame, 0.1% potassium (as chloride) should be added to sample and standard solutions to suppress this interference.

GOLEB [424] [425] has determined the $^{235}U/^{238}U$ ratio employing a cooled hollow cathode lamp and a flameless atomization cell.

8.56. Vanadium

Vanadium can be determined in a nitrous oxide/acetylene flame at the triplet 318.3/318.4/318.5 nm with a reciprocal sensitivity of about 2 μg/ml 1% and a detection limit of 0.05 μg/ml. CARTWRIGHT [214] has examined numerous other resonance lines, a number of which are compiled in table 34.

Table 34. V resonance lines

Wavelength nm	Energy level (*K*)	Reciprocal sensitivity (μg/ml 1%)
318.3	137–31541	2
318.4	323–31722	
318.5	553–31937	
370.4	2425–29418	5
437.9	2425–25254	5
438.5	2311–25112	6
370.5	2311–29296	10
257.4	553–39391	20
251.7	323–40039	35
253.0	553–40064	35
250.8	137–40001	40

CAPACHO-DELGADO and MANNING [207] found an increase in the vanadium signal in the presence of phosphoric acid. GOECKE [422] discovered that iron in concentrations over 1000 μg/ml caused interferences, so he removed it from strong hydrochloric acid solutions by extraction with isopropyl ether. The high concentration of hydrochloric acid then had to be removed by evaporation. GOECKE found further that 200 μg/ml aluminium increased the vanadium signal by a third, so he added this cation to all sample and standard solutions and was thereby able, with optimization of the nitrous oxide/acetylene flame, to obtain a reciprocal sensitivity of 0.2 μg/ml 1%.

WEST and co-workers account these increases in the absorbance to a hindrance of the lateral diffusion in the flame. The additives delay the atomization of vanadium and thereby reduce the time available for lateral diffusion of the atoms in the flame. The atoms under study are concentrated in the centre of the flame and hence in the beam path [1306].

Approximately 1×10^{-10} g vanadium can be determined in the graphite tube furnace. Although it can be thermally pre-treated up to about 1600°C without losses, the determination is not interference free, so that frequently the method of additions must be employed [1101].

Higher concentrations of nitric acid strongly reduce the vanadium signal, sulphuric acid only a little. Many interferences can be expected due to carbide formation, especially in the presence of finely divided carbon, which results from the thermal decomposition of biological materials.

8.57. Yttrium

Yttrium can be determined in a nitrous oxide/acetylene flame at the 410.2 nm resonance line with a reciprocal sensitivity of 1.8 μg/ml 1% and a detection limit of about 0.1 μg/ml. The resonance lines at 407.7 nm, 412.8 nm and 414.3 nm exhibit only slightly lower sensitivity. To suppress ionization interferences, 0.1–0.2% potassium (as chloride) should be added to sample and standard solutions.

8.58. Zinc

Zinc is one of the metals most frequently determined by atomic absorption spectroscopy. A reciprocal sensitivity of 0.02 μg/ml 1% and a detection limit of about 0.001 μg/ml can be achieved in an air/acetylene flame at the 213.8 nm resonance line. An improved signal-to-noise ratio and a detection limit better than 0.001 μg/ml are obtained in an air/hydrogen flame.

In the early years of atomic absorption spectroscopy, DAVID [267] and ALLAN [41] engaged in thorough studies on the determination of zinc and found that atomic absorption spectroscopy was superior to all other methods for the determination of this element.

The determination of zinc appears to be free of interferences in an air/acetylene flame as SPRAGUE and SLAVIN [1171] established for the analysis of biological materials. PLATTE and MARCY [989] verified this by finding no influence on the absorption of 1 μg/ml zinc by 1000 μg/ml sulphate, phosphate, nitrate, nitrite, bicarbonate, silicate, EDTA and nine other cations. The interferences quoted by GIDLEY and JONES [414] [415] can be traced to the use of an unsuitable lamp and an unsuitable burner.

High zinc concentrations can be advantageously determined at the 307.6 nm resonance line; the reciprocal sensitivity at this line is about 100 μg/ml 1%. Unnecessary dilutions can be avoided and high concentrations can be precisely determined.

In principle, even less than 1×10^{-13} g zinc can be determined in the graphite tube furnace, so that this element exhibits the highest sensitivity. Because of the extremely high contamination risk and the omnipresence of zinc, even in pure water, it has not yet really been possible to obtain this value. CLARK, DAGNALL and WEST systematically investigated the determination of zinc in the graphite tube furnace and found that the majority of ions interferred when they were added in great excess. This is easily understood because zinc can only be thermally pre-treated to about 400 °C without losses. The following did not interfere in the determination of zinc in a 100 fold excess: Na, K, Al, Sn, Cu, Ba, Sb, sulphuric acid, nitric acid and acetic acid. Since the absorption of zinc was found to be independent of organic solvents, these authors recommended an extraction with diethyl dithiocarbonate in MIBK to avoid interferences [235].

8.59. Zirconium

Zirconium can be determined in a nitrous oxide/acetylene flame at the 360.1 nm resonance line with a reciprocal sensitivity of 10 μg/ml 1% and a detection limit of about 1 μg/ml. Further resonance lines are compiled in table 35.

Table 35. Zr resonance lines

Wavelength nm	Energy level (*K*)	Reciprocal sensitivity (μg/ml 1%)
360.1	1241–29002	15
354.8	570–28750	30
301.2	570–33764	35
298.5	0–33487	40
303.0	1241–34240	40
362.4	570–28157	40

Zirconium is ionized by about 10% in the nitrous oxide/acetylene flame [794]; to suppress this interference, 0.1% potassium (as chloride) should be added to sample and standard solutions [1152].

AMOS and WILLIS [48] found that the absorption of zirconium is increased considerably in the presence of fluoric acid and high iron concentrations; they therefore added 2% fluoric acid to all solutions. SLAVIN and co-workers [1152] verified this and also found that hydrochloric acid had a similar influence. 10% hydrochloric acid increased the signal of a zirconium standard solution by 400% compared to the acid-free standard. BOND [141] obtained the strongest effect by adding 0.1 M aluminium fluoride solution which gave an eight fold signal. BOND and co-workers [142] [143] found that the absorption of zirconium is increased in the presence of many nitrogen containing compounds, which can react as LEWIS bases, and from this they developed a procedure for the determination of ammonium traces. TYLER [1247] extracted zirconium as the cupferon complex with a benzene/isopentanol (1 : 1) mixture and so eliminated a number of interferences. FOSTER [383] found that a thenoyltrifluoroacetone (TTA)/xylene mixture was ideally suited for the extraction of zirconium and that it increased the signal by a factor of at least five.

9. Specific Applications

Because atomic absorption spectroscopy is specific and has great freedom from interferences, its range of applications is very wide. Although the analytical procedures are largely similar, the various problems posed by different branches of analytical chemistry justify a separate treatment of each area. Emphasis will be placed on the actual application of atomic absorption spectroscopy, while sample preparation, such as solution or digestion of solid substances, can be assumed as largely known and will therefore be only briefly touched upon. This is in particular the case when the preparation of samples does not differ from known wet chemical procedures; only special preparation techniques will be described more fully. The chemical and physical interferences already mentioned under the individual elements will also be assumed as known and only special interferences will receive fuller treatment.

9.1. Medicine, Biochemistry, Toxicology

One of the most important applications of atomic absorption spectroscopy is in the routine clinical-chemical laboratory for the determination of calcium and magnesium in various body fluids. The analysis of sodium and potassium [956], which has been largely carried out in simple flame photometers up to the present, is also being increasingly taken over by atomic absorption. Recently, analysis of elements, such as iron, copper and zinc, in serum is becoming increasingly routine, because they cannot be analysed with sufficient speed and reliability by other methods.

The determination of calcium and magnesium in serum and urine was first described by WILLIS in 1960 in various publications and it is now one of the most widely distributed applications of atomic absorption spectroscopy. A survey of the extensive literature which has been published on the determination of these elements in biological materials is given in table 36.

The simplest procedure, which is employed most frequently, is the direct analysis of the 1:20 to 1:50 diluted sample using a three-slot burner head and an air/acetylene flame. To suppress phosphate interference 1% sodium EDTA, 0.5% lanthanum in hydrochloric acid, or 0.25% strontium must be added to sample and standard solutions. Lanthanum may only be added to diluted serum as otherwise the protein will coagulate. For this reason EDTA is occasionally preferred. Although protein does not interfere in the determination of calcium and magnesium when using a good nebulizer-burner system at the high dilution and in the presence of lanthanum, various authors have found better reproducibility with deproteinized samples [1081] [1121]. It must be taken into consideration, however, that trichloroacetic acid and hydrochloric acid influence the determination of these two elements, i.e. the standard solutions must contain the same concentration of the corresponding acid.

Numerous comparative measurements with conventional procedures, such as complexometric titrations, flame photometry, fluorimetry, etc., have shown that for the determination of calcium and magnesium in the most varying body fluids, atomic absorption spectroscopy is at least as good, and with respect to precision and speed is usually superior ([438] [467] [1041] [1081] [1144] [1263]).

Nevertheless, HANSEN has drawn attention to the great care required in the determination of calcium, since this element is relatively sensitive to changes in the composition of the flame. When all parameters of the nebulizer-burner system have been optimized, a precision of better than 1% for calcium and 0.6% for magnesium can be obtained [465].

Table 36. Bibliographical survey: Application of atomic absorption spectroscopy for the determination of calcium and magnesium in biological materials (reference numbers)

Element	Serum	Urine	Other body fluids, tissues, etc.
Ca	416 420 484 583 671 873 956 958 998 1041 1081 1113 1245 1323 1324 1384	416 496 673 873 1081 1144 1244 1245 1263 1328 1384	121 283 416 462 504 673 873 910 951
Mg	279 416 420 438 467 505 533 542 552 554 672 782 873 956 958 1081 1113 1175 1322 1325	279 416 496 505 533 554 782 873 1081 1175 1244 1328	121 229 279 283 416 442 462 463 476 505 542 545 554 782 873 951 969

The determination of *sodium* and *potassium* in serum has been described by WILLIS [1326], HERRMANN and LANG [504] and PASCHEN [956]. It is best carried out on a 1:50 dilution of the sample in an air/hydrogen flame, and sodium should be measured at the secondary line at 330.2 nm to avoid too high dilutions.

If an air/acetylene flame is employed, the sodium content in the potassium standard solution must be approximately matched to that in the samples. Even an air/propane flame, in which there is no observable ionization, can be used for the determination of these two elements.

The micro method described by PASCHEN [956] for the determination of sodium, potassium, calcium and magnesium in a single serum dilution is particularly elegant. 100 µl serum is diluted 1:100, or better 1:50 [957], with 0.25% strontium chloride solution and analysed directly for the four elements. The addition of strontium eliminates both the influence of phosphate on calcium and the ionization of sodium and potassium in the air/acetylene flame. The coefficients of variation for the determination lie between 0.3% and 0.5%.

GRUNBAUM and PACE have described a micro method for the determination of sodium, potassium, calcium and magnesium in urine [446]. 0.1 ml urine is diluted 1:100 under addition of lanthanum and cesium and aspirated directly into the flame. HAZEBROUCQ controlled the sodium, potassium, calcium and magnesium contents of solutions used for extrarenal dialysis (artificial kidney machines) and found that atomic absorption is superior to flame photometry [484]. The precision for sodium is 1%.

However, not only serum and urine can be determined by atomic absorption; HANKIEWICZ for example, analysed gastric juices for magnesium by diluting them 1:10 with water and aspirating directly into the flame [463].
Flameless atomic absorption is not particularly suitable for the determination of sodium, potassium, calcium and magnesium in serum since very high dilutions are necessary and, further, the expected precision and accuracy cannot be achieved [836] because of the high risk of contamination.
The elements iron, copper and zinc are present in similar concentrations of about 1 µg/ml (= 100 µg%) in serum and therefore count as trace elements. A survey of the literature published on the determination of these three elements in various body fluids and tissues is given in table 37.

Table 37. Bibliographical survey: Application of atomic absorption spectroscopy for the determination of copper, iron and zinc in biological materials

Element	Serum, Plasma	Urine	Other body fluids, tissues, etc.
Cu	110 135 294 493 506 529 836 952 1035 1150 1171 1204 1301	110 1150 1204	110 230 462 476 929 1204
Fe	112 135 419 493 836 927 930 1035 1171 1281 1301 1362 1370 1385	112 135 1161 1380	135 462 476 507 930 1370 1381 1382
Zn	280 455 482 493 784 836 952 958 1000 1030 1035 1171	280 784 1000	121 230 280 462 476 929 951

While tissue samples, for obvious reasons, must be wet or dry ashed before the analysis, iron, copper and zinc can be directly determined in serum with good atomic absorption spectrophotometers. Zinc is the easiest to determine because of its high sensitivity [1035]; it can be directly analysed against aqueous standard solutions without interferences in 1:5 diluted serum and undiluted or 1:2 diluted urine [1030]. To prevent blockage of the burner slot, a three-slot burner head is used with advantage with the air/acetylene flame. DAWSON and WALKER [280] preferred higher dilutions and scale expansion. The determination of copper and iron in serum presents greater problems because of the low sensitivities of these two elements. Owing to the relatively high viscosity of blood serum, it cannot be compared with aqueous standard solutions; on the other hand, higher dilutions considerably worsen the sensitivity and precision of the determination. For this reason, various authors have recommended deproteinization [110] [1218] or extraction [952] [1385], which however is somewhat time consuming for routine determinations. OLSEN and HAMLIN [930] have recommended a simple deproteinization with which 95% of the haemoglobin iron can be removed at the same time.

Deproteinization with trichloroacetic acid is by no means without problems as established by FIELDING and RYALL [366], since it can have a considerable influence on the subsequent analysis. The precipitation of protein reduces the liquid phase by up to about 13% since the protein occupies about the same volume before and after the precipitation. Furthermore, the precipitate has a positive surface charge which can lead to adsorption of negatively charged ions, such as iron complexes, from the supernatant liquid especially in the presence of EDTA.

On the other hand, a direct determination is also not without problems. While the determination of copper usually presents no difficulties, various points must be observed for the determination of iron. The viscosity problem can be very reliably solved by diluting the serum 1:2 with water and matching the standard solutions with a commercially available synthetic plasma expander [1035]. Nevertheless, this procedure can only be employed for iron when a reliable haemolysis can be carried out after the sample has been taken. Haemolytic sera can only be determined with reasonable reliability after deproteinization. Furthermore, the direct method is not suitable for the determination of iron in *serum*, since 40–70 µg Fe/100 ml too much are always found compared to other (e.g. colorimetric) procedures (figure 73). In plasma, the determined iron values agree well with other methods [493].

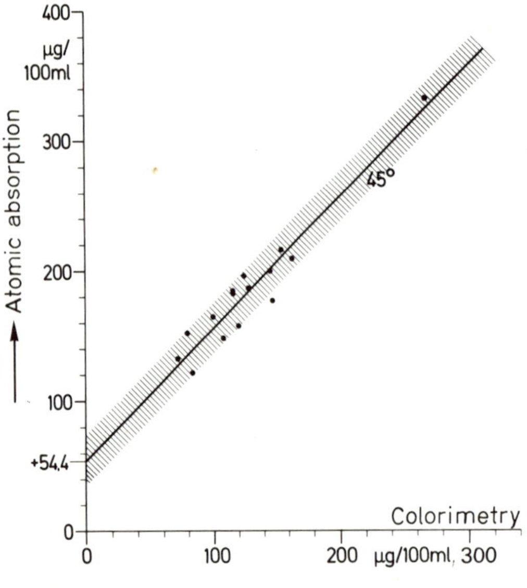

Figure 73. For the determination of iron by atomic absorption in serum which has not been previously treated, 40–70 µg Fe/100 ml too much are usually found compared to colorimetric procedures or deproteinized sera.

The determination of iron and copper in urine offers no great problems. The direct determination of acidified (2 ml urine + 0.1 ml conc. sulphuric acid), undiluted urine has the necessary precision, sensitivity and specificity for clinical chemistry. There are no interferences due to protein and inorganic components, and the agreement with other, more time-consuming procedures is good [1161].

The greatest disadvantage for the determination of iron, copper and zinc in serum or plasma is the high sample requirement of 1 to 2 ml per determination. Here, the graphite tube furnace is an interesting alternative offering the same precision of 3–5%, but only requiring 1 to 2 µl per determination. The disadvantage of this micro procedure is the markedly greater time requirement for each determination.

The remarks made for the determination of iron in the flame naturally also apply here, i.e. a direct determination can only be made on plasma but not on serum. The majority of authors therefore prefer a micro deproteinization procedure recommended by OLSEN and co-workers [927] which gives perfect results even with only 20 µl serum [1362].

Copper in serum, and iron and copper in urine can be determined either directly or after simple dilution with water in the graphite tube furnace [112] [1301]. The agreement with other methods is good, as shown in table 38 for the determination of copper in serum.

Table 38. Cu in serum
Comparative measurements: Flameless atomic absorption with a Heated Graphite Atomizer HGA-70, colorimetry, and flame atomic absorption

Serum	Colorimetry	Flame AA	Flameless AA
1	105	104	101
2	96	94	90
3	109	110	111
4	87	87	80
5	75	77	75
6	111	111	107
7	123	121	119
8	114	115	110
9	124	127	128
10	184	179	183
11	113	107	110
12	118	117	123
Mean	113	112	111

The determination of zinc in serum or urine in the graphite tube furnace suffers from the high volatility of this element. An attempt should be made to facilitate the thermal decomposition by the addition of nitric acid [482] [784] or lanthanum salts, or the sample should be externally ashed [784].

Atomic absorption also serves ideally for the determination of the iron binding capacity [930] [1370] [1385] and iron in haemoglobin [507] [1381] [1382].

A micro procedure has also been worked out for determining the iron binding capacity using the graphite tube furnace [112] [1362] which gives very good agreement with the macro method of OLSON and HAMLIN [930]. The sample requirement here is also around 20 µl serum.

Even though virtually nothing was known about the other trace elements in the human body a few years ago, the publication of a number of review articles [278] [293] [1098] [1201] shows the suddenly awakened interest in metallic elements which are often present in concentrations under 1 µg/ml (100 µg%) in body fluids. There are at least three reasons for making a thorough investigation of these trace elements: Concentrations which are too low, due to dietry insufficiencies or to metabolic disturbances, can lead to deficiency diseases. Concentrations which are too high because of excessive environmental intake or metabolic disturbances can produce toxic effects. Abnormal concentrations of trace elements can also be caused by certain illnesses, so that a way is perhaps offered for a better understanding of these illnesses.

Table 39. Bibliographical survey: Application of atomic absorption spectroscopy for the determination of essential and non-essential trace elements in biological materials

Element	References
Al	389 707 708
B	65
Co	294 521 1192 1329
Cr	199 200 201 277 355 429 970 983 1059 1085 1383
Mn	24 96 230 429 462 787 941 1301
Mo	983
Ni	294 521 523 918 970 1086 1136 1200 1201 1203
Rb	1207
Se	1117
Sr	96 246 272 294 297 876 1287

For a long time, the greatest obstacle for a routine determination of these elements was the large sample requirement and the necessity of a more of less time consuming sample preparation for determinations in the flame. Flameless atomic absorption in the graphite tube furnace has given the biochemist and the medical practitioner the means of collecting the quantity of data necessary for making a statistically reliable prognosis for the first time.

With trace elements, a differentiation is made between essential, non-essential, and toxic elements. Here it must be taken into consideration that an essential element can be thoroughly toxic when it is present in excess, and that a toxic element may only become dangerous for humans after a certain concentration. It must further be considered that an element regarded as indifferent today may only be so because its importance for the human body has not yet been recognized. Table 39 gives a literature survey on the determination of essential and indifferent trace elements in biological materials; and in table 40, the normal values for essential elements are given together with their effects on the human body.
KRISHNAN and co-workers made a detailed examination of the aluminium content in the brain in connection with neurofibrillitic degeneration [707] [708]. The brain tissue was wet ashed with nitric acid and the aluminium content determined in a nitrous oxide/acetylene flame. FUCHS and co-workers determined aluminium in serum with the graphite tube furnace [389]. By using the temperature programme they were able to thermally decompose 25 µl serum directly and determine the aluminium, free of interferences.

Chromium can only be determined with sufficient sensitivity in the flame after previous dry [983] or wet ashing and extraction [355]. The graphite tube furnace, on the other hand, offers sufficient sensitivity for a direct determination; DAVIDSON and SECREST found that the various oxidation states of chromium exhibited the same sensitivity [277]. ROSS and co-workers found that treatment with nitric acid for the determination of chromium in serum and urine brought no alterations and they thus decided on a direct determination [429] [1059], like other authors [970] [1085].

Table 40. Essential trace elements: Normal values in human serum or plasma and their effect

Element	Normal value µg/l	Effect in the body
Aluminium	30–40	Influence on cell mitosis; increased values with neurofibrillitic degeneration
Chromium	20–30	Essential for glucose metabolism; deficiency with arterial sclerosis and diabetes
Cobalt	0.1–0.2	Vitamin B_{12} coupling; excess with cardiomyopathy
Maganese	8–18	Essential for glucose and lipid metabolism and a number of other important biochemical processes
Molybdenum	4	Essential for xanthine oxidase and aldehyde oxidase
Nickel	2.1–2.6	Deficiency with liver cirrhosis and chronic renal insufficiency; increased values with myocardic disorders, etc.
Selenium	10	Connected with vitamin E
Strontium	50	Hardens bones and teeth, possibly connected with osteoporosis
Vanadium	10	Possibly essential for fatty acid metabolism

Cobalt is the only essential element whose concentration is too low even for a direct determination in the graphite tube furnace. SULLIVAN and co-workers have described the determination of cobalt in tissue after wet ashing [1192] and DELVES determined cobalt in oxidized blood with the flame after extraction in MIBK.

Manganese is present in blood serum in relatively high concentrations so that a direct determination in the flame can be made in a good atomic absorption spectrophotometer after 1:2 dilution with water [787]. For the determination of manganese in urine, an extraction is however to be preferred [24] [941]. Manganese in serum and other body fluids can be easily determined in the graphite tube furnace. The results are in agreement with those from ashed [96] or extracted [429] samples.

PIERCE and CHOLAK determined molybdenum in blood and urine in a flame after ashing and extraction with APDC/MIBK [983].

Nickel can be determined in serum, whole blood or urine in a flame after acid treatment or deproteinization and subsequent extraction with APDC/MIBK [294] [521] [918] [1086] [1200]. Nickel can be determined directly in serum in the graphite tube furnace employing 50 µl samples [970] [1136], or with addition of lanthanum in nitric acid solution [1203].

The use of a temperature programme for the thermal pre-treatment of large serum volumes is indispensable. If this is not possible, an extraction stage usually cannot be avoided.
Strontium can be determined in urine in a flame after dry ashing and addition of lanthanum [1287], or directly after 1:2 dilution with a hydrochloric acid lanthanum solution [297]. BEK and co-workers determined strontium in serum directly with the graphite tube furnace and obtained good conformity with values from ashed samples [96].
Besides these essential trace elements, other "indifferent" elements such as gold, lithium or rubidium are determined in serum and urine since they are introduced into the body therapeutically. The interest here is not the normal concentrations, which are very low, but solely the increased content in blood or serum and the amount excreted in the urine.
Gold is administered for the treatment of articular rheumatism. Gold in serum can be determined directly in a flame after 1:2 to 1:4 dilution. The viscosity of the standard solutions should be matched to that of the samples [302] [303] [313] [762]. For determination in blood, Triton X-100 is advantageously added for haemolysis [477]. Urine is treated with trichloroacetic acid, nitric acid/perchloric acid [314], or potassium permanganate [69] and the gold then extracted with MIBK. For all clinical requirements, the flame gives sufficient sensitivity and precision.
Lithium is administered for the treatment of manic psychosis. Lithium in serum can be easily determined in the flame after 1:10 dilution [466] [744] [745] [1009]; in urine, greater dilutions may be necessary [1383]. FRAZER and co-workers determined lithium in erythrocytes after 1:50 dilution with water [385]. The graphite tube furnace is suitable for the determination of lower lithium concentrations or microanalyses [1301].
Rubidium is also employed in the therapy of psychoses. Like lithium, it is analysed in serum, plasma, whole blood and urine employing 1:10 dilution with water for serum or blood and in undiluted urine [1207].
Besides the metallic elements, various non-metals or anions in body fluids are determined indirectly by atomic absorption. Of especial mention is the method described by BARTELS for the determination of chloride in serum in which an excess of silver nitrate solution is added, and after centrifugation, the silver in the supernatant liquid is determined directly [82]. WOLLIN and also ROE and co-workers determined sulphate in serum and urine by first oxidizing with H_2O_2 (SCHÖNINGER) or potassium chlorate (BENEDICT) and then adding barium, and after centrifugation, determining the barium in the supernatant liquid [1042] [1350]. ZAUGG and KNOX described a method for the determination of phosphate in which ammonium molybdate is added after deproteinization, the free heteropoly acid is shaken with octan-2-ol and molybdenum determined in the organic extract [1372] [1373].
PARSONS and co-workers modified this procedure for automated analyses and the simultaneous determination of calcium in 0.2 ml plasma. The method is suitable for 0.2–20 mg P/100 ml and 1–20 mg Ca/100 ml and has a coefficient of variation of about 1% [954].
MANNING and FERNANDEZ described the direct determination of phosphorus in urine and liver tissue in the graphite tube furnace [804]. They employed the absorption of a PO_x radical which can be observed at a lead ion line at 220.3 nm.
The determination of metals of toxicological interest was described early in the history of atomic absorption. In 1962, WILLIS reported the determination of bismuth, cadmium, lead and mercury in urine after extraction with an organic solvent. Similarly, most other toxic

metals are extracted with APDC in MIBK over a wide pH range. A literature survey on the various publications in this area is compiled in table 41.
The fact that the "normal values" of most toxic elements cannot be detected by direct aspiration so that an enrichment by means of extraction is always required, somewhat hindered the expansion of atomic absorption spectroscopy in this field. It is therefore hardly surprising that the considerably more sensitive flameless procedures were rapidly applied in this sector. The first successful attempt was the procedure of BRANDENBERGER and BADER [152–155] for the determination of mercury, later followed the boat technique [616] which can be applied

Table 41. Bibliographical survey: Application of atomic absorption spectroscopy for the determination of toxic elements in biological materials

Element	References
Ag	178 199 200 201
As	298
Be	138
Bi	299 644 1329
Cd	111 112 199 200 201 218 247 294 319 429 483 738 739 750 784 1058 1329
Hg	111 146 564 565 566 703 711 735 753 754 786 847 863 864 871 874 920 926 1249 1329
Pb	46 53 85 109 113 199 200 201 216 217 219 292 294 308 318 333 357 359 480 482 483 492 508 509 523 591 592 593 616 618 685 695 696 697 712 783 784 869 983 1052 1056 1106 1112 1301 1327 1329 1386
Se	1117 1130
Te	644 1117
Tl	111 112 248 482 784 1080 1120

to virtually all elements of toxicological interest. Urine undiluted and whole blood diluted 1:10 are introduced into a tantalum boat, dried near the flame, lightly ashed and then inserted into the flame of the atomic absorption spectrophotometer. The method is rapid and simple and permits routine detection of normal values and toxically increased values without special sample preparation. The technique has been described in detail for lead in urine [616], lead in whole blood [618] and thallium in various biological materials [248]. Thereafter followed the DELVES system, which was specially developed for the determination of lead in whole blood, and also the flameless mercury system. The conclusive breakthrough of atomic absorption into toxicology came with the introduction of the graphite tube furnace which can be universally employed for all the toxic elements mentioned here [1301]. In the following, only the determination of the three most important elements, lead, cadmium and mercury, will be examined in greater detail.
The fact that lead can only be determined in whole blood or urine after ashing and extraction [869] [985] has already been mentioned. KOPITO and co-workers, however, further emphasize that despite all the advances, this determination remains relatively difficult and contains many

sources of error; it must be carried out with the greatest care and requires experience [696]. SCHLEBUSCH and NIEHOFF go a step further by maintaining that despite extraction and the allied enrichment, the flame does not give sufficient sensitivity for the determination of normal values (14–16 μg Pb/100 ml) [1089].

The DELVES procedure [292] for the determination of lead in whole blood has found wide distribution and application (see page 42). The results correspond well with conventional flame procedures [509] and colorimetry [359]. Besides the original method, in which the blood sample is dried in a nickel crucible and treated with hydrogen peroxide, numerous modifications have been described. EDIGER and COLEMAN merely dried the blood sample on a hot-plate and then introduced it directly into the flame [318], while BARTHEL and co-workers replaced hydrogen peroxide by nitric acid [85] and ROSE and WILDEN by *aqua regia* [1056] and were thus able to improve the precision of the determination. JOSELOW and BOGDEN extracted the lead with APDC/MIBK before carrying out the analysis in the DELVES system [592].

An interesting variation has been described by CERNIK and SAYERS, who placed a droplet of blood on filter paper, and after drying, punched out a standard disc and analysed it in the DELVES system [219]. Even under very different conditions the blood spread on the filter paper so that the lead values showed no significant variations [217]. JOSELOW and SINGH found that the precision of this procedure could be markedly improved if the cup remained in the holder during a measurement series [593]; nevertheless, the paper disc method appears to be less precise than that with liquid blood.

Because of its high volatility, the determination of lead in whole blood or urine in the graphite tube furnace is very difficult and in all cases requires the use of a background corrector [199–201] [1301]. The analysis of urine can be noticeably simplified by a relatively simple pre-treatment of the sample with Triton X-100 [712] or with nitric acid [482]. Other authors prefer digestion with HNO_3/H_2O_2 [492] or even ashing in a high frequency oxygen plasma [783], or extraction [308]. EVENSON and PENDERGAST found that the determination of lead in erythrocytes is much more reliable and informative than in whole blood [333]. They diluted the erythrocytes 1:2 with Triton X-100 and determined the lead directly in the graphite tube furnace with a coefficient of variation of 3.5%.

For the determination of cadmium in blood, serum or urine with the flame, wet ashing with subsequent extraction with APDC/MIBK is most suitable [294] [738] [739]. Although a direct determination of cadmium in the DELVES system appears to be possible [218] [319], previous extraction is preferred for precision determinations [750]. For determinations in the graphite tube furnace, extraction [112] or cold ashing [784] is of advantage because of the high volatility. Nevertheless, with careful thermal pre-treatment and atomization at lower temperatures (e.g. 1300 °C), direct determinations, which are in good agreement with extraction procedures, appear to be possible [429] [1058].

In principle, even though mercury can be determined in the flame after extraction and respective enrichment [111] [864] or by the boat technique [863], nowadays the cold vaporization procedure after reduction with tin(II) chloride described by POLUEKTOV and co-workers is almost exclusively used [1341]. The sole procedural differences are in the preparation of the sample. Urine is usually allowed to stand for several hours with sulphuric acid/potassium permanganate [711] [754]. The same procedure can also be used for blood, plasma [711] and hair [920]. BOUCHARD recommended digestion in chromic acid because of the great speed [146] and LINDSTEDT and SKARE decomposed blood with nitric acid/perchloric acid at 70 °C overnight [754]. MAGOS and CERNIK determined mercury in urine, blood and tissue without previous ashing [786]; they were only able to determine the inorganic mercury, while after ashing all mercury can be determined. A good review on the other flameless methods for the determination of mercury has been published by MANNING [797].

Finally, the analysis of tissue, etc., will be briefly discussed, for although this presents no especial problems, prior digestion is generally required. First, a number of new digestion methods will be mentioned which offer various advantages over conventional dry or wet (with various acid mixtures) procedures.

TÖLG and co-workers have examined a modified "cold combustion" with activated oxygen for sample amounts below 1 g and trace elements in the ng range. In tests on radionucleides the losses even for such volatile elements as selenium and mercury were below 2% [620]. The same authors also examined the pressure digestion of tissue (up to 0.5 g) with nitric acid in a PTFE vessel; the losses here were also under 2% [700]. Similar digestions have been described by PAUSS [966] and ADRIAN [19]. The latter employed relatively large sample amounts (up to 5 g), low temperatures (< 100 °C) and long digestion periods.

JACKSON and co-workers described the digestion of tissue with a quarternary ammonium hydroxide (Soluene-100) overnight at room temperature [558]. MURTHY and co-workers found the digestion with tetramethyl ammonium hydroxide more rapid and protective than usual wet ashing [897], and the agreement with other procedures was good. LUYTEN and co-workers found from a comparison between this procedure and the pressure digestion of ADRIAN that brain tissue could be digested much better with quarternary ammonium hydroxide (possibly because of the high lipid content), while fish tissue, for example, could be better digested with acid under pressure [744].

The analysis of tissue can usually be carried out in the flame when sufficient material is available. KAWAMURA and co-workers determined barium in bones [629] and SPOONER and CRASSWELLER determined sodium, potassium, calcium and magnesium in kidney stones [1166]. HELSBY determined strontium [499], copper and molybdenum [498], and HUNTER and co-workers lead [543] in teeth. HARRISON and TYREE determined calcium, magnesium, copper, iron and zinc in fingernails [474].

For the determination of the lowest concentrations and especially for the microanalysis of the smallest tissue samples, the graphite tube furnace is most suitable. Furthermore, solid tissue samples can be analysed directly in the furnace, so that in certain cases the total analysis time can be reduced considerably. MAES and co-workers determined magnesium, iron, copper, zinc, manganese, lead and silver in 2 mm long hair sections by introducing them directly into the graphite tube [785] and BARNETT and KAHN determined copper in fingernails [77] in good agreement with conventional methods.

However, for reasons of homogenity, or if a large number of elements are to be determined, a prior digestion is frequently to be recommended even for determinations in the graphite tube furnace. Thus SCHULZ-BALDES determined copper in small tissue samples after digestion in nitric acid [1099] while SCHRAMEL determined eight trace elements in the international biological standard of BOWEN after wet ashing with sulphuric acid/hydrogen peroxide [1096] in the graphite tube furnace.

9.2. Agricultural and Foodstuffs Chemistry

Agricultural chemistry encompasses three main branches; the analysis of soil and soil extracts, fertilizers, and plants. Besides the main elements sodium, potassium, calcium and magnesium, the trace elements copper, manganese, zinc, iron, molybdenum and boron are of especial interest because they play an important role in the life cycle of plants. If these elements are not contained in the soil in sufficient quantity, damage to plants or poor harvests can result; the soil should be enriched with fertilizer. On the other hand, over-fertilization can also lead to damage, so that a regular control is opportune. The aluminium content of soils is also frequently examined nowadays since it is of importance for regeneration and the acid content of the soil. A literature survey on the individual elements which have been analysed in soil extracts, fertilizers and plants is compiled in table 42.
Atomic absorption spectroscopy is very suitable for the direct analysis of soil extracts since the often complex and concentrated matrix rarely gives greater problems. For trace analyses,

Table 42. Bibliographical survey: Application of atomic absorption spectroscopy in agricultural chemistry

Element	Soil and soil extracts	Fertilizers, feeding stuffs	Plants
Al	173 720 968 1365		
B	464 979	461 980 1288	859 963
Ca	173 269 968		117 268 274 418 1259
Cd	612		
Co			1127
Cr	173		
Cu	39 64 173 464 1260	461 490 846	22 39 418 1040 1259
Fe	37 64 173 968 1365	461 846	22 37 273 274 418 1040 1111 1259
K	173 269 968 1221	490 989 1230	274 418
Li	173		961
Mg	269 441 968 1321	846	117 274 418 1259
Mn	64 173 464 535	461 846	22 174 273 274 418 1040 1111 1259
Mo	464 1250	271 530	503
Na	269 968	490	274 418
Ni	173		
Pb	612		532 612 1111
Rb	173		
SO_4	401 1261		
Se		553	1017
Si	721 1365		
Sn			1319
Sr	173 272		272 1242
Zn	41 64 173 464	461 846	22 41 174 274 418 1040 1259

the extract can be directly aspirated with ammonium acetate or other extracting agents using a 3-slot burner head. The results correspond to neutron activation [173] and frequently are better than with conventional procedures [464]. The freedom from interferences and speed of atomic absorption permit easier automation when large numbers of samples must be handled [532]. Boron is the only element which frequently cannot be determined directly with the required sensitivity in soil and soil extracts. A recently described extraction procedure with an organic solvent brings an increase in the sensitivity by a factor of about 25 and is particularly useful for this element [979] [1288].

The analysis of fertilizers usually requires a digestion; for the analysis of the most important trace elements a single fuming off with concentrated hydrochloric acid and solution of the residue in diluted hydrochloric acid is often sufficient. At least for manganese, copper, zinc and iron the results obtained agree well with those from potassium hydrogen sulphate melts or fluoric acid/nitric acid digestions [461]. McBRIDE [846] applied a very critical test to the determination of copper, iron, manganese, magnesium and zinc in fertilizers by atomic absorption spectroscopy [1363] and additionally carried out comparitive measurements. The results were very satisfactory and the procedure was adopted as the official method of the AOAC.

MANNING and FERNANDEZ reported the direct determination of phosphorus in fertilizers with the graphite tube furnace [804]. They employed the absorption of a PO_x radical at a lead ion line at 220.3 nm which is specific for phosphorus. IHNAT determined selenium in feeding stuffs, to which it is added as an essential element, by flameless atomization [553] and recommended prior digestion in nitric acid/perchloric acid. JUNG and CLARKE determined cadmium in fungicides in the flame at the secondary line at 326.1 nm [599].

The analysis of plant materials practically always requires ashing of the dried material; atomic absorption spectroscopy offers the advantage that the ash, mostly dissolved in dilute hydrochloric acid, can be analysed directly. Only very rarely are further enrichment steps such as extraction in an organic solvent required.

Recently, a number of authors examined the relationship between sample preparation and analytical results for plant analysis. GIRON found no significant differences between dry and wet ashing with nitric acid/perchloric acid for the determination of eight elements, only the precision was better with dry ashing [418]. VARJU found that pre-treatment of the plant samples with nitric acid before the dry ashing accelerated this and increased the solubility of the ash. Copper and iron as their nitrates are not bound to silicic acid and can therefore also be quantitatively determined. An extraction of plant materials with 6 N hydrochloric acid at 110°C in a pressure vessel dissolves iron, copper, manganese and zinc quantitatively, but calcium and magnesium to less than 50% [1259].

ROBLES and LACHICA compared wet ashing (with nitric acid/perchloric acid) with three simple extraction procedures employing nitric acid/hydrochloric acid, hydrochloric acid alone and hydrochloric acid/fluoric acid respectively to remove the silicic acid. The last procedure gave the highest values for iron, copper, manganese and zinc [1040]. AGUILAR and co-workers examined the relationship between the conventional analysis of leaves and the analysis of the sap, and found good agreement for copper, manganese and zinc, while the results for iron were up to ten times higher for the conventional method [22].

Besides the conventional elements, pronounced trace elements or toxic elements are also being increasingly determined in plants. SIMMONS extracted cobalt from plant material [1127] and WILLIAMS determined tin in wood [1319]. KAHN and co-workers used the boat technique for the direct determination of lead in dried leaves [612]. SEGAR and co-workers determined lead, iron and manganese from aqueous suspensions of leaves employing the graphite tube furnace and obtained good agreement with other procedures [1111]. HENNING and JACKSON

Table 43. Bibliographical survey: Application of atomic absorption spectroscopy in foodstuffs chemistry

Element	Drinks	Fruit, vegetables, grain	Fish, meat	Oils, fats	Milk, milk products
Al		525	129		
B		526	526		
Ca	386 594	594 951	121 489 951 1129	23 382 448	166 722
Cd		751	49 129 527		
Cl^-	405				
Co	387		598		722
Cr	387 594	594	49 129		
Cu	180 386 387 594 1184 1258 1378	39 594	49 129 203 489 490 491 674 741 1129	1143	15 722 802 1331
Fe	386 387 594 1003 1258	273 525 594	129 489 674 1129		
Hg		531 576 948	57 146 528 531 576 754 895 934 949 1225 1229 1251		
K	386 512 594	273 594	121 490 1226	1001	166 722
Li	114 594	594			
Mg	387 549 1258	36 273 594 951	121 489 674 951	23 448	116 722
Mn	387 594	174 273 594	489 674	1143	722
Mo					1046
Na	386 512 594	273 594	121 490 867	130 1001	166 722
Ni	180 387			1005 1143	
P			654	1143	654 804
Pb	1055 1377 1378	352	129 144 527 590 1055 1179		369 456 724 803
Rb	594	594 1185			
Sn	1003		276 1054 1129		1093
Sr		270 594 1242	275 1242		722 802 1242
Zn	387 594 1378	174 267 273 594 951	49 121 129 489 674 951 1129 1246	23	

determined molybdenum in plants with the graphite tube furnace after dry ashing from hydrochloric acid solution in good agreement with colorimetry [503].

Nowadays, the analysis of foodstuffs is of great importance. Both the environmental influence during growth, such as pesticides, feeding stuffs, air pollution, etc., and possible contamination during processing must be taken into consideration. A further factor of importance, which is being increasingly examined, is the influence of packaging materials (tin plate, plastic). Corresponding to this wide range of analytical requirements, the number of elements to be determined in foodstuffs is also relatively large. A literature survey on the various applications of atomic absorption spectroscopy in this sector is compiled in table 43.

The required sample preparation varies according to the nature of the sample. While fruit and vegetables, and also fish and meat products always require either wet or dry ashing before analysis, other products can frequently be determined directly or with only a minimum of preparation by atomic absorption. Of especial mention in this connection are several newer digestion procedures which frequently bring a considerable saving of time and can also be employed for easily volatile elements. BOUCHARD used chromic acid to digest fish and meat for subsequent analysis of mercury. The digestion was complete after $^1/_2$–$1^1/_2$ hours at 25 °C [146]. Various digestions in PTFE autoclaves under pressure and increased temperature are particularly interesting since they usually only require about half an hour. PAUS described the digestion of fish and seaweed for the subsequent analysis of lead, cadmium, copper, iron, mercury and zinc [966]. HOLAK found good agreement with the official AOAC method for mercury in fish [528] and TÖLG and co-workers, by employing radionucleides, established that losses of even such volatile elements as selenium or mercury were less than 2% [700]. ADRIAN employed larger sample amounts (5 g) and lower temperatures, but at the cost of time; he allowed the samples to stand overnight at room temperature and then heated them to 90 °C for three hours [19]. TÖLG described a modified "cold combustion" with activated oxygen which also worked without losses [620]. These new techniques are of especial interest in connection with the vastly increased importance attached to the determination of such toxic elements as lead, cadmium and mercury. Drinks, with the exception of fruit juices or concentrates, which still contain fruit pulp, can be aspirated directly into the flame. Liqueurs and concentrates must be diluted since they have very high viscosities, and with drinks containing carbon dioxide, this should be driven off before the analysis, most suitably by warming. Drinks containing high or changeable sugar or alcohol contents should be analysed according to the method of additions if possible, since a high alcohol content can influence the individual elements to varying degrees. VARJU has reported that a direct determination is also possible with alcoholic drinks, if the alcohol content is brought to 50% and is not allowed to vary more than 5% [1258].

Milk can be aspirated directly, without dilution, into the flame; occasionally however, the elements under study are not present in sufficient concentrations so that they must be extracted after wet ashing or determined from the ashes after dry ashing. BROOKS and co-workers preferred deproteinization of milk with trichloroacetic acid to be able to analyse the main elements free of interferences [166]. Milk can also be directly analysed very well in the graphite tube furnace. LAGATHU and DESIRANT determined iron, cobalt, copper, manganese and strontium [722] with coefficients of variation of 2–5%; MANNING and FERNANDEZ determined copper and strontium [802] and lead [803] directly, employing a background corrector. Similarly,

almost all milk products must be wet or dry ashed before analysis. A sole exception is the determination of copper in butter and other fats; this is most conveniently carried out via a simple extraction of the dissolved of melted sample with nitric acid [15]. PRICE and co-workers have described a similar extraction for nickel in edible fats which is also carried out with 10% nitric acid in the presence of carbon tetrachloride [1005].
Liquid fats and oils are diluted 1:5 to 1:10 with an organic solvent such as MIBK and aspirated directly into the flame if the concentration of the element under study is sufficient. Calibration can be made against oil soluble standard metallo-organic substances in a similar manner to the analysis of mineral oils (see section 9.6). For the lowest metal traces, oils and fats must be ashed if a flameless procedure is not employed. Although ashing appears to be necessary for very exact analysis of trace elements in foodstuffs, simple heating with 1+1 hydrochloric acid and subsequent filtration has brought very satisfactory results [1129]. This procedure appears very attractive for long measurement series since it permits a rapid determination of trace elements in foodstuffs with sufficient accuracy in combination with atomic absorption spectroscopy.

9.3. Water and Air Analysis

Very little has been reported up to the present on the direct determination of metal traces in air. WHITE [1317] described an atomic absorption spectrophotometer which permitted the direct analysis of air for lead and cadmium vapours. EDWARDS [322] determined silver in air with a detection limit of 3 $\mu g/m^3$ by conducting the air to be analysed directly into the primary air supply of a commercial atomic absorption spectrophotometer. THILLIEZ [1224] also used atomic absorption spectroscopy for the determination of tetramethyl and tetraethyl lead in the neighbourhood of a manufacturing plant for these substances and LINDNER determined mercury in room air by simply drawing it through an absorption cell mounted in an atomic absorption spectrophotometer [752].
In general, however, the analysis of air is carried out indirectly via filters or precipitation (rain or dust, etc.) which frequently make preparation necessary [120]. SACHDEV, ROBINSON and WEST [1077] reported the determination of manganese, iron, cobalt and nickel in air and water with a modified atomic absorption spectrophotometer. KETTNER [640] compared the determination of lead in dust by the dithizone method with atomic absorption and found the latter to be at least as accurate and, at the same time, more rapid and reliable. Various authors also determined lead [181] [295] [548] [559] [1094] [1227], cadmium [1094] [1227], copper [534] [1094], iron [534] [559] [1227], zinc [559] [1094] and other elements [1227] by extracting the filter with acid, or ashing the filter either at low temperature or with oxygen plasma and then taking up the residue in acid and aspirating directly into the flame.
Flameless procedures are also being increasingly employed for the determination of trace elements in air since the high sensitivity and low sample requirements make relatively short sampling times possible. VIJAN and WOOD determined arsenic by reducing it to arsine and passing this through a quartz tube heated to 800°C [1271]. OMANG [932] and JANSSEN and DAMS determined lead [572] and cadmium [573] in air with the graphite tube furnace and found excellent sensitivity and got accurate results.

Mercury again assumes a special position here and has been determined by a number of authors employing the various low temperature procedures [70] [71] [703].
ROSE and BOLTZ [1057] indirectly determined the sulphur dioxide content in air via lead sulphate and the analysis of lead by atomic absorption spectroscopy.
Atomic absorption spectroscopy is ideally suited to the determination of water of all types; river and lake water, drinking water, sea-water, waste water, boiler and feed waters, etc.
Apart from the ionization interferences which are naturally expected for the determination of the alkali metals, no marked chemical interferences occur in the analysis of water for either the main elements or the trace elements. As long as the concentrations of the elements to be determined are sufficiently large, or the sensitivity of flame atomic absorption is adequate, the metal salts which are dissolved in the water can be determined directly and without preparation. For natural waters this is principally valid for sodium, potassium, calcium and

Table 44. Bibliographical survey: Application of atomic absorption spectroscopy in water analysis

Element	References
Ag	255 1109 1308
Al	373 1302
Au	1387
Be	510
Ca	108 125 136 192 242 371 374 386 510 989
Cd	72 177 510 718 770 771 965 967 988 1015 1109 1168
Cl^-	404
Co	162 168 177 182 321 337 375 510 597 815 916 965 1077 1109 1168
Cr	177 225 289 321 361 510 815 865 883 916 989 1302
Cu	72 136 162 167 177 192 320 321 337 361 371 374 386 434 510 597 704 718 815 900 916 965 967 988 989 1015 1109 1168 1237 1302
Fe	136 162 168 177 192 320 321 337 386 510 588 596 704 916 967 988 989 1077 1109 1111 1168 1300 1302
Hg	372 621 798 926 1235 1302
K	125 192 371 374 386
Li	56 371 374
Mg	125 136 192 371 374 989
Mn	72 136 177 320 321 361 371 374 596 597 815 916 928 967 989 1077 1109 1111 1168 1300 1302
Mo	195 289 815 899
Na	125 192 371 374 386 510
Ni	136 162 168 177 182 321 337 375 588 597 815 916 965 989 1077 1109 1168
PO_4^{3-}	1371
Pb	72 136 162 168 177 320 361 375 597 916 965 967 988 1109 1111 1168 1302
Rb	125
Se	699
Sr	125 371 374
V	815 901 1109
Zn	136 162 168 177 337 361 371 374 421 434 597 916 967 988 989 1015 1109

magnesium, and also frequently for zinc. The most important trace elements, copper, iron, cobalt, nickel, manganese and chromium, on the other hand, usually cannot be determined with the required sensitivity by the direct method. Frequently an enrichment procedure or the use of a graphite tube furnace is necessary. A bibliographical survey on the use of atomic absorption spectroscopy in water analysis is compiled in table 44.

The most frequently employed enrichment procedure in water analysis for trace elements is complexing the metal ions with ammonium pyrrolidine dithiocarbamate (APDC) and subsequently extracting into methyl isobutyl ketone (MIBK). This extraction system has the advantage that relatively stable chelates are formed with numerous metals over a wide pH range, so that with a single extraction a large number of trace elements can be simultaneously extracted. Many authors using this system have successfully extracted and subsequently analysed copper, zinc, iron, cobalt, nickel, lead, manganese and other elements. JOYNER and FINLEY [596] complexed iron and manganese in sea-water with sodium diethyl dithiocarbamate and extracted the complex with MIBK. CHAU, SIM and WONG [228] found that chromium can best be extracted as the acetyl acetonate complex, and DELAUGHTER [289] employed diphenyl dithiocarbazone for chromium and dithiol for molybdenum with extraction into MIBK.
In contrast, other authors found that these elements could be optimally extracted with APDC/MIBK. MIDGETT and FISHMAN [865] ascertained that chromium must be oxidized to the hexavalent state before extraction with this system. Nevertheless, the extraction of trace elements with organic solvents is not completely without problems. OLSEN and SOMMERFELD draw attention to the relative instability of various metal chelates.

Manganese in particular begins to precipitate directly after extraction. These authors therefore evaporated the extract to dryness and dissolved the residue in 0.1 N hydrochloric acid/acetone (1:1) [928]. KOIRTYOHANN and WEN found that the sensitivity of various extracted trace elements was dependent upon the pH of the aqueous solution and they eliminated this dependence by adding 0.2 N perchloric acid to the aqueous solution [691]. FISHMAN extracted aluminium from water with oxine/MIBK at pH 8, the extraction must be finished within three minutes however, otherwise magnesium oxinate comes out of solution and takes the aluminium with it [373]. HICKS and co-workers draw attention to a possible interference owing to the presence of water-soluble organic solvents in waste water [510].
MULFORD [894] described the APDC/MIBK extraction system in detail and BURRELL [184] treated the technique of solvent extraction in combination with atomic absorption spectroscopy for the determination of trace elements in sea-water. Comprehensive instructions and methodology on the application of atomic absorption in water analysis have been published by FISHMAN and DOWNS [374], BREWER, SPENCER and SMITH [162], EDIGER [317] and WELZ [1298].

JOYNER and co-workers [597] have published a comprehensive report on the various procedures for sample enrichment which are of interest for the trace analysis of sea-water. The authors deal critically with evaporation, co-precipitation, solvent extraction and other enrichment procedures, and give a comprehensive bibliographical survey on this topic. Various authors have reported on the application of atomic absorption spectroscopy to water analysis in "Trace Inorganics in Water", vol. 73, in the series "Advances in Chemistry". KAHN [609] wrote an introduction to atomic absorption spectroscopy, FISHMAN and MIDGETT reported on the extraction of cobalt, nickel and lead from natural waters [375].

BOETTNER and GRUNDER routinely determined numerous trace elements by atomic absorption [136] and quoted values for the precision of the individual determinations. PLATTE [988] drew especial attention to the simplicity, speed and specificity of atomic absorption for the analysis of various industrial waters. He quoted extraction procedures for iron, copper, lead, zinc and cadmium. McFARREN and LISHKA [849] made a comparative summary of the results of various studies on the determination of trace elements, and ELLIS and DEMERS [326] described the application of atomic fluorescence spectroscopy to water analysis.

Flameless atomic absorption has found particularly wide application in water analysis. This method often saves an extraction stage and then, as a direct procedure, offers greater speed and simplicity for the same precision and better accuracy. Particularly the last point is of importance, and it should always be borne in mind that at every preparative stage the risk of contamination and losses increases, especially in the trace range.

The determination of mercury can be carried out by the low temperature procedure of POLUEKTOV and co-workers [996] with a detection limit of 0.2 µg/l, which is sufficient for most interests. MANNING has drawn attention to the risk of molecular absorption from organic solvents such as acetone, etc., with waste water analysis [798]. In cases of doubt, the determination of mercury should always be checked with a background corrector.

The analysis of high purity water can be carried out with high sensitivity and completely free of problems in the graphite tube furnace. For this purpose, PICKFORD and ROSSI have even developed an automatic sampling system for flameless atomic absorption which operates with better precision than manual dispensing [981] [982]. CRAGIN and HERRON determined calcium in polar ice in good agreement with the flame and found that the graphite tube furnace gives not only better detection limits but also better precision [242].

The most varying trace elements can be determined directly in natural inland waters without sample preparation, but not without interferences, however [1294]. Usually the inorganic main components of the water effect a certain reduction of the signal [361] [1302], which appears to remain constant above a given total salt concentration (compare page 102). Therefore, it is possible to eliminate the interference by matching the matrix (e.g. making the standard solutions up in tap water [1302]). Other authors recommend the method of additions [72] [321] with which several elements such as chromium and nickel hardly exhibit matrix influences [321] [883]. However, PAUS preferred extraction of the trace elements in APDC/MIBK before determination in the graphite tube furnace [965].

SEGAR and GONZALEZ were the first to make a comprehensive examination of the analysis of sea-water with the graphite tube furnace and found that volatile elements such as lead, cadmium, silver and zinc could not be determined. More difficultly atomized elements such as iron, cobalt, manganese, nickel and vanadium could be freed sufficiently from the sodium chloride matrix, but owing to co-vaporization the sensitivity is in part strongly reduced [1107] [1109]. Only iron was still sensitive enough for a direct determination; for this reason the authors recommended extraction of the trace elements with APDC/MIBK before determination in the graphite tube furnace [1111]. This method has also been successfully used by other authors [704] [967]. An especial advantage of the graphite tube furnace for sea-water analysis is that the extraction can be performed on a micro scale [704] and the determination carried out "on site" so that samples do not need to be transported [185].

MUZZARELLI and ROCCHETTI determined copper [900], molybdenum [899] and vanadium [901] in sea-water after treating the water with persulphate and running it down a Chitosan® column to remove sodium chloride. LUNDGREN and co-workers successfully determined cadmium in sea-water with a modified graphite tube furnace having a very rapid rate of temperature rise and exact temperature control. They atomized cadmium at 820 °C, a temperature at which sodium chloride is not noticeably volatile, and obtained a detection limit of 0.03 μg/l [770] [771].
EDIGER and co-workers recently reported that the direct analysis of sea-water is greatly facilitated through the addition of ammonium nitrate. The reaction

$$NaCl + NH_4NO_3 \longrightarrow NaNO_3 + NH_4Cl$$

takes place and both the end products, sodium nitrate and ammonium chloride, can be separated at temperatures around 400 °C, while sodium chloride is first noticeably volatile at temperatures above 1100 °C [320]. The determinations of lead, iron, copper and manganese were successfully carried out with detection limits around 1 μg/l.
The analysis of sediments has increased greatly in importance in connection with environmental control. Usually, a determined sediment fraction is digested in a known acid mixture and after dilution is analysed in the flame [50] [177] [380]. Naturally the graphite tube furnace can also be employed, especially if the concentrations are relatively low [151]. Under certain circumstances, sediments in suspension can be directly analysed in the graphite tube furnace [380].

9.4. Geochemistry

Because of its high specificity and freedom from interferences, which make separation of the main components even with trace analyses superfluous, atomic absorption spectroscopy has found rapid acceptance in the geochemical laboratory. It not only replaces conventional wet chemical analyses, but also stands on an equal footing with other more time-consuming procedures such as X-ray fluorescence or even neutron activation analysis.
At the start of the 1960's, the elements calcium, magnesium, sodium, potassium, iron, manganese, copper and zinc in various rock samples and silver and gold in ores were principally analysed. After the introduction of the nitrous oxide/acetylene flame, elements such as aluminium, silicon and titanium were added in increasing measure. Thus, with atomic absorption spectroscopy it is possible to analyse the most important metallic elements of interest in rocks usually with sufficient sensitivity and very good precision. A survey of literature published up to the present on the application of atomic absorption to the geochemical sector is compiled in table 45. Geochemistry is also the first specialized area to which atomic absorption spectroscopy was applied and for which a comprehensive monograph [2] containing much useful information was published.
The most difficult part of a rock or ore analysis is frequently the solution procedure. This is particularly the case when the sample to be analysed consists of silicate material. A number

MUZZARELLI and ROCCHETTI determined copper [900], molybdenum [899] and vanadium [901] in sea-water after treating the water with persulphate and running it down a Chitosan® column to remove sodium chloride. LUNDGREN and co-workers successfully determined cadmium in sea-water with a modified graphite tube furnace having a very rapid rate of temperature rise and exact temperature control. They atomized cadmium at 820 °C, a temperature at which sodium chloride is not noticeably volatile, and obtained a detection limit of 0.03 µg/l [770] [771].

EDIGER and co-workers recently reported that the direct analysis of sea-water is greatly facilitated through the addition of ammonium nitrate. The reaction

$$NaCl + NH_4NO_3 \longrightarrow NaNO_3 + NH_4Cl$$

takes place and both the end products, sodium nitrate and ammonium chloride, can be separated at temperatures around 400 °C, while sodium chloride is first noticeably volatile at temperatures above 1100 °C [320]. The determinations of lead, iron, copper and manganese were successfully carried out with detection limits around 1 µg/l.

The analysis of sediments has increased greatly in importance in connection with environmental control. Usually, a determined sediment fraction is digested in a known acid mixture and after dilution is analysed in the flame [50] [177] [380]. Naturally the graphite tube furnace can also be employed, especially if the concentrations are relatively low [151]. Under certain circumstances, sediments in suspension can be directly analysed in the graphite tube furnace [380].

9.4. Geochemistry

Because of its high specificity and freedom from interferences, which make separation of the main components even with trace analyses superfluous, atomic absorption spectroscopy has found rapid acceptance in the geochemical laboratory. It not only replaces conventional wet chemical analyses, but also stands on an equal footing with other more time-consuming procedures such as X-ray fluorescence or even neutron activation analysis.

At the start of the 1960's, the elements calcium, magnesium, sodium, potassium, iron, manganese, copper and zinc in various rock samples and silver and gold in ores were principally analysed. After the introduction of the nitrous oxide/acetylene flame, elements such as aluminium, silicon and titanium were added in increasing measure. Thus, with atomic absorption spectroscopy it is possible to analyse the most important metallic elements of interest in rocks usually with sufficient sensitivity and very good precision. A survey of literature published up to the present on the application of atomic absorption to the geochemical sector is compiled in table 45. Geochemistry is also the first specialized area to which atomic absorption spectroscopy was applied and for which a comprehensive monograph [2] containing much useful information was published.

The most difficult part of a rock or ore analysis is frequently the solution procedure. This is particularly the case when the sample to be analysed consists of silicate material. A number

BOETTNER and GRUNDER routinely determined numerous trace elements by atomic absorption [136] and quoted values for the precision of the individual determinations. PLATTE [988] drew especial attention to the simplicity, speed and specificity of atomic absorption for the analysis of various industrial waters. He quoted extraction procedures for iron, copper, lead, zinc and cadmium. McFARREN and LISHKA [849] made a comparative summary of the results of various studies on the determination of trace elements, and ELLIS and DEMERS [326] described the application of atomic fluorescence spectroscopy to water analysis.

Flameless atomic absorption has found particularly wide application in water analysis. This method often saves an extraction stage and then, as a direct procedure, offers greater speed and simplicity for the same precision and better accuracy. Particularly the last point is of importance, and it should always be borne in mind that at every preparative stage the risk of contamination and losses increases, especially in the trace range.

The determination of mercury can be carried out by the low temperature procedure of POLUEKTOV and co-workers [996] with a detection limit of 0.2 μg/l, which is sufficient for most interests. MANNING has drawn attention to the risk of molecular absorption from organic solvents such as acetone, etc., with waste water analysis [798]. In cases of doubt, the determination of mercury should always be checked with a background corrector.

The analysis of high purity water can be carried out with high sensitivity and completely free of problems in the graphite tube furnace. For this purpose, PICKFORD and ROSSI have even developed an automatic sampling system for flameless atomic absorption which operates with better precision than manual dispensing [981] [982]. CRAGIN and HERRON determined calcium in polar ice in good agreement with the flame and found that the graphite tube furnace gives not only better detection limits but also better precision [242].

The most varying trace elements can be determined directly in natural inland waters without sample preparation, but not without interferences, however [1294]. Usually the inorganic main components of the water effect a certain reduction of the signal [361] [1302], which appears to remain constant above a given total salt concentration (compare page 102). Therefore, it is possible to eliminate the interference by matching the matrix (e.g. making the standard solutions up in tap water [1302]). Other authors recommend the method of additions [72] [321] with which several elements such as chromium and nickel hardly exhibit matrix influences [321] [883]. However, PAUS preferred extraction of the trace elements in APDC/MIBK before determination in the graphite tube furnace [965].

SEGAR and GONZALEZ were the first to make a comprehensive examination of the analysis of sea-water with the graphite tube furnace and found that volatile elements such as lead, cadmium, silver and zinc could not be determined. More difficultly atomized elements such as iron, cobalt, manganese, nickel and vanadium could be freed sufficiently from the sodium chloride matrix, but owing to co-vaporization the sensitivity is in part strongly reduced [1107] [1109]. Only iron was still sensitive enough for a direct determination; for this reason the authors recommended extraction of the trace elements with APDC/MIBK before determination in the graphite tube furnace [1111]. This method has also been successfully used by other authors [704] [967]. An especial advantage of the graphite tube furnace for sea-water analysis is that the extraction can be performed on a micro scale [704] and the determination carried out "on site" so that samples do not need to be transported [185].

SUHR and INGAMELLS have recommended a rapid and simple procedure employing a lithium metaborate ($LiBO_2$) digestion with which silicon as well as other metallic elements in silicate rocks and minerals can be digested and maintained in solution. This procedure was originally recommended for emission spectroscopy [1191] and colorimetry [553], and was only later applied to atomic absorption spectroscopy [855] [1190].

Figure 74. Autoclave-2 for the digestion of difficultly dissolved samples under pressure and at increased temperature in a closed PTFE vessel (Bodenseewerk Perkin-Elmer & Co. GmbH).

According to the original directions of SUHR and co-workers, 0.100 g of a finely powdered sample are mixed with 0.500 g $LiBO_2$ and heated in a baked-out graphite crucible in a muffle furnace at 900–1000 °C for 10–15 minutes. The clear melt is agitated in the hot crucible, so that small pellets are formed which do not adhere to the walls, it is then rapidly poured into 40 ml nitric acid contained in a 200 ml PTFE or polyethylene beaker. The beaker is covered and the contents stirred with a magnetic stirrer to obtain complete solution. From this solution—with further dilution if required—silicon, aluminium, magnesium, calcium, potassium, sodium, manganese, chromium, strontium, titanium, copper, zinc and other elements can be determined. YULE and SWANSON [1366] checked this procedure for nine elements with various standard rocks and found good agreement with the certified values. Good agreement was also found with wet chemical procedures (which take five times as long) for both the main component (up to 80% SiO_2) and for the trace elements. GOVINDARAJU and L'HOMEL have semi-automated this procedure for the subsequent determination of silicon in silicate rocks, either directly or after passage through a cation exchanger [437]. The latter method gives a precision of better than 1% and excellent accuracy. INGAMELLS has published a good survey on the use of lithium metaborate digestions for geological samples [556].

VAN LOON and PARISSIS modified the lithium metaborate digestion independently from SUHR and adapted it especially for subsequent atomic absorption analysis [1257]. Silicon, aluminium, calcium, iron, potassium, magnesium, manganese, sodium and titanium were analysed in 12 standard silicate rocks. Exact procedures for the preparation required for precise and interference-free atomic absorption analyses are given for every element. The results obtained compare very well with values from three other analytical procedures. BOAR and INGRAM [131] also gave detailed procedures for the analysis of silicate rocks with this digestion.

LUECKE and co-workers have described the use of the boat technique for the determination of very small concentrations of lead, zinc and silver in soil samples for geochemical prospecting. They draw especial attention to the speed of the analysis and the high sensitivity [329] [776].

FRATTA likewise used the boat technique for the determination of thallium in silicate rocks in the ppb range [384].

GOVINDARAJU has described a method for the direct introduction of solid samples into the flame [435] [436]. The powdered rock sample is mixed with an ionization buffer (e.g. sodium carbonate) and then placed onto a threaded iron rod wetted with acetone which is then directly introduced into the flame. Lithium, rubidium, cesium, lead and zinc can be determined by this method with detection limits of around 1 ppm.

The graphite tube furnace has also received a great deal of attention for the determination of trace elements in rock samples. The possibility of analysing solid samples directly appears to be of particular interest since the frequently time-consuming digestion procedures can be avoided, and at the same time, the highest sensitivity can be obtained. Various authors reported the direct analysis of easily atomized elements such as lead, cadmium, mercury, silver, thallium, bismuth and zinc [496] [730] [768] [1294] [1295], and also cesium and rubidium [731]. With these elements, good separation can be achieved by selective vaporization by atomizing at relatively low temperatures (e.g. 1400–1700 °C). At these temperatures the rock samples are usually molten but do not have a very large vapour pressure, so that only slight non-specific light losses occur; nevertheless, the element under study is atomized rapidly enough from the molten rock sample. With a good background corrector this procedure can even be used at considerably higher temperatures, so that copper, for example, can be directly analysed in solid samples [1295].

Besides the above-mentioned advantages, the direct analysis of solid samples also has a number of marked disadvantages. Since the maximum usable quantity is in the order of 1 mg, the grain size should not exceed 1 µm to ensure that the dispensed amount is representative of the whole sample. Nevertheless, the coefficient of variation for the direct analysis of solids is typically 10–20%, which is not always sufficient for the requirements. Calibration also occasionally causes difficulties and the time required to make a precise weighing in the region of 1 mg should not be underestimated. However, the procedure is excellent for a rapid, semi-quantitative examination of a sample for trace elements.

The application of atomic absorption spectroscopy to geochemical analyses usually brings no special problems. The tendency to use hotter flames is recommended, with the nitrous oxide/acetylene flame for the determination of the alkaline-earths, for example [764]. This can be recommended more, when the concentrations of silicon and aluminium are higher, or if sulphuric acid has been used for the digestion. On the other hand, the use of a 3-slot burner head for trace analyses is recommended, because of the high total salt content. Under certain circumstances the method of additions must be employed for the determination of very low concentrations in complex or concentrated matrixes in order to avoid errors: occasionally simple extraction procedures, such as the APDC/MIBK system, are employed for enrichment.

Attention is again drawn to the fact that background absorption can occur when concentrated solutions are aspirated directly into the flame, especially at short wavelengths and if refractory materials are present [211]. This leads to false measurements if it is not correspondingly corrected. The risk is particularly high in flames of lower temperature and in the presence of high silicon, aluminium or calcium contents. However, with attention to these interferences, very rapid and reliable results can be obtained by atomic absorption spectroscopy, as the numerous comparative measurements which have been made on standard rocks and in inter-laboratory tests show.

DALL'AGLIO [264] compared atomic absorption with optical emission spectrography and LANGMYHR and PAUS [729] carried out an inter-laboratory test in which atomic absorption spectroscopy stood in very good agreement with X-ray spectrography, emission spectrography, conventional wet chemical techniques and other methods. It is therefore hardly surprising that atomic absorption spectroscopy has found rapid acceptance in the routine geochemical laboratory [748].

Various authors have employed the graphite tube furnace for the determination of trace elements in acid digestions. SCHWEIZER used a modified fluoric acid/perchloric acid digestion, repeatedly fumed off with hydrochloric acid at the end for the determination of the main and subsidiary components in carbonaceous rock materials in the flame, and the trace elements chromium, cobalt, copper, molybdenum, nickel and vanadium in the graphite tube furnace [1101]. Chromium and cobalt could be determined against aqueous standard solutions, but for the other elements the method of additions had to be employed; the coefficients of variation were between 2.5% and 4%. Other authors used the normal fluoric acid/perchloric acid digestions directly for the analysis of beryllium [1123], vanadium [234], cadmium, nickel, zinc [245], aluminium, iron, manganese and titanium [495] [496] and cobalt [768]. In some cases the perchloric acid appears to reduce the non-specific absorption [245], but it reduces the signal for cobalt very strongly [768], and also interferes with a number of other elements, so that a separation appears quite desirable [234] [1123]. To reduce the interferences further, several authors extracted such elements as tellurium [91] or thallium [496] [1124] from the sample solution before determination in the graphite tube furnace.

The elements silver and gold hold a special position in geochemical analysis. A docimastic analysis, which is usually employed for the determination of these elements, is difficult and very time-consuming; atomic absorption spectroscopy offers speed and sensitivity. Only the solution of the sample occasionally causes difficulties and frequently requires certain modifications for each raw material. FIXMAN and BOUGHTON [376] dissolved silver-containing minerals in nitric acid under the addition of hydrogen peroxide. RAWLING and co-workers [1025] [1026] dissolved the sample in hydrochloric acid/nitric acid and maintained the silver in solution by always using 6 N hydrochloric acid.

RUBEŠKA and co-workers [1072] give three digestion procedures for sulphidic silver ores employing (1) nitric acid/tartaric acid under addition of mercury(II) nitrate, or (2) nitric acid/sulphuric acid, or (3) sulphuric acid, under certain circumstances with the addition of tartaric acid. HUFFMANN and co-workers [538] found, on the other hand, that simply heating the sample with concentrated nitric acid for about 15 minutes brings all silver into solution. The authors examined 44 ore samples with contents between 1 and 8000 ppm silver and found no mentionable differences between the values obtained from the simple nitric acid extraction and those from complete solution of the sample. The agreement with conventional methods of determination was also good. SPITZER and TESIK determined silver in copper ironstone [494] and in calcine [1165] from 7 N hydrochloric acid solution, from which the iron had previously been extracted. This corresponded well with docimastic analyses.

The determination of gold does not appear to be completely interference-free and acid solutions especially cannot be compared to cyanide ones. In general however, gold is not determined directly, but is extracted because the natural concentrations are too low for a direct analysis and the requirements for accuracy are high. BUTLER and co-workers [198] complexed gold with dimethylglyoxime and extracted with MIBK. STRELOW and co-workers examined the problem of gold analysis in detail [1182] and found that gold can be quantitatively extracted directly from 3 N hydrochloric acid with MIBK without complexing agent. With a mixture of aqueous solution to MIBK in the ratio 10:1, 99% of the gold can be extracted in one stage. SPITZER and TESIK [1165] used the same procedure to extract gold from calcine. The co-extracted iron was removed from the organic phase in a second stage. These authors, like TINDALL [1231] [1232], found good agreement with values obtained by docimastic analyses. GREAVES [439] extracted gold with MIBK in the presence of hydrobromic acid instead of hydrochloric acid and obtained satisfactory results.

9.5. Metallurgy and Plating

In the analytical laboratories of iron and steel works, direct reading spectrographs (known as quantometers) are used nowadays principally for production control. Alongside these, atomic absorption spectroscopy has also secured its position, so that metallurgy now belongs to one of the main fields of application. A literature survey on the determinations carried out on irons, steels and alloys is compiled in table 46.

For iron and the majority of steels, the solution procedure presents no major problems; frequently they can be dissolved in a hydrochloric acid/nitric acid mixture which is ideally suited for subsequent atomic absorption analysis. BEYER [119] warmed about 500 mg sample with 10 ml 1+1 hydrochloric acid and then added concentrated nitric acid drop by drop until all soluble material had been taken up. KINSON [645] [646] and BELCHER [100] found that tungsten steels can be better dissolved in phosphoric acid/fluoric acid [547] and SCHILLER employed the same acid mixture to dissolve niobium in low alloy steels [1088]. HEADRIDGE and SOWERBUTTS used a pressure digestion in a PTFE autoclave for the determination of total aluminium in steel [488].

The analytical procedure is usually adjusted to suit the nature of the analysis to be carried out and the required accuracy. Interferences of a chemical nature are rarely observed in steel analyses. In the presence of higher silicon contents, interferences occur in the analysis of manganese, but these can be avoided by adding calcium and using a 3-slot burner head. Chromium is influenced by iron in the air/acetylene flame so that either the iron content must be matched in the standard solutions or a nitrous oxide/acetylene flame must be employed. KÖNIG, SCHMITZ and THIEMANN [680] found that the determination of copper, manganese and aluminium is influenced both by the acid content and the iron concentration, so they matched the standard solutions correspondingly. KNIGHT and PYZYNA [675] solved this problem for the analysis of nine elements in tool steel by calibrating with standard steels of the NBS, they found very good agreement with emission spectrography and X-ray fluorescence. With the latter, phase problems which lead to errors can occur, as examples have shown. HUSLER used a nitrous oxide/acetylene flame for the determination of chromium, manganese, molybdenum, vanadium and tungsten in tool steel. After the addition of iron as the main element

Table 46. Bibliographical survey: Application of atomic absorption spectroscopy for the analysis of iron and steel

Element	References
Ag	522
Al	47 145 175 488 679 680 791
As	513
Ca	487 743 1220
Cd	919
Co	675 851 852 1169
Cr	119 547 675 680 743 945 947 1045 1053 1299
Cu	119 646 675 680 743 1169
Fe	743 919
Mg	98 119 851 919 1169 1220
Mn	100 119 537 547 675 680 743 919 947 1169
Mo	271 547 663 675 887 990 1045
Nb	1088
Ni	119 236 645 675 743 919 947 1169 1299
Pb	328 743 851 919
Se	972
Si	656 675 845 1004
Sn	903
Te	75 819
Ti	486 905
V	207 457 547 675
W	547
Zn	632 743

to the standard solutions and potassium as ionization buffer, the determinations in this flame were free of interferences [547].

The use of standard steels for calibration has proved of especial value for high precision analyses. If high accuracy is necessary, such as in the determination of the main alloying constituents in chrome-nickel steels, small interferences can have a marked effect. As WELZ and SEBESTYEN [1299] were able to show, even with matching of the acid and iron contents, an accuracy of only 1% could be achieved for chromium and nickel. With standard steels for calibration and employing baseline suppression and scale expansion, an accuracy of 0.3% relative can be achieved. An amazing fact is that the standard steels do not particularly have to resemble the sample in their composition; no special selection of standard steels must be made.

Magnesium, which has considerable effect on the properties of cast iron, can only be determined with great difficulty by conventional methods. Atomic absorption spectroscopy, because of its simplicity and high sensitivity offers the ideal method for this element. BELCHER and BRAY [98] developed a procedure in 1962 which is still frequently used nowadays as the standard method. 1 g sample is dissolved in 30 ml 1+1 hydrochloric acid and oxidized with 5 ml nitric acid. The solution is evaporated to dryness, mildly heated for a further five minutes, taken up in 10 ml hydrochloric acid and then diluted with strontium chloride solution. The use of a nitrous oxide/acetylene flame offers advantages for both magnesium and calcium [1220]. The determination of molybdenum in steel is preferably carried out in a nitrous oxide/acetylene flame since only slight interferences are to be expected. KIRKBRIGHT [663] only added the corresponding amount of iron to the standard solutions and otherwise found no interferences. Nevertheless, POLLOCK [990] frequently found values that were too low for the determination of molybdenum in slags, even in the nitrous oxide/acetylene flame, and was only able to obtain the theoretical values by adding 1000 µg/ml aluminium. In an air/acetylene flame, which must be rich in fuel gas, numerous interferences that can be eliminated by the addition of either a large excess of aluminium [271] or 2% ammonium chloride [887] occur.

The determination of aluminium [175] [488] [791], vanadium [207] [457], titanium [486], niobium [1088] and other refractory elements in steel can be carried out in a nitrous oxide/acetylene flame with no interferences worth mentioning, only the iron concentration should be matched. BOSCH and co-workers [145] have reported a rapid determination of acid soluble aluminium in steel in the important concentration range 0.002–0.060% Al. According to this procedure, 20 samples can be analysed in 11 minutes with a standard deviation of about ±0.001% Al. COLLIN and co-workers analysed the various non-metallic phases in steel which have considerable influence on the physical and mechanical properties [239]. The steel was dissolved in fluoric acid and the residue (sulphides, carbides, nitrides, oxides, aluminosilicates of calcium, iron, magnesium and manganese) was fused with sodium carbonate/sodium tetraborate. After solution in sulphuric or hydrochloric acid and dilution, the individual elements were determined in the flame.

MCAULIFFE [845] has described a procedure for the direct determination of silicon in cast iron. 0.50 g sample are warmed with 25 ml 3 N sulphuric acid until solution is complete. Then 10 ml 12% ammonium persulphate solution are added and the solution is heated to boiling for about one minute until it is clear. After cooling, the solution is diluted, filtered and aspirated directly. The values obtained by this procedure agree very well with those of other methods and with certificate values. KIRKBRIGHT and co-workers [656] determined silicon in low alloyed steels by atomic fluorescence spectroscopy and found no interferences.

HILL [513] has described the determination of arsenic in steels, ores, etc., in an argon/hydrogen diffusion flame following a simple distillation procedure. The agreement of values obtained in this way with certificate values is excellent.

HOFTON and HUBBARD determined silver in steel after extraction of the dithizonate in MIBK [522]. This represents a considerable simplification compared to conventional methods. PETERSON determined selenium in stainless steel in the range 0.2–0.5% free of interferences by dissolving the sample in hydrochloric acid/nitric acid and then treating it with sulphuric acid/phosphoric acid [972].

MARČEC, KINSON and BELCHER [819] determined traces of tellurium in steels and iron by precipitating with tin(II) chloride and extracting the diethyl dithiocarbamate complex with amyl acetate. The subsequent atomic absorption analysis is interference-free. BARNETT and KAHN [75] described a procedure for the determination of tellurium in steel without prior extraction. 2 g steel are dissolved in 25 ml of an acid mixture consisting of two parts concentrated nitric acid and one part 1+1 sulphuric acid and then diluted to 100 ml. Since a marked non-specific light loss occurs with this solution at the 214.3 nm resonance line, this background absorption must be corrected to prevent errors. The authors employed a deuterium background corrector (see page 109). KAHN [606] also gives a good survey on the application of atomic absorption spectroscopy in the metallurgical laboratory.

Atomic absorption spectroscopy is equally suitable for the determination of trace elements and main components in alloys. Only the techniques and the accuracy requirements are different. SATTUR [1079] has examined numerous non-ferrous alloys routinely by atomic absorption spectroscopy. He determined copper, iron, manganese and nickel in copper alloys in which the copper was present to more than 80% in some cases. He similarly determined aluminium, lead, tin and zinc containing alloys. GIDLEY and JONES [414] [415] analysed copper alloys for zinc, ELWELL and GIDLEY for lead [328] and manganese [7], and CAPACHO-DELGADO and MANNING [206] [208] determined tin in brass and lead alloys using an air/acetylene flame because chemical interferences were stronger in an air/hydrogen flame. SPRAGUE, MANNING and SLAVIN [1167] determined selenium and tellurium in various copper alloys. NAKAHARA and co-workers determined beryllium in copper alloys with a nitrous oxide/acetylene flame and found no interferences other than those from palladium and silicon [904].

Similar to the analysis of steel, the main components in alloys can only be determined with the required accuracy when baseline suppression and scale expansion are employed. The use of a less sensitive resonance line to avoid excessive dilutions [1299], and to make comparative analyses against standard alloys of similar composition are also frequently recommended.

LEWIS, OTT and SINE [747] have recommended atomic absorption spectroscopy as the standard method for the determination of lead in nickel alloys. They determined lead in concentrations of 0.0001% downwards and found no chemical interferences. DYCK [316] determined chromium, magnesium and manganese in nickel alloys and found that atomic absorption was rapid and reliable. ANDREW and NICHOLS [54] determined magnesium in high purity nickel and in nickel alloys with a standard deviation of 0.00005%. BURKE found that the concentrations of antimony, lead, bismuth and tin in nickel are too low for a precise direct determination so he employed co-precipitation with manganese(IV) oxide for enrichment [179]. WELCHER and co-workers dissolved nickel alloys in nitric acid/fluoric acid for the determination of lead, selenium, tellurium, thallium and bismuth since hydrochloric acid completely suppresses the signals for bismuth and thallium and partially for lead, and sulphuric acid influences the sensitivity of tellurium [1291]. The agreement with emission analyses and X-ray fluorescence was good. ARENDT described two procedures for the determination of niobium in nickel alloys [60] and NEWLAND extracted bismuth from nickel alloys with MIBK before the determination [911].

WELCHER and KRIEGE [1289] developed a special procedure for the precision determination of Al, Cr, Co, Fe, Mo, Nb, Ta, Ti, V and W in high temperature nickel alloys. The agreement with values obtained by the tedious wet chemical method was good. Furthermore, the accuracy and precision were checked by repeated determinations on a standard alloy. CARPER [212] determined the three main components in iron-nickel-cobalt alloys. The composition of the alloy is particularly critical with respect to its magnetic properties. 150 mg sample are dissolved in 5 ml *aqua regia* and diluted to 500 ml for the determination of cobalt. The solution is further diluted 1:4 for iron and 1:20 for nickel. The measurement is against aqueous standard solutions, which in each case contain the other two alloy components in the corresponding concentrations. The coefficient of variation for cobalt (about 3%) is 1.5% relative, for iron (about 17%) 0.28% relative and for nickel (about 80%) 0.21% relative. The agreement with classical chemical analysis, which is more time consuming, is excellent. No chemical interferences were observed when the standard solutions contained Ni—Fe—Co in the ratio 80:17:3. WELCHER and KRIEGE determined the main elements in high temperature cobalt alloys after solution in hydrochloric acid/nitric acid/fluoric acid. The standards did not have to be matched exactly to the sample solution; only slight interferences occurred and the agreement with certificate values was good [1290].

QUARRELL and co-workers determined antimony and tin in lead alloys used for cable screening [1011]. They dissolved the sample in a mixture of fluoroboric acid and hydrogen peroxide and determined antimony in the air/acetylene flame and tin in the nitrous oxide/acetylene flame directly from this solution. BELL has drawn attention to the necessity of a rapid determination of the components in lead alloys in the electrical industry and also that lead and tin are not soluble in the same medium [104]. Mixtures of fluoroboric acid, hydrogen peroxide and di-sodium EDTA dissolve lead alloys rapidly at room temperature, and after dilution, the solutions can be aspirated directly into the flame. SKERRY and CHAPMAN dissolved tin-lead solders in hydrochloric acid/nitric acid by heating the solder in the form of fine turnings to 195°C and adding to the acid under stirring [1131]. HWANG and SANDONATO determined numerous trace elements in a tin-lead solder after dissolving in fluoroboric acid/nitric acid [550].

BELL [102] [103] investigated the application of atomic absorption spectroscopy for the analysis of aluminium alloys in detail and found that Cd, Cr, Cu, Fe, Mg, Mn, Ni and Zn could be determined free of interferences. The sample is simply dissolved in 1+1 hydrochloric acid, with the addition of a few drops of hydrogen peroxide if required, and correspondingly diluted. For the determination of alkaline-earth metals the addition of lanthanum or the use of a nitrous oxide/acetylene flame is recommended. WILSON [1339] determined silver traces in aluminium alloys and compared the results with the conventional method of VOLHARD [1274]; the agreement was very good. MANSELL and co-workers [816] likewise found good agreement with wet chemical procedures for the determination of Ca, Cu, Mn and Zn in magnesium alloys. Since the majority of alloys also contain aluminium, they added lanthanum to the samples. Besides other elements, WILSON [1340] determined chromium and zirconium in aluminium alloys and found good agreement with titrimetric or gravimetric values. ELROD and EZELL [327] made a more comprehensive study on the determination of chromium in aluminium and aluminium oxide materials and found that potassium peroxydisulphate eliminated all the interferences owing to larger amounts of aluminium, iron or titanium in the air/acetylene flame. The samples were dissolved in 1+1 sulphuric acid and then sufficient potassium peroxydisulphate was added before dilution so that the end solution contained 1%. The determination could be made against aqueous chromium standard solutions. The agreement with certificate values was very good and the day-to-day precision of atomic absorption was better than the diphenylcarbazide method by about a factor of three.

CAMPELL [204] determined silicon in aluminium alloys; he dissolved 0.5 g sample in 10 ml 20% sodium hydroxide solution with subsequent oxidation of the silicon with hydrogen peroxide. The solution was acidified, made up to 100 ml and directly aspirated. The solution was further diluted with higher silicon contents. After the standard solutions had been matched with the main components, an agreement of 1% relative could be achieved with wet chemical procedures. NAKAHARA and co-workers determined beryllium in aluminium and magnesium alloys in a nitrous oxide/acetylene flame and found no interferences other than those caused by palladium and silicon [904].

BARTON, JOHNSON and SHEPHERD [86] determined calcium in calcium-lead alloys with a standard deviation of 0.3% relative and found by recovery assays a mean of 100%.

MYERS [538] determined aluminium, vanadium and iron in titanium alloys, and iron in zirconium alloys. The samples were dissolved in fluoric acid and oxidized with a little nitric acid. The mean deviation over several months was $\pm 0.10\%$ for aluminium and vanadium and $\pm 0.002\%$ for iron in the titanium alloys. CAPACHO-DELGARDO and MANNING [206] [208] determined tin in zirconium by dissolving the metal in a mixture of HBF_4 and HCL.

SCHLEWITZ and SHIELDS determined chromium, copper, iron, nickel and tin in zirconium alloys and found that the matrix must be matched for an exact analysis [1090]. For the determination of tantanlum in zirconium alloys, they recommended an extraction with MIBK [1091].

THORMAHLEN and FRANK determined silicon in niobium alloys which they dissolved in sulphuric acid/hydrogen peroxide [1228].

LIEBER determined the main components in silver alloys after dissolving them in hydrochloric acid/nitric acid [749], and SCHAEFER and VOMHOF determined gold in jewelry alloys after solution in *aqua regia* in good agreement with gravimetry and X-ray fluorescence, and with high precision [1084].

Besides the analysis of alloys, atomic absorption spectroscopy is also suitable for the determination of traces in high purity metals. In general, two precautionary measures must be taken. First, slight background effects can occur when high total salt concentrations are aspirated, they must be taken into account; and second, the method of additions should be employed to avoid the influences of viscosity (physical interferences). Further, attention must be drawn to the risk of aspirating high concentrations of copper and silver into a mixing chamber burner with acetylene as the fuel gas since acetylide formation can take place, leading to an explosion in the burner system. In such cases, the removal of the main component or dilution is recommended if a fuel gas other than acetylene cannot be used.

McCRACKEN and co-workers [848] have reported the determination of traces of tin, cadmium and zinc in copper after previous separation of the copper by means of ion exchangers, and GOODWIN determined silver in copper without separation and found that only nitrose gases interfere and must therefore be boiled off before the analysis [432].

NEUMANN determined chromium, iron, cobalt, manganese and nickel in high purity molybdenum and tungsten and found no interferences [908], while for the determination of calcium, either the addition of cesium and strontium was necessary [907] or, as for the determination of aluminium, the use of a nitrous oxide/acetylene flame [909]. In all cases prior separation was not required.

SCARBOROUGH and co-workers [1083] have described a procedure for the determination of chromium, cobalt, iron and nickel in concentrations under 10 ppm in high purity sodium metal. The trace elements under study are precipitated together with added lanthanum at pH 9 and thus separated from sodium. The precipitate is dissolved in hydrochloric acid and determined by atomic absorption.

PANDAY and GANGULY [947] determined chromium, manganese and nickel in bismuth and tellurium metals and found marked background absorption which they eliminated by removing the main components. ROTH, BOHL and SELLERS [1061] determined traces of tellurium in high purity bismuth by extracting with MIBK from 6 N hydrochloric acid solution and aspirating the organic layer directly. The method is accurate and rapid; the effectiveness of the extraction is 98–99%.

The determination of trace noble metals in other noble metals does not appear to be completely free of interferences; various authors use additions of 0.5% copper and 0.5% cadmium [1114], 0.5–1% lithium [485] [1234] or 1% sodium and 1% copper [57] to eliminate various interferences. ADRIAENSSENS and VERBEEK found that in a 2% potassium cyanide solution, most of the interelemental interferences observed with noble metals in acid solutions did not occur [17] and they determined gold, palladium, platinum and silver without interferences and with good precision in cupellation granules [18]. KALLMANN and HOBART also determined gold, palladium and silver in noble metal beads from smelting processes in cyanide solution with an accuracy of 1% [623]. The graphite tube furnace also appears to be ideally suited to noble metal analyses since interelemental interferences hardly occur [16]. Various authors have drawn attention to the risk of background absorption during noble metal analyses in the flame which must be taken into consideration [324] [687]. Two good reviews on the analysis of noble metals have been published by SEN GUPTA [1116] and BEAMISH and VAN LOON [90].

Of interest in connection with the analysis of high purity metals is the enrichment procedure for trace metals on activated carbon or mercury described by JACKWERTH and co-workers. Activated carbon can act as a trace adsorber for enrichment of numerous metals from high purity aluminium, silver, tungsten [563], cadmium or magnesium [562]. For the subsequent determination, the activated carbon suspended in diluted nitric acid can be aspirated directly into the flame.

When two metals are in electrical contact, the more base of the two determines the potential and will therefore alone be attacked by oxidizing agents. If a piece of metal coated with a layer of mercury is dissolved in acid, the trace impurities are enriched while the main component passes into the amalgam layer. The "end point" of the solution can be clearly identified potentiometrically. This enrichment procedure has been successfully employed for the determination of numerous trace elements in high purity aluminium [519], cadmium [561] and zinc [560].

The determination of trace elements in uranium by atomic absorption spectroscopy has been frequently reported. TAKAHASHI and URUNO [1215] determined magnesium after extraction of the main bulk of uranium; HUMPHREY carried out the same determination but without separation of the uranium [541]. He employed the method of additions to avoid interferences. JURISK [600] determined nickel, iron and aluminium in good agreement with colorimetric procedures. SCARBOROUGH [1082] analysed uranium alloys for molybdenum, ruthenium, palladium and rhodium; when the corresponding quantity of uranium had been added to the standard solutions, the four elements did not mutually interfere. SHEPHERD and JOHNSON determined iron and nickel in uranium after separation of the uranium on an ion exchanger, either directly or according to the method of additions. The latter method permitted very accurate determination of the smallest contents.

WALKER and VITA [1280] determined Al, Ca, Cd, Co, Cr, Cu, Fe, Mg, Mn, Na, Ni, K, Pb and Zn in uranium and uranium compounds. The sample is dissolved in 6–8 N nitric acid and the uranium is selectively separated with tributylphosphate. The aqueous layer is evaporated to dryness and the residue dissolved in 0.2 N hydrochloric acid and directly aspirated. Atomic absorption spectroscopy is approximately as accurate as colorimetry and much more accurate than emission spectroscopy. PAGLIAI and POZZI separated uranium by reversed phase separation chromatography (solid phase: KEL-F, stationary phase: TBP) and determined ten elements without problems by atomic absorption [946].

PICHOTIN and CHASSEUR [975] determined magnesium, calcium, lithium and potassium in plutonium metal after extraction of the plutonium with tri-n-octylphosphine and dibutyl-phosphoric acid.

Plating baths are mostly suitable for direct analysis by atomic absorption spectroscopy after dilution; however, since the determinations are mainly simple, little has been published on this application of the method. SHAFTO [1119] reported the determination of copper, iron, lead and zinc in nickel baths, and WHITTINGTON and WILLIS [1318] analysed various plating baths for trace elements. KOMETAIN [694] determined the concentrations of silver, cobalt, copper, iron, nickel and lead in gold baths; he further employed atomic absorption spectroscopy for the determination of the composition of plating baths, diffusion data and the thickness of the electroplated coatings, etc. KRAFT [701] described the application of atomic absorption for determining cobalt in gold baths, silver, copper and iron in chromium, nickel and tin baths, and copper in copper cyanide baths.

VAN LOON [1253] determined iridium in a noble metal multicomponent concentrate with satisfactory precision, and EZELL [336] reported the indirect determination of chloride in liquid industrial wastes via precipitation with silver nitrate. EZELL found that the precision with atomic absorption is better than with titrimetric methods. KUNISHI and OHNO determined sulphate in a rhodium bath. They exploited the suppressing effect of sulphate on the signal for iron and its elimination by lanthanum for the determination of sulphate [717].

KAPETAN [624] reported the determination of cobalt and impurities in gold baths. Before the introduction of atomic absorption, cobalt was determined polarographically, and copper, iron and lead by optical emission. The replacement of both these methods by atomic absorption saves a considerable amount of time. The trace elements are determined directly from the undiluted bath, and for cobalt the bath is diluted 1:20.

9.6. Petrochemistry

The application of atomic absorption spectroscopy ranges from the analysis of crude oil through the control of additives and impurities in the most varying intermediate and end products to the determination of wear metals in used lubricating oil. Atomic absorption offers the great advantage that the samples can be diluted virtually without exception with a suitable solvent, and then directly aspirated into the flame. The calibration substance must frequently be chosen with great care, otherwise false measurements can occur. For the majority of oil analyses, oil soluble metal organic salts, mostly cyclohexane-butyrates, etc., which are offered by various manufacturers, are suitable.

The determination of nickel, which is a strong catalyst poison, and vanadium, which leads to corrosion problems, in crude oil has been described by various authors after the introduction of the graphite tube furnace [222] [923] [933]. While the majority introduce the sample into the graphite tube either directly or after dilution and then carry out the thermal pre-treatment, Eskamani and co-workers preferred ashing with a magnesium sulphonate at 650 °C before introducing the sample into the graphite tube [332].

The determination of catalyst poisons in crude oil and distillation residues was one of the first applications of atomic absorption in petrochemistry. Barras [79] [80] examined oil sump for nickel, copper and iron and found that the results with atomic absorption compared favourably with other analytical procedures. Kerber [633] determined nickel in gas oil diluted 1:10 and examined a large number of solvents for their suitability and combustion properties in the flame. He found that numerous aliphatic hydrocarbons, alcohols and ketones did not dissolve the sample completely, and that although aromatic hydrocarbons were good solvents, they burnt with yellow, sooty flames which lead to poor results. p-Xylene was finally established as a good solvent with acceptable combustion properties. Trent and Slavin [1243] likewise determined nickel in gas oil and similar substances and found that in the trace range the stray light must be corrected since values which are too high are otherwise obtained. These authors also employed p-xylene as solvent, but diluted 1:5, and used the method of additions.

Moore and co-workers [878] used dioxan to dilute oil samples. Other authors [127] had difficulty with this solvent as it has a tendency to freeze on the nebulizer jet. Robinson [1037] determined sodium in gas oil and found no interferences from potassium. Capacho-Delgado and Manning [207] determined vanadium in gas oil with a nitrous oxide/acetylene flame. By employing a 1:3 dilution and the method of additions, they were able to detect 0.05 µg/ml vanadium in the oil.

Frequently, barium, calcium and zinc salts are employed as additives in lubricating oils. Since they are always present in relatively high concentrations, their determination by atomic absorption presents no difficulties; the sample must only be correspondingly diluted. Mostyn and Cunningham [889] determined zinc in lubricating oils with an air/acetylene flame and n-heptane as the diluting solvent.

For these analyses, Slavin [12] recommended MIBK as solvent and the use of a nitrous oxide/acetylene flame for the determination of calcium and barium additives. The determination is free of chemical interferences in this flame; only 0.2% sodium should be added to sample and standard solutions to suppress ionization interferences.

Binding and Gawlick [127] examined various solvents for the analysis of barium, calcium and zinc additives. n-Hexane gave an unsteady display and cyclohexane froze behind the nebulizer jet. Barium additives were either poorly soluble or insoluble in toluene and MIBK. n-Heptane proved to be suitable, but the authors recommended iso-octane as the ideal solvent for this application. Barium and calcium were determined in a nitrous oxide/acetylene flame and the sodium salt of petroleum sulphonic acid was added to suppress ionization. The authors made a detailed investigation of the emission of the nitrous oxide/acetylene flame in the regions of the calcium and barium resonance lines and found that increasing the lamp current improved the signal-to-noise ratio for the barium determination considerably. The authors employed two nebulizers to be able to work with only a single dilution of the oil samples. One nebulizer adjusted for optimum performance served for the less sensitive barium and a second nebulizer having only about $^{1}/_{10}{}^{th}$ of the aspiration performance served for the more sensitive calcium and zinc.

PETERSON and KAHN diluted the samples with xylene and also determined these three elements in a nitrous oxide/acetylene flame, but reduced the sensitivity for zinc by turning the burner head through 90° [974]. SALMAMA employed less sensitive resonance lines for all three elements and was also able to work with a single dilution [1078].

SUPP determined antimony in additives by dissolving them in MIBK under addition of 2-ethylhexanoic acid and aspirating directly [1205]. He found good agreement with the wet chemical method and no interferences from zinc or barium. Antimony dialkyl-dithiocarbamate and dialkylphosphorodithionate give oils excellent anti-wear, high pressure and anti-oxidation properties. The author also determined lead in lead diamyldithiocarbamate, a much used additive for lubricating oils and greases, by this method [1206]. NORWITZ and GORDON determined lithium in sebacate-based lubricants which, because of their high and low temperature properties, gain continuously in importance. Such lubricants typically contain 8% lithium stearate and 1.5% barium petroleum sulphonate. The sample is treated with hydrochloric acid, extracted with ether and the aqueous phase evaporated off with perchloric acid and then analysed by atomic absorption [921].

MOSTYN and CUNNINGHAM [889] determined molybdenum, which is in the form of the disulphide in greases, after prior ashing in a quartz crucible. Unpublished investigations in the author's laboratory have shown that molybdenum disulphide suspensions in oil can be directly determined in a nitrous oxide/acetylene flame, but that the height of the signal is strongly dependent upon the particle size. Further, after dilution and during aspiration the suspension must be magnetically stirred since separation otherwise occurs. KÄGLER [601] [602] determined lithium, sodium and calcium in greases, and sodium in heating oil and found that atomic absorption spectroscopy is considerably simpler than X-ray fluorescence and flame emission spectroscopy. 10–200 mg of the grease are homogenized with 4–8 ml butanol, made up to 100–200 g with 1 N hydrochloric acid and well shaken. Lithium, sodium and calcium are determined directly from the HCl extract. Heating oil is diluted with either toluene or a toluene/methanol mixture (8 : 2) and aspirated directly. For trace analyses of sodium or vanadium in light fraction heating oils, the author's own researches have shown that this can be directly aspirated either undiluted or diluted 1 : 2.

KÄGLER has made a detailed examination of the determination of sodium in various mineral oil products with respect to instrument parameters and solvent influences [603].

A relatively difficult analytical problem is the determination of traces in light fraction fuel oils used for gas turbines since they must satisfy stringent purity requirements. Particularly vanadium must be reliably determined in concentrations below 0.1 µg/g. The graphite tube furnace is particularly suitable for these analyses; nevertheless, it must be fitted with a graphite tube which prevents the oil from running during the thermal pre-treatment. Also, a temperature programme with a slow rate of temperature increase is important for the removal of the matrix. When these requirements are met, 0.01 µg/g vanadium can be determined in fuel oil [1296].

ESKAMANI and co-workers determined selenium in petrol and other petrochemical products in a graphite tube furnace after the samples had been digested in a Parr bomb [331].
MILLER and co-workers [866] determined silicon in traces of silicone grease and found that atomic absorption spectroscopy was ideal.
In 1961, ROBINSON [1038] reported the determination of lead in petrol by atomic absorption and found it rapid, interference-free and selective. He diluted petrol with iso-octane to obtain a more favourable measurement range. Later, a number of authors found differences in the absorbance for lead tetraethyl and lead tetramethyl which brought difficulties, especially with mixtures. TRENT [1239] examined the determination and found that lead tetramethyl in heptane, MIBK or toluene gave a signal two to three times higher than lead tetraethyl. The length of the nebulizer aspirating capillary was also not without importance. Finally, TRENT employed a 25 cm long capillary and iso-octane as diluent since no differences between the two additives could then be observed. Lead tetraethyl required 30 seconds to give a stable signal, while the full signal height is obtained immediately with lead tetramethyl. MOSTYN and CUNNINGHAM [889] therefore used an accurately measured aspiration time of 15 seconds to avoid waiting until equilibrium had been reached. With this procedure the authors achieved a coefficient of variation of 1.6%.
KASHIKI and co-workers found that lead tetramethyl, lead tetraethyl and lead alkyls could be determined by calibration against standard lead alkyls when 3 mg iodine were added to 1 ml petrol sample and diluted to 50 ml with MIBK before aspiration [628]. Later, the same authors described this procedure for the graphite tube furnace to facilitate determinations at the lower concentration range (0.02–0.1 μg Pb/ml). Also in this case, all the lead compounds examined exhibited the same sensitivity after the addition of iodine [627].
KOLB and co-workers [692] used a combination of gas chromatography/atomic absorption spectroscopy for the separation and element specific determination of both the lead additives in petrol. In this case atomic absorption served as a highly specific detector for gas chromatography by only responding to the metallic elements.
Later, SEGAR described this procedure for the graphite tube furnace. He conducted the eluate from the gas chromatograph directly into a continuously heated graphite tube at 1500–1700 °C and was able to achieve good separation of five lead alkyls [1108] [1110].
MOORE and co-workers [878] determined nickel and copper and MOSTYN and CUNNINGHAM [889] determined zinc in petrol by aspirating it undiluted. BARTELS and WILSON [84] examined jet fuels for manganese which is added to these fuels as methyl-cyclopentadienyl-manganesetricarbonyl to eliminate the typical vapour trails. When undiluted fuel is aspirated, additives in the range 0.02–0.3 weight percent can be directly determined by atomic absorption. Higher concentrations can be determined after dilution.
Within the last few years the determination of wear metals in lubricating oils has become an important application of atomic absorption spectroscopy. The rapidity and high precision of these determinations permit continuous monitoring of engines, gears and other machine parts. Frequently, the element specific examination of a lubricating oil enables early recognition and often localization of incipient damage to be made.
Usually the lubricating oil is diluted 1:10 with MIBK and aspirated directly. Calibration is mainly against oil-soluble standards (cyclohexanebutyrates, etc.) or against naphthenates [853]. SPRAGUE and SLAVIN [1138] [1173] developed an automated procedure for the determina-

tion of Fe, Ni, Cr, Pb, Cu, Ag, Sn, Mg and Al in aircraft lubricating oils. With a sample frequency of 14 seconds a reproducibility of 3% could be achieved. BURROWS and co-workers [187] compared atomic absorption spectroscopy and colorimetry for the examination of machine oils which were ashed before the determination. The agreement was good. Similar results could also be obtained by atomic absorption without ashing, however. This is because suspended metallic components in lubricating oils, as well as dissolved ones, can be determined by atomic absorption so long as the particles are not too large.

KRISS and BARTELS thoroughly examined the effectiveness of the nebulizer/burner system for the determination of suspended particles in oil. For iron particles of 3.0 µm and 1.2 µm, the recovery was only in the order of a few percent [83] [709]. Nevertheless, the authors state that such particle sizes are hardly present in genuine oil. Further, they found that the effectiveness of the atomization is independent of the flame and that problems with the reproducibility come from the oil and not from the flame [710]. The authors have reported a procedure with which the large particles can also be determined by atomic absorption. The oil is diluted with 14 ml MIBK and 5 ml of a mixture consisting of 80% methanol, 5% water and 15% concentrated hydrochloric acid. For the 1.2 µm particles the recovery is 100% after having stood an hour, but for the 3 µm particles the waiting time is 16 hours [709].

With lubricating oil analyses, it is usually not the absolute determination of the metal concentration which is of importance, but only typical increases in the relative concentration which can indicate incipient damage or damage that has already occurred. A widely distributed monitoring network for a heavy contractors plant, for example, has been developed according to this system and has proved of great use [1279]. KAHN and co-workers have also shown that no statistical deviation can be observed between atomic absorption and emission for oil analyses [615].

The analysis of wear metals in lubricating oil in the graphite tube furnace has been described by two groups of workers [165] [233]. However, this does not appear to be a typical application since it is neither a trace nor a micro determination and the flame is in all cases more rapid.

9.7. Glass, Ceramics, Cement

With glass and more especially with ceramic materials, it is not the actual analysis which is the problem, but the solution of the samples. Glass is normally dissolved in fluoric acid and the silicon removed since it is usually not of interest. The subsequent analysis by atomic absorption is very simple since the total salt concentration is lower. Ceramics are frequently dissolved in fluoric acid, also often in combination with perchloric acid, etc. Occasionally small PTFE autoclaves can be of use [964]. Failing this, alkaline fusions must be used. However, under certain circumstances, these can make the atomic absorption analysis difficult. ADAMS and PASSMORE [13] [14] determined magnesium, calcium and strontium in glass and ceramics and also the alkali metals in ceramics. The chemical interferences which are expected for the alkaline-earth elements in the air/acetylene flame were eliminated with a mixture of lanthanum and EDTA. JONES [584] determined Li, Na, K, Ca, Mg, Ba, Cu, Mn, Fe, Co, Ni, Pb and Zn in various glasses and found good agreement with both certified values and photometric results. PASSMORE and ADAMS determined iron, zinc [959] and copper [960] in numerous glasses and found good agreement with colorimetric results. SLAVIN and co-workers [1148] determined arsenic in glass, and BOBER and MILLS [132] analysed lead glasses for lead in good agreement with chemical spectrochemical procedures. CHARLIER and MAYENCE determined chromium, cobalt, copper, manganese and nickel in glasses against matched standard solutions without interferences after fuming off with fluoric acid/perchloric acid and taking the residue up in 5% hydrochloric acid. The detection limits were around 0.5% metal in glass [227].

Various authors have examined the application of the graphite tube furnace for glass analyses. FULLER dissolved high purity quartz in fluoric acid, volatilized off the silicon tetrafluoride and the acid in the graphite tube and determined iron and copper down to 0.01 µg/g in good agreement with colorimetry [392]. He also determined chromium, iron, cobalt, copper, manganese, nickel and vanadium in glasses by fuming them off to dryness with fluoric acid/perchloric acid, dissolving the residue in warm perchloric acid and extracting the trace elements with MIBK for subsequent determination in the graphite tube furnace [396]. WOOLLEY determined iron and copper in the high purity glasses required for the production of fibre optics in the graphite tube furnace [1357]. The sample preparation is of particular interest: The sample is pulverized by an electro-hydraulic pressure wave which is generated in the water contained in the sample by a spark. The sample is then dissolved in a PTFE autoclave in which the silicon is separated.

A determination carried out frequently nowadays in connection with environmental control is the testing of glazed ceramic surfaces for their release of toxic heavy metals. The vessel is allowed to stand in 4% acetic acid for half an hour and the metals of interest then determined. KRINITZ and FRANCO increased the induction period to 24 hours and found an asymptotic increase in the metal concentration. In an interdisciplinary test, coefficients of variation of 4–5% were obtained for lead and cadmium [706].

MUNTZ determined yttrium in hafnium and zirconium oxides in good agreement with other methods after the samples had been dissolved in fluoric acid/sulphuric acid [896]. VAN LOON [1252] has described a procedure for the determination of aluminium in materials with high silicon contents. He found good agreement with certified values. BELCHER [97] dissolved tungsten carbide in orthophosphoric acid/hydrochloric acid under addition of nitric acid and determined iron in these solutions in good agreement with X-ray fluorescence.

The determination of secondary ingredients and trace elements in cement was carried out at an early stage by atomic absorption. CAPACHO-DELGADO and MANNING [209] determined Fe, Mg, Mn, K, Li, Na, Sr, Al and Ti in cement against aqueous standard solutions in which the calcium concentration had been matched. For Mg, Sr, Al and Ti the nitrous oxide/acetylene flame was employed. MONTAGUT-BUSCAS and co-workers [875] determined cobalt after extraction with MIBK in cement to be used for the construction of reactors.

While the routine accuracy of atomic absorption of approximately 1–3% is sufficient for the determination of the secondary ingredients in cement, calcium must be determined with the greatest accuracy. CROW, HIME and CONNOLLY [244] found after a comprehensive study that it is possible to achieve an accuracy of 0.2% by this method when using standard cements for calibration. The authors give complete methodology for the analysis of all the important elements in cement by atomic absorption.

TENOUTASSE determined ten elements including calcium in cement after a digestion in fluoric acid/hydrochloric acid in a PTFE autoclave. No foreign ions (except boric acid) were added and all ions could be determined from one solution [1222]. FIERENS and DEGRE have described an automatic procedure for the determination of 11 elements in cement, raw materials and clinker products [367].

9.8. Plastics, Textiles, Paper

SLAVIN [1141] reported the determination of manganese in plastics after prior wet ashing with concentrated sulphuric acid. To avoid matrix influences, he employed the method of additions. FARMER [339] determined iron in nylon 66; he dry ashed the sample, dissolved the residue in hydrochloric acid and extracted with APDC/MIBK. DRUCKMAN [311] analysed polypropylene for titanium, aluminium and iron and found that atomic absorption spectroscopy had decisive advantages over earlier procedures. A sample of the polymer is ashed at 800 °C and taken up in molten sodium carbonate. This is dissolved in 3 N sulphuric acid and aspirated directly. This procedure can also be extended to other plastics (polyethylene, polystyrene, PVC) and other metals such as magnesium, copper and zinc. PARALUSZ [950] determined silicon in various polymers and made a comprehensive study of various standards and hollow cathode lamps; the most varying analytical methods were also compared and atomic absorption spectroscopy proved to be especially suitable.

OLIVIER [924] employed atomic absorption for the determination of 20 metallic elements in a multitude of polymers and he gives detailed methodology for the sample preparation. Soluble polymers are dissolved in a solvent suitable for atomic absorption spectroscopy, such as MIBK, DMF [925], cyclohexanone, formic acid, etc., and aspirated directly. Insoluble polymers are digested in sulphuric acid/hydrogen peroxide; volatile metals are precipitated twice with subsequent enrichment. The author emphasizes that atomic absorption spectroscopy permits rapid and accurate determination of metals in polymers both in the ppm and % ranges.

PRICE determined antimony, lead, iron, cobalt, copper, manganese, tin and zinc in raw materials and in polymers such as rayon, nylon and polyesters directly in the flame rapidly and accurately after solution in sulphuric acid/hydrogen peroxide [1002]. KERBER determined gold in polyester fibres [634], aluminium and iron in polypropylene, and copper and iron in fluorinated hydrocarbon polymers [637] by directly introducing the samples into the graphite tube furnace; he obtained good agreement with flame atomic absorption, colorimetry and emission.

DELLA MONICA and MCDOWELL determined chromium in leather and found that compared to alkaline fusions, atomic absorption is reliable and brings a considerable saving in time [291].

HARTLEY and INGLIS [478] described the determination of aluminium in wool by atomic absorption. The sample is dissolved in boiling hydrochloric acid and the extract is aspirated directly into the flame. The detection limit is 0.02% aluminium in the wool. SIMONIAN [1128] determined copper in textiles using the same procedure and found good agreement with colorimetry and electrolysis. OLIVIER [924] decomposed wool in 5% sodium hydroxide solution or in concentrated hydrochloric acid, and cotton in 72% sulphuric acid and determined numerous elements in the resulting solutions.

BETHGE and RÅDESTRÖM [118] employed atomic absorption to determine copper, iron, manganese, calcium, magnesium and sodium in cellulose and found the accuracy and precision were comparable to those obtained by conventional methods. If a large number of samples are to be determined, or several metals in one sample, atomic absorption spectroscopy saves a great deal of time. The samples were dry ashed, taken up in 6 N hydrochloric acid and aspirated directly. ANT-WUORINEN and VISAPÄÄ [59] have described a procedure for the determination of Na, K, Mg, Ca, Fe, Cu and Zn in various cellulose samples. The influence of various methods of pre-treatment was examined in detail and the procedure was checked with X-ray fluorescence analysis. PARALUSZ [950] investigated the content of organic silicon compounds in paper by extracting them with petrol, evaporating the solvent, taking up the residue in a minimum of MIBK and aspirating directly.

LANGMYHR and co-workers determined cadmium, copper, lead and manganese in paper and paper pulp directly in the graphite tube furnace by introducing the solid or slurried sample and employing a background corrector [732]. KERBER and co-workers found good agreement with flame atomic absorption for the determination of iron, copper, manganese and silicon in paper in the graphite tube furnace [637].

9.9. Miscellaneous Industrial Products

Atomic absorption spectroscopy is suitable for numerous analytical problems in industry, and because of its simplicity, sensitivity and specificity it can be universally applied. Frequently, however, these numerous "secondary tasks" of atomic absorption are either not reported or only reported in passing.

A relatively frequent application of atomic absorption is the analysis of traces in ultra pure chemicals. AGAZZI [20] determined zinc in hydrogen peroxide to which it is added as a stabilizer. The range of interest is 2–50 mg/l hydrogen peroxide. LUNDY and WATJE [772] examined fuming nitric acid for Al, Fe, Cr, Ni and Zn. The sample was evaporated almost to dryness and then taken up in water. The results with atomic absorption were in satisfactory agreement with the time-consuming colorimetric method. YANAGISAWA and co-workers [1360] have described a procedure for the determination of calcium in phosphoric acid and its salts. Calcium is complexed with hydroquinoline and extracted with 3-methylbutan-1-ol. The extract is aspirated directly and the calcium concentration determined by the method of additions. BEDROSIAN and LERNER [94] determined Na, K, Ca, Mg, Fe, Mn, Cu, Ni and Cr in elemental boron and obtained a relative standard deviation of 4% or better. Atomic absorption compared well with emission and the relative error of the latter was frequently the greater. SPITZER [1146] analysed zinc oxide for Cu, Pb, Fe, Mn, Ni and Ca. The results agreed well with those for chemical analysis. After the standard solutions had been matched to the zinc concentration (40 g/l), all the elements could be determined free of interferences. This method is reliable and convenient.

HARRINGTON determined ruthenium in hydrated ruthenium chloride and titanium in tetra-ortho-butyltitanate in good agreement with gravimetric methods [470]. OLIVIER examined caustic soda for mercury using a flameless procedure [926]. REVERSAT determined calcium, chromium, iron, copper, magnesium and nickel in single crystals of lithium hydroxide in the graphite tube furnace [1033].

The analysis of catalysts on support materials without prior solution has been reported by various authors. KASHIKI and OSHIMA aspirated a suspension of aluminium oxide catalysts in methanol directly into the flame and determined cobalt and molybdenum in good agreement with a digestion method [626]. COUDERT and VERGNAUD determined platinum on activated carbon or aluminium oxide by introducing the powder evenly into the gas mixture with a screw feed, so guaranteeing pneumatic transport to the flame [240]. JANOUŠKOVÁ and co-workers also determined platinum on aluminium oxide, but carried out the determination in the graphite tube furnace and only found slight interferences [570].

DALRYMPLE and KENNER [265] determined calcium in various pharmaceutical preparations and found good agreement between atomic absorption and complexometric titrations. Zinc in crystalline insulin and insulin preparations was determined by SPIELHOLTZ and TORALBALLA [1162] by atomic absorption spectroscopy and various other procedures. Atomic absorption spectroscopy is more rapid and simpler for the same or better precision of the results.

LEATON [736] has reported the application of atomic absorption spectroscopy for the analysis of pharmaceutical products. As, Ca, Co, Cu, Fe, Mg, K and Hg were determined in solutions, powders and tablets with good accuracy. To reduce sample preparation to a minimum, the single substances were only dissolved in acid and aspirated directly into the flame.

KARAKHANIS and ANFINSEN determined aluminium in pharmaceuticals by dissolving or leaching the samples with concentrated hydrochloric acid and aspirating directly after dilution. The method was free of interferences and in good agreement with complicated procedures [625]. TARLIN and BATSCHELDER determined iron in tablets and capsules in agreement with colorimetric, volumetric and gravimetric procedures [1216], and MOODY and TAYLOR determined zinc in numerous pharmaceutical preparations with good accuracy and precision [877].

A good review on the determination of 15 elements in the most varying pharmaceutical products has been published by SMITH [1157].

SEARLE and co-workers [1103] employed atomic absorption spectroscopy to determine lead in old paints and obtained a standard deviation of 0.05% lead with this method, while with gravimetry the standard deviation was 0.3% with a five times higher sample requirement. HENN used the DELVES system for the same determination and found good agreement between this procedure and prior ashing [502].

MOTEN determined chromium with a detection limit of 5 ppm in triphenylmethane paint additives [890]. BRANDT has reported the universal application of atomic absorption to the analysis of raw materials, pigments, polymers and additives [156] and EIDER found that the main advantage of atomic absorption is the possibility of directly analysing metallo-organic compounds and complex mixtures without prior separation. This signifies a considerable saving of time compared to wet chemical analyses for the same precision [323].

FULLER employed the graphite tube furnace for the determination of aluminium, iron, copper and manganese in titanium dioxide pigments [392]. The sample was dissolved in fluoric acid and introduced directly into the graphite tube; the agreement with flame atomic absorption and X-ray fluorescence was very good.

MATZ [841] determined Pb, Mg, Fe, K, Na, Zn, Cu and Cd in rubber seals by atomic absorption; the samples were dry ashed at 650°C, the ash dissolved in 6 N hydrochloric acid and then aspirated directly into the flame. The determination is rapid, simple and interference free.

DEILY [287] employed atomic absorption spectroscopy for the determination of aluminium in aluminium trialkyls. The sample is hydrolysed in a 1% solution of hydrochloric acid in ethyleneglycol-monomethylether and the solution is aspirated directly into a nitrous oxide/acetylene flame.

PERKINS [971] determined sodium in phosphorescent compounds by dissolving about 250 mg in hydrochloric acid and diluting to 10 ml. Atomic absorption was superior to flame photometry for this application. SCOTT [1102] examined yttrium phosphors for europium by either dissolving the sample in hydrochloric acid or digesting with potassium carbonate. Europium activated yttrium orthovanadate, yttrium oxide and yttrium oxysulphide find application in the colour television industry and a rapid and simple method had to be found. Atomic absorption spectroscopy completely meets all the requirements.

DITTRICH and LIESCH determined chromium in coke used for making electrodes; the samples were ashed, the ash fused with potassium bisulphate and the melt dissolved in 1 N sulphuric acid. This procedure can be employed for the range 0.1–2 ppm Cr in coke [306]. HARTSTEIN and co-workers determined 10 elements in coal after a pressure digestion, first in fuming nitric acid ($2^1/_2$ h at 150°C) and then fluoric acid (15 min. at 150°C), by aspirating directly into the flame [479].

BELCHER and BROOKS [99] determined Na, K, Mg, Ca, and Sr in coal ash in an air/acetylene flame. The samples were dissolved in fluoric acid and diluted with a lanthanum solution to suppress chemical interferences. OBERMILLER and FREEDMAN [922] dissolved coal ash in perchloric acid and fluoric acid and determined Na, K, Mg, Ca and Fe. Lanthanum was also added to the sample solutions for the alkaline-earth metals. The results with atomic absorption for Na, K, and Fe agreed well with those of chemical methods; for Ca and Fe other values were obtained with complexometric titrations. It could, however, be shown that the atomic absorption values were the correct ones. WOOLLEY [1356] has described the application of atomic absorption spectroscopy in the electronics industry for the analysis of various raw materials and finished products. Synthetic quartz crystals and semi-conductor materials, glasses, fluorescent lamp materials, plating baths, magnetic materials, thermistors, alloys and plastics are all analysed.

Bibliography

Monographs and Books

[1] Analytical Methods for Atomic Absorption Spectrophotometry. 7th edition. Perkin-Elmer Corporation, Norwalk, Connecticut 1973.
[2] Angino, E. E., and Billings, G. K.: Atomic Absorption Spectrometry in Geology. Elsevier Publishing Company, Amsterdam/London/New York 1967.
[3] Atomic Absorption Spectroscopy. ASTM Special Technical Publication 443, Amer. Soc. Test. Mater., Philadelphia 1969.
[4] Christian, G. D., and Feldman, F. J.: Atomic Absorption Spectroscopy: Applications in Agriculture, Biology, and Medicine. Interscience Publishers, New York/London/Sydney 1970.
[5] Dean, J. A., and Rains, T. C.: Flame Emission and Atomic Absorption Spectrometry, Vol. 1, Theory. Marcel Dekker Inc., New York/London 1969.
[6] Dean, J. A., and Rains, T. C.: Flame Emission and Atomic Absorption Spectrometry, Vol. 2, Components and Techniques. Marcel Dekker Inc., New York 1971.
[7] Elwell, W. T., and Gidley, J. A. F.: Atomic Absorption Spectrophotometry, 2nd edition. Pergamon Press, Oxford/London/Edinburgh/New York/Toronto/Sydney/Paris/Braunschweig 1966.
[8] L'vov, B. V.: Atomic Absorption Spectrochemical Analysis. English translation: Adam Hilger, London 1970.
[9] Price, W. J.: Analytical Atomic Absorption Spectrometry. Heyden & Son Ltd., London/New York/Rheine 1972.
[10] Ramirez-Muñoz, J.: Atomic Absorption Spectroscopy. Elsevier Publishing Company, Amsterdam/London/New York 1968.
[11] Rubeška, I., and Moldan, B.: Atomic Absorption Spectrophotometry. English translation: Iliffe Books Ltd., London 1969.
[12] Slavin, W.: Atomic Absorption Spectroscopy. Interscience Publishers, New York/London/Sydney 1968.

Miscellaneous References

[13] Adams, P. B.: in Standard Methods of Chemical Analysis, Vol. III B, 6th edition. Edited by F. J. Welcher, Van Nostrand, Princeton, N. J. 1966.
[14] Adams, P. B., and Passmore, W. O.: Anal. Chem. **38,** 630 (1966).
[15] Adda, J., Rousselet, F., and Mocquat, C.: Rev. Latière Française **231,** 227 (1966).
[16] Adriaenssens, E., and Knoop, P.: Anal. Chim. Acta **68,** 37 (1973).
[17] Adriaenssens, E., and Verbeek, F.: At. Absorption Newslett. **12,** 57 (1973).
[18] Adriaenssens, E., and Verbeek, F.: At. Absorption Newslett. **13,** 41 (1974).
[19] Adrian, W. J.: At. Absorption Newslett. **10,** 96 (1971).
[20] Agazzi, E. J.: Anal. Chem. **37,** 364 (1965).
[21] Agazzi, E. J.: Anal. Chem. **39,** 233 (1967).
[22] Aguilar, A., Jaime, S., and Lachica, M.: 3rd CISAFA, Paris 1971, p. 545.
[23] Aitzetmüller, K.: 3rd CISAFA, Paris 1971, p. 487.
[24] Ajemian, R. S., and Whitman, N. E.: Ann. Ind. Hyg. Assoc. J. **30,** 52 (1969).
[25] Al Ani, M. J., Dagnall, R. M., and West, T. S.: Analyst **92,** 597 (1967).
[26] Aldous, K. M., Browner, R. F., Dagnall, R. M., and West, T. S.: Anal. Chem. **42,** 939 (1970).
[27] Aldous, K. M., Dagnall, R. M., and West, T. S.: Anal. Chim. Acta **44,** 457 (1969).
[28] Aldous, K. M., Mitchell, D. G., and Jackson, K. W.: 4th ICAS, Toronto 1973.
[29] Alkemade, C. Th. J.: Anal. Chem. **38,** 1252 (1966).
[30] Alkemade, C. Th. J.: Appl. Optics **7,** 1261 (1968).
[31] Alkemade, C. Th. J.: in [5], p. 101.

[32] Alkemade, C. Th. J., Hoomayers, H. P., and Lijnse, P. L.: Spectrochim. Acta **27B,** 149 (1972).
[33] Alkemade, C. Th. J., and Milatz, J. M. W.: Appl. Sci. Res. Sect. B **4,** 289 (1955).
[34] Alkemade, C. Th. J., and Milatz, J. M. W.: J. Opt. Soc. Am. **45,** 583 (1955).
[35] Alkemade, C. Th. J., and Voorhuis, M. H.: Z. Anal. Chem. **163,** 91 (1958).
[36] Allan, J. E.: Analyst **83,** 466 (1958).
[37] Allan, J. E.: Spectrochim. Acta **10,** 800 (1959).
[38] Allan, J. E.: Nature **187,** 1110 (1960).
[39] Allan, J. E.: Spectrochim. Acta **17,** 459 (1961).
[40] Allan, J. E.: Spectrochim. Acta **17,** 467 (1961).
[41] Allan, J. E.: Analyst **86,** 530 (1961).
[42] Allan, J. E.: Spectrochim. Acta **18,** 259 (1962).
[43] Allan, J. E.: 4th Australian Spectry. Conf. 1963.
[44] Althaus, E.: Analysentechn. Berichte **3,** 1964 (Bodenseewerk Perkin-Elmer & Co. GmbH, Überlingen).
[45] Althaus, E.: N. Jahrb. Miner. Mh. **9,** 259 (1966).
[46] Amos, M. D., Bennett, P. A., Brodie, K. G., Lung, P. W. Y., and Matoušek, J. P.: Anal. Chem. **43,** 211 (1971).
[47] Amos, M. D., and Thomas, P. E.: Anal. Chim. Acta **32,** 139 (1965).
[48] Amos, M. D., and Willis, J. B.: Spectrochim. Acta **22,** 1325 (1966).
[49] Anderson, J.: At. Absorption Newslett. **11,** 88 (1972).
[50] Anderson, J.: At. Absorption Newslett. **13,** 31 (1974).
[51] Anderson, R. G., Maines, I. S., and West, T. S.: IAASC, Sheffield 1969.
[52] Anderson, R. G., Maines, I. S., and West, T. S.: Int. Sympos. Microchem., Graz 1970.
[53] Anderson, W. N., Broughton, P. M. G., Dawson, J. B., and Fisher, G. W.: Clin. Chim. Acta **50,** 129 (1974).
[54] Andrew, T. R., and Nichols, P. N. R.: Analyst **87,** 25 (1962).
[55] Andrew, T. R., and Nichols, P. N. R.: Analyst **92,** 156 (1967).
[56] Angino, E. E., and Billings, G. K.: Geochim. Cosmochim. Acta **30,** 153 (1966).
[57] Antonacopoulos, N.: Chem. Mikrobiol. Technol. Lebensm. **3,** 8 (1974).
[58] Antonetti, A., Amiel, C., Mulher, R., and Rousselet, F.: 3rd CISAFA, Paris 1971, p. 589.
[59] Ant-Wuorinen, O., and Visapää, A.: Paperi Ja Puu **48,** 649 (1966).
[60] Arendt, D. H.: At. Absorption Newslett. **11,** 63 (1972).
[61] Armstrong, F. A. J., and Uthe, J. F.: At. Absorption Newslett. **10,** 101 (1971).
[62] Atkinson, R. J., Chapman, G. D., and Krause, L.: J. Opt. Soc. Am. **55,** 1269 (1965).
[63] Atwell, M. G., and Hebert, J. Y.: Appl. Spectry. **23,** 480 (1969).
[64] Bachler, W.: Die Bodenkultur **20,** 17 (1969).
[65] Bader, H., and Brandenberger, H.: At. Absorption Newslett. **7,** 1 (1968).
[66] Bailey, N. T., and Wood, S. J.: Anal. Chim. Acta **69,** 19 (1974).
[67] Baker, M. R., and Vallee, B. L.: J. Opt. Soc. Am. **45,** 775 (1955).
[68] Baker, R. A., Hartshorne, D. J., and Wilshire, A. G.: At. Absorption Newslett. **8,** 21 (1969).
[69] Balazs, N. D. H., Pole, D. J., and Masarei, J. R.: Clin. Chim. Acta **40,** 213 (1972).
[70] Ballard, A. E., Stewart, D. W., Kamm, W. O., and Zuehlke, C. W.: Anal. Chem. **26,** 921 (1954).
[71] Ballard, A. E., and Thornton, C. W. D.: Ind. Eng. Chem. Anal. Ed. **13,** 893 (1941).
[72] Barnard, W. M., and Fishman, M. J.: At. Absorption Newslett. **12,** 118 (1973).
[73] Barnett, W. B.: Anal. Chem. **44,** 695 (1972).
[74] Barnett, W. B.: At. Absorption Newslet. **12,** 142 (1973).
[75] Barnett, W. B., and Kahn, H. L.: At. Absorption Newslett. **8,** 21 (1969).
[76] Barnett, W. B., and Kahn, H. L.: 3rd CISAFA, Paris, 1971.
[77] Barnett, W. B., and Kahn, H. L.: Clin. Chem. **18,** 923 (1972).
[78] Barnett, W. B., Kahn, H. L., and Manning, D. C.: At. Absorption Newslett. **8,** 46 (1969).
[79] Barras, R. C.: Jarrell-Ash Newslett. **13,** (1962).
[80] Barras, R. C., and Helwig, J. D.: Am. Petrol. Inst. Abstr. Refining Lit. **5** (1963).
[81] Barringer, A. R.: Inst. Mining & Metal. Bull. 714, **75B,** 120 (1966).

[82] Bartels, H.: At. Absorption Newslett. **6,** 132 (1967).
[83] Bartels, T. T., and Slater, M. P.: At. Absorption Newslett. **9,** 75 (1970).
[84] Bartels, T. T., and Wilson, C. E.: At. Absorption Newslett. **8,** 3 (1969).
[85] Barthel, W. F., Smrek, A. L., Angel, G. P., Liddle, J. A., Landrigan, P. J., Gehlbach, S. H., and Chisolm, J. J.: J. Ass. Offic. Agr. Chem. **56,** 1252 (1973).
[86] Barton, H. N., Johnson, A. J., and Shepherd, G. A.: Dow Chem. Corp. RFP-**978** (1967).
[87] Baudin, G., Bonne, R., Chaput, M., and Feve, L.: 3rd CISAFA, Paris 1971, p. 853.
[88] Baudin, G., Chaput, M., and Feve, L.: Spectrochim. Acta **26 B,** 425 (1971).
[89] Bazhov, A. S.: J. Anal. Chem. USSR **23,** 1446 (1968).
[90] Beamish, F. E., and Van Loon, J. C.: Recent Advances in the Analytical Chemistry of the Noble Metals. Pergamon Press, Oxford/New York/Toronto/Sydney/Brauchschweig 1972.
[91] Beaty, R. D.: At. Absorption Newslett. **13,** 38 (1974).
[92] Beaty, R. D.: At. Absorption Newslett. **13,** 44 (1974).
[93] Beccaluva, L., and Venturelli, G.: At. Absorption Newslett. **10,** 50 (1971).
[94] Bedrosian, A. J., and Lerner, M. W.: Anal. Chem. **40,** 1104 (1968).
[95] Beer, A.: Ann. Physik **86,** 78 (1852).
[96] Bek, F., Janoušková, J., and Moldan, B.: At. Absorption Newslett. **13,** 47 (1974).
[97] Belcher, C. B.: Anal. Chim. Acta **29,** 340 (1963).
[98] Belcher, C. B., and Bray, H. M.: Anal. Chim. Acta **26,** 322 (1962).
[99] Belcher, C. B., and Brooks, K. A.: Anal. Chim. Acta **29,** 202 (1963).
[100] Belcher, C. B., and Kinson, K.: Anal. Chim. Acta **30,** 483 (1964).
[101] Belcher, R., Dagnall, R. M., and West, T. S.: Talanta **11,** 1257 (1964).
[102] Bell, G. F.: At. Absorption Newslett. **5,** 73 (1966).
[103] Bell, G. F.: At. Absorption Newslett. **6,** 18 (1967).
[104] Bell, H. F.: Anal. Chem. **45,** 2296 (1973).
[105] Bell, W. E., Bloom, A. L., and Lynch, J.: Rev. Sci. Int. **32,** 688 (1961).
[106] Belt, C. B.: At. Absorption Newslett. **3,** 23 (1964).
[107] Belt, C. B.: Econ. Geol. **59,** 240 (1964).
[108] Bentley, E. M., and Lee, G. F.: Environ. Sci. Technol. **1,** 721 (1967).
[109] Berman, E.: At. Absorption Newslett. **3,** 111 (1964).
[110] Berman, E.: At. Absorption Newslett. **4,** 296 (1965).
[111] Berman, E.: At. Absorption Newslett. **6,** 57 (1967).
[112] Berman, E.: 4th ICAS, Toronto 1973.
[113] Berman, E., Valavanis, V., and Dubin, A.: Clin. Chem. **14,** 239 (1968).
[114] Bermejo-Martinez, F., and Baluja-Santos, C.: CSI XVII, Florence 1973, p. 577.
[115] Bernas, B.: Anal. Chem. **40,** 1682 (1968).
[116] Berry, J. W., Chappell, D. G., and Barnes, R. B.: Ind. Eng. Chem. Anal. Ed. **18,** 19 (1946).
[117] Berry, W. L., and Johnson, C. M.: Appl. Spectry. **20,** 209 (1966).
[118] Bethge, P. O., and Rådeström, R.: Svk. Papperstidn. **69,** 772 (1966).
[119] Beyer, M.: At. Absorption Newslett. **4,** 212 (1965).
[120] Beyer, M.: At. Absorption Newslett. **8,** 23 (1969).
[121] Bianchi, C. P.: Cell Calcium. Butterworth, London 1969.
[122] Billings, G. K.: At. Absorption Newslett. **4,** 312 (1965).
[123] Billings, G. K.: At. Absorption Newslett. **4,** 357 (1965).
[124] Billings, G. K., and Adams, J. A. S.: At. Absorption Newslett. **3,** 65 (1964).
[125] Billings, G. K., and Harriss, R. C.: Texas J. Sci. **17,** 129 (1965).
[126] Billings, G. K., Ragland, P. C., and Harriss, R. C.: Texas J. Sci. **18,** 277 (1966).
[127] Binding, U., and Gawlick, H.: Analysentechn. Berichte **18** (1969).
[128] Binnewies, M., and Schäfer, H.: Z. Anorg. Allg. Chem. **395,** 77 (1973).
[129] Bishop, J. R., Dunn, J. W., and Hill, W. H.: IAASC, Sheffield, 1969.
[130] Black, L. T.: J. Am. Oil Chem. Soc. **47,** 313 (1970).
[131] Boar, P. L., and Ingram, L. K.: Analyst **95,** 124 (1970).
[132] Bober, A., and Mills, A. L.: Appl. Spectry. **22,** 62 (1968).
[133] Bode, H., and Fabian, H.: Z. Anal. Chem. **162,** 328 (1958).

[134] Bode, H., and Fabian, H.: Z. Anal. Chem. **163,** 187 (1958).

[135] Böhmer, M., Auer, E., and Bartels, H.: Ärztl. Lab. **13,** 258 (1967).

[136] Boettner, E. A., and Grunder, F. I.: in Trace Inorganics in Water. Amer. Chem. Soc. Publ., Advan. Chem. Ser. **73,** 236 (1968).

[137] Bokowski, D. L.: At. Absorption Newslett. **6,** 97 (1967).

[138] Bokowski, D. L.: Am. Indust. Hygn. Assn. J. **29,** 474 (1968).

[139] Boling, A. E.: Spectrochim. Acta **22,** 425 (1966).

[140] Boling, F. A.: Spectrochim. Acta **23 B,** 495 (1968).

[141] Bond, A. M.: Anal. Chem. **42,** 932 (1970).

[142] Bond, A. M., and O'Donnell, T. A.: Anal. Chem. **40,** 560 (1968).

[143] Bond, A. M., and Willis, J. B.: Anal. Chem. **40,** 2087 (1968).

[144] Boppel, B.: Z. Anal. Chem. **268,** 114 (1974).

[145] Bosch, H., Büchel, E., and Lohau, K. H.: Analysentechn. Berichte **20** (1970).

[146] Bouchard, A.: At. Absorption Newslett. **12,** 115 (1973).

[147] Bouguer, P.: Essai d'Optique sur la gradation de la lumière. Paris 1729.

[148] Bowman, J. A.: Anal. Chim. Acta **42,** 285 (1968).

[149] Bowman, J. A., Sullivan, J. V., and Walsh, A.: Spectrochim. Acta **22,** 205 (1966).

[150] Bowman, J. A., and Willis, J. B.: Anal. Chem. **39,** 1210 (1967).

[151] Brady, D. V., Montalvo, J. G., Glowacki, G., and Pisciotta, A.: Anal. Chim. Acta **70,** 448 (1974).

[152] Brandenberger, H., and Bader, H.: Helv. Chim. Acta **50,** 1409 (1967).

[153] Brandenberger, H., and Bader, H.: At. Absorption Newslett. **6,** 101 (1967).

[154] Brandenberger, H., and Bader, H.: At. Absorption Newslett. **7,** 53 (1968).

[155] Brandenberger, H., and Bader, H.: Chimia **21,** 597 (1967).

[156] Brandt, J.: Am. Paint J. **57,** 28 (1973).

[157] Bratzel, M. P., and Chakrabarti, Ch. L.: Anal. Chim. Acta **63,** 1 (1973).

[158] Bratzel, M. P., Chakrabarti, Ch. L., and Sturgeon, R. E.: Anal. Chem. **44,** 372 (1972).

[159] Bratzel, M. P., Dagnall, R. M., and Winefordner, J. D.: Anal. Chem. **41,** 713 (1969).

[160] Bratzel, M. P., Dagnall, R. M., and Winefordner, J. D.: Anal. Chem. **41,** 1527 (1969).

[161] Bratzel, M. P. Winefordner, J. D., and Dagnall, R. M.: IAASC, Sheffield 1969.

[162] Brewer, P. G., Spencer, D. W., and Smith, C. L.: in [3], 70 (1969).

[163] Brimhall, W. H.: Anal. Chem. **41,** 1349 (1969).

[164] Brimhall, W. H., and Adams, J.: Geochim. Cosmochim. Acta **33,** 1308 (1969).

[165] Brodie, K. G., and Matoušek, J. P.: Anal. Chem. **43,** 1557 (1971).

[166] Brooks, I. B., Luster, G. A., and Easterly, D. G.: At. Absorption Newslett. **9,** 93 (1970).

[167] Brooks, R. R., Presley, P. J., and Kaplan, I. R.: Anal. Chim. Acta **38,** 321 (1967).

[168] Brooks, R. R., Presley, P. J., and Kaplan, I. R.: Talanta **14,** 809 (1967).

[169] Browner, R. F., Dagnall, R. M., and West, T. S.: Anal. Chim. Acta **45,** 163 (1969).

[170] Browner, R. F., Dagnall, R. M., and West, T. S.: IAASC, Sheffield 1969.

[171] Browner, R. F., Dagnall, R. M., and West, T. S.: Talanta **16,** 75 (1969).

[172] Browner, R. F., Patel, B. M., and Winefordner, J. D.: CSI XVII, Florence 1973, p. 227.

[173] Brunelle, R. L., Hoffman, C. M., Snow, K. B., and Pro, M. J.: J. Ass. Offic. Agr. Chem. **52,** 911 (1969).

[174] Buchanan, J. R., and Muraoka, T. T.: At. Absorption Newslett. **3,** 79 (1964).

[175] Buck, L.: Chimie Analytique **47,** 10 (1965).

[176] Buell, B. E.: in [5], p. 267.

[177] Bukenberger, U., Lodemann, C. K. W., and Loeschke, J.: Oberrhein. Geol. Abh. **21,** 43 (1972).

[178] Buneaux, F., and Fabiani, P.: Ann. Biol. Clin. **28,** 273 (1970).

[179] Burke, K. E.: Anal. Chem. **42,** 1536 (1970).

[180] Burke, K. E., and Albright, C. H.: J. Ass. Offic. Agr. Chem. **53,** 531 (1970).

[181] Burnham, C. D., Moore, C. E., Kowalski, T., and Krasniewski, J.: Appl. Spectry. **24,** 4 (1970).

[182] Burrell, D. C.: At. Absorption Newslett. **4,** 309 (1965).

[183] Burrell, D. C.: At. Absorption Newslett. **4,** 328 (1965).

[184] Burrell, D. C.: Anal. Chim. Acta **38,** 447 (1967).

[185] Burrell, D. C.: 3rd CISAFA, Paris 1971, p. 409.
[186] Burrell, D. C., and Wood, G. G.: Anal. Chim. Acta **48,** 45 (1969).
[187] Burrows, J. A., Heerdt, J. C., and Willis, J. B.: Anal. Chem. **37,** 579 (1965).
[188] Busch, K. W., and Morrison, G. H.: Anal. Chem. **45,** 712A (1973).
[189] Busch, K. W., Morrison, G. H., and Feldman, M.: 4th ICAS, Toronto, 1973.
[190] Butler, L. R. P.: J. S. Afr. Inst. Mining Met. **62,** 786 (1962).
[191] Butler, L. R. P.: At. Absorption Newslett. **5,** 99 (1966).
[192] Butler, L. R. P., and Brink, D.: S. Afr. Ind. Chem. **17,** 152 (1963).
[193] Butler, L. R. P., and Brink, D.: IAASC, Sheffield, 1969.
[194] Butler, L. R. P., and Brink, D.: in [6], p. 21.
[195] Butler, L. R. P., and Mathews, P. M.: Anal. Chim. Acta **36,** 319 (1966).
[196] Butler, L. R. P., and Schroeder, W. W.: IAASC, Sheffield 1969.
[197] Butler, L. R. P., and Strasheim, A.: Spectrochim. Acta **21,** 1207 (1965).
[198] Butler, L. R. P., Strasheim, A., Strelow, F. W. E., Mathews, P., and Feast, E. C.: CSI XII, Exeter 1965.
[199] Buttgereit, G.: Analysentechn. Berichte **26** (1972).
[200] Buttgereit, G.: Arbeitsmed. Sozialmed. Arbeitshyg. **10,** 286 (1972).
[201] Buttgereit, G.: Z. Anal. Chem. **267,** 81 (1973).
[202] Calkins, R. C.: Appl. Spectry. **20,** 146 (1966).
[203] Cameron, A. G., and Hackett, D. R.: J. Sci. Fd. Agric. **21,** 535 (1970).
[204] Campbell, D. E.: XVI Anachem Conf., Detroit 1968.
[205] Campbell, W. C., Ottaway, J. M., and Strong, B.: 4th ICAS, Toronto 1973.
[206] Capacho-Delgado, L., and Manning, D. C.: At. Absorption Newslett. **4,** 317 (1965).
[207] Capacho-Delgado, L., and Manning, D. C.: At. Absorption Newslett. **5,** 1 (1966).
[208] Capacho-Delgado, L., and Manning, D. C.: Spectrochim. Acta **22,** 1505 (1966).
[209] Capacho-Delgado, L., and Manning, D. C.: Analyst **92,** 553 (1967).
[210] Capacho-Delgado, L., and Sprague, S.: At. Absorption Newslett. **4,** 363 (1965).
[211] Carlson, G. G., and Van Loon, J. C.: At. Absorption Newslett. **9,** 90 (1970).
[212] Carper, J. L.: At. Absorption Newslett. **9,** 48 (1970).
[213] Cartwright, J. S., and Manning, D. C.: At. Absorption Newslett. **5,** 114 (1966).
[214] Cartwright, J. S., Sebens, C., and Manning, D. C.: At. Absorption Newslett. **5,** 91 (1966).
[215] Cartwright, J. S., Sebens, C., and Slavin, W.: At. Absorption Newslett. **5,** 22 (1966).
[216] Cernik, A. A.: 4th ICAS, Toronto 1973.
[217] Cernik, A. A.: At. Absorption Newslett. **12,** 42 (1973).
[218] Cernik, A. A.: At. Absorption Newslett. **12,** 163 (1973).
[219] Cernik, A. A., and Sayers, M. H. P.: Brit. J. Industr. Med. **28,** 392 (1971).
[220] Chakrabarti, Ch. L.: Appl. Spectry. **21,** 160 (1967).
[221] Chakrabarti, Ch. L.: Anal. Chim. Acta **42,** 379 (1968).
[222] Chakrabarti, Ch. L., and Hall, G.: Spectry. Letters **6,** 385 (1973).
[223] Chakrabarti, Ch. L., Lyles, G. R., and Dowling, F. B.: Anal. Chim. Acta **29,** 489 (1963).
[224] Chakrabarti, Ch. L., Robinson, J. W., and West, P. W.: Anal. Chim. Acta **34,** 269 (1966).
[225] Chao, T. T., Fishman, M. J., and Ball, J. W.: Anal. Chim. Acta **47,** 189 (1969).
[226] Chapman-Andresen, C., and Christensen, S.: Compt. Rend. Trav. Lab. Carlsberg **38,** No. 2, 19 (1970).
[227] Charlier, H., and Mayence, R.: 3rd CISAFA, Paris 1971, p. 799.
[228] Chau, Y-K., Sim, S-S., and Wong, Y-H.: Anal. Chim. Acta **43,** 13 (1968).
[229] Cheek, D. B., Graystone, J. E., Willis, J. B., and Holt, A. B.: Clin. Sci. **23,** 169 (1962).
[230] Cheek, D. B., Powell, G. K., Reba, R., and Feldman, M.: Bull. Johns Hopkins Hosp. **118,** 338 (1966).
[231] Christensen, S.: At. Absorption Newslett. **11,** 51 (1972).
[232] Christian, G. D., and Feldman, F. J.: Anal. Chim. Acta **40,** 173 (1968).
[233] Chuang, F. S., and Winefordner, J. D.: Appl. Spectry. **28,** 215 (1974).
[234] Cioni, R., Innocenti, F., and Mazzuoli, R.: At. Absorption Newslett. **11,** 102 (1972).
[235] Clark, D., Dagnall, R. M., and West, T. S.: Anal. Chim. Acta **63,** 11 (1973).

[236] Clarke, W. E.: IAASC, Sheffield, 1969.
[237] Clinton, O. E.: Spectrochim. Acta **16,** 985 (1960).
[238] Cobb, W. D., and Harrison, T. S.: Joint Symp. acc. Meth. Anal. Maj. Const., London, 1970.
[239] Collin, J., Sire, J., and Merklen, J.: 3rd CISAFA, Paris 1971, p. 713.
[240] Coudert, M., and Vergnaud, J. M.: 3rd CISAFA, Paris 1971, p. 757.
[241] Cowley, T. G., Fassel, V. A., and Kniseley, R. N.: Spectrochim. Acta **23 B,** 771 (1968).
[242] Cragin, J. H., and Herron, M. M.: At. Absorption Newslett. **12,** 37 (1973).
[243] Creeser, M. S., and West, T. S.: Spectrochim. Acta **25 B,** 61 (1970).
[244] Crow, R. F., Hime, W. G., and Connolly, J. D.: Res. J. Develop. Lab. Portland Cement Ass. **9,** No. 2, 60 (1967).
[245] Cruz, R., and Van Loon, J. C.: 4th ICAS, Toronto 1973.
[246] Curnow, D. H., Gutteridge, D. H., and Horgan, E. D.: At. Absorption Newslett. **7,** 45 (1968).
[247] Curry, A. S., and Knott, A. R.: Clin. Chim. Acta **30,** 115 (1970).
[248] Curry, A. S., Read, J. F., and Knott, A. R.: Analyst **94,** 744 (1969).
[249] Dagnall, R. M.: II ČS. Conf. Flame Spectry., Zvikov 1973.
[250] Dagnall, R. M., Kirkbright, G. F., West, T. S., and Wood, R.: Anal. Chim. Acta **47,** 407 (1969).
[251] Dagnall, R. M., Taylor, M. R. G., and West, T. S.: Spectry. Letters **1,** 397 (1968).
[252] Dagnall, R. M., Thompson, K. C., and West, T. S.: Anal. Chim. Acta **36,** 269 (1966).
[253] Dagnall, R. M., Thompson, K. C., and West, T. S.: At. Absorption Newlett. **6,** 117 (1967).
[254] Dagnall, R. M., Thompson, K. C., and West, T. S.: Analyst **92,** 506 (1967).
[255] Dagnall, R. M., Thompson, K. C., and West, T. S.: Talanta **14,** 551 (1967).
[256] Dagnall, R. M., Thompson, K. C., and West, T. S.: Talanta **14,** 557 (1967).
[257] Dagnall, R. M., Thompson, K. C., and West, T. S.: Talanta **14,** 1151 (1967).
[258] Dagnall, R. M., Thompson, K. C., and West, T. S.: Talanta **14,** 1467 (1967).
[259] Dagnall, R. M., Thomspon, K. C., and West, T. S.: Analyst **93,** 72 (1968).
[260] Dagnall, R. M., Thompson, K. C., and West, T. S.: Analyst **93,** 153 (1968).
[261] Dagnall, R. M., Thompson, K. C., and West, T. S.: Analyst **94,** 643 (1969).
[262] Dagnall, R. M., and West, T. S.: Talanta **11,** 1553 (1964).
[263] Dagnall, R. M., and West, T. S.: Appl. Opt. **7,** 1287 (1968).
[264] Dall'Aglio, M., Gragnani, R., and Visibelli, D.: Rend. Soc. Ital. Mineral. Petrol. Cia. Soc. Mines. Italy **24,** 188 (1968).
[265] Dalrymple, B. A., and Kenner, G. T.: J. Pharm. Sci. **58,** 604 (1969).
[266] Dalton, E. F., and Malanoski, A. J.: At. Absorption Newslett. **10,** 92 (1971).
[267] David, D. J.: Analyst **83,** 655 (1958).
[268] David, D. J.: Analyst **84,** 536 (1959).
[269] David, D. J.: Analyst **85,** 459 (1960).
[270] David, D. J.: Nature **187,** 1109 (1960).
[271] David, D. J.: Analyst **86,** 730 (1961).
[272] David, D. J.: Analyst **87,** 576 (1972).
[273] David, D. J.: At. Absorption Newslett. **1,** 45 (1962).
[274] David, D. J.: in Modern Methods of Plant Analysis, Vol. 5. Springer Verlag, Berlin/Göttingen/Heidelberg 1972.
[275] David, D. J.: Analyst **89,** 747 (1964).
[276] David, D. J.: Spectrochim. Acta **20,** 1185 (1964).
[277] Davidson, I. W. F., and Secrest, W. L.: Anal. Chem. **44,** 1808 (1972).
[278] Dawson, J. B.: Proc. Soc. Anal. Chem. **1970,** 195.
[279] Dawson, J. B., and Heaton, F. W.: Biochem. J. **80,** 99 (1961).
[280] Dawson, J. B., and Walker, B. E.: Chlin. Chim. Acta **26,** 465 (1969).
[281] Dean, J. A.: Analyst **85,** 621 (1960).
[282] Dean, J. A., and Carnes, W. J.: Anal. Chem. **34,** 192 (1962).
[283] Decker, C. F., Aras, A., and Decker, L. R.: Anal. Biochem. **8,** 344 (1964).
[284] De Galan, L.: Spectry. Letters **3,** 123 (1970).
[285] De Galan, L., and Winefordner, J. D.: J. Quant. Spectrosc. Radiat. Transfer **7,** 251 (1967).

[286] De Galan, L., and Samaey, G. F.: Spectrochim. Acta **24 B,** 679 (1969).
[287] Deily, J. R.: At. Absorption Newslett. **5,** 119 (1966).
[288] Deily, J. R.: At. Absorption Newslett. **6,** 65 (1967).
[289] Delaughter, B.: At. Absorption Newslett. **4,** 273 (1965).
[290] Delfino, J. J., Bortleson, G. C., and Lee, G. F.: Environm. Sci. Technol. **3,** 1189 (1969).
[291] Della Monica, E. S., and McDowell, P. E.: J. Am. Leather Chem. Ass. **1971,** 21.
[292] Delves, H. T.: Analyst **95,** 431 (1970).
[293] Delves, H. T.: At. Absorption Newslett. **12,** 50 (1973).
[294] Delves, H. T., Shepherd, G., and Vinter, P.: Analyst **96,** 260 (1971).
[295] Den Tonkelaar, W. A. M., and Bikker, M. A.: Atmospher. Environm. **1971,** 353.
[296] Derschau, H. A. V., and Prugger, H.: Z. Anal. Chem. **247,** 8 (1969).
[297] Descube, J., Roques, N., Rousselet, F., and Girard, M. L.: Ann. Biol. Clin. **35,** 1011 (1967).
[298] Devoto, G.: Boll. Soc. Ital. Biol. Sper. **44,** 425 (1968).
[299] Devoto, G.: Boll. Soc. Ital. Biol. Sper. **44,** 1253 (1968).
[300] Dickinson, G. W., and Fassel, V. A.: Anal. Chem. **41,** 1021 (1969).
[301] Dickson, R. E., and Johnson, C. M.: Appl. Spectry. **20,** 214 (1966).
[302] Dietz, A. A., and Rubinstein, H. M.: Clin. Chem. **15,** 787 (1969).
[303] Dietz, A. A., and Rubinstein, H. M.: Ann. Rheum. Dis. **32,** 124 (1973).
[304] Dinnin, J. I., and Helz, A. W.: Anal. Chem. **39,** 1489 (1967).
[305] Dinnin, J. I.: Anal. Chem. **39,** 1491 (1967).
[306] Dittrich, K., and Liesch, G.: Talanta **20,** 691 (1973).
[307] Dodson, R., Forney, F., and Swift, E.: J. Am. Chem. Soc. **58,** 2573 (1936).
[308] Döllefeld, E.: Ärztl. Lab. **17,** 369 (1971).
[309] Donega, H. M., and Burgess, T. E.: Anal. Chem. **42,** 1521 (1970).
[310] Dowling, F. B., Chakrabarti, Ch. L., and Lyles, G. R.: Anal. Chim. Acta **28,** 392 (1963).
[311] Druckman, D.: At. Absorption Newslett. **6,** 113 (1967).
[312] D'Silva, A. P., Kniseley, R. N., and Fassel, V. A.: Anal. Chem. **36,** 1287 (1964).
[313] Dunckley, J. V.: Clin. Chem. **17,** 992 (1971).
[314] Dunckley, J. V.: Clin. Chem. **19,** 1081 (1973).
[315] Dunk, R., Mostyn, R. A., and Hoare, H. C.: At. Absorption Newslett. **8,** 79 (1969).
[316] Dyck, R.: At. Absorption Newslett. **4,** 170 (1965).
[317] Ediger, R. D.: At. Absorption Newslett. **12,** 151 (1973).
[318] Ediger, R. D., and Coleman, R. L.: At. Absorption Newslett. **11,** 33 (1972).
[319] Ediger, R. D., and Coleman, R. L.: At. Absorption Newslett. **12,** 3 (1973).
[320] Ediger, R. D., Peterson, G. E., and Kerber, J. D.: Pittsburgh Conf. Anal. Chem. Appl. Spectry. 1974.
[321] Edmunds, W. M., Giddings, D. R., and Morgan-Jones, M.: At. Absorption Newslett. **12,** 45 (1973).
[322] Edwards, H. W.: Anal. Chem. **41,** 1172 (1969).
[323] Eider, N. G.: Appl. Spectry. **25,** 313 (1971).
[324] Elliott, E. V., and Stever, K. R.: At. Absorption Newslett. **12,** 60 (1973).
[325] Ellis, D. W., and Demers, D. R.: Anal. Chem. **38,** 1943 (1966).
[326] Ellis, D. W., and Demers, D. R.: in Trace Inorganics in Water. Amer. Chem. Soc. Publ., Advan, Chem. Ser. **73,** 326 (1968).
[327] Elrod, B. B., and Ezell, J. B.: At. Absorption Newslett. **8,** 129 (1969).
[328] Elwell, W. T., and Gidley, J. A. F.: Anal. Chim. Acta **24,** 71 (1961).
[329] Emmermann, R., and Lücke, W.: Z. Anal. Chem. **248,** 325 (1969).
[330] Erinc, G., and Mangee, R. J.: Anal. Chim. Acta **31,** 197 (1964).
[331] Eskamani, A., Strecker, H. A., and Vigler, M. S.: 4th ICAS, Toronto 1973.
[332] Eskamani, A., Vigler, M. S., Strecker, H. A., and Anthony, N. R.: 4th ICAS, Toronto 1973.
[333] Evenson, M. A., and Pendergast, D. D.: Clin. Chem. **20,** 163 (1974).
[334] Everson, R. T., and Schrenk, W. G.: 4th ICAS, Toronto 1973.
[335] Ezell, J. B.: At. Absorption Newslett. **5,** 122 (1966).
[336] Ezell, J. B.: At. Absorption Newslett. **6,** 84 (1967).

[337] Fabricand, B. P., Sawyer, R. R., Ungar, S. G., and Adler, S.: Geochim. Cosmochim. Acta **26,** 1023 (1962).
[338] Fallgatter, K., Svoboda, V., and Winefordner, J. D.: Appl. Spectry. **25,** 347 (1971).
[339] Farmer, M. H.: At. Absorption Newslett. **6,** 121 (1967).
[340] Farrar, B.: At. Absorption Newslett. **4,** 325 (1965).
[341] Farrar, B.: At. Absorption Newslett. **5,** 62 (1966).
[342] Fassel, V. A.: IAASC, Sheffield, 1969.
[343] Fassel, V. A.: CSI XVI, Heidelberg 1971, p. 63.
[344] Fassel, V. A., and Becker, D. A.: Anal. Chem. **41,** 1522 (1969).
[345] Fassel, V. A., and Golightly, D. W.: Anal. Chem. **39,** 466 (1967).
[346] Fassel, V. A., and Mossotti, V. G.: Anal. Chem. **35,** 252 (1963).
[347] Fassel, V. A., Mossotti, V. G., Grossman, W. E. L., and Kniseley, R. N.: XII CSI, Exeter, 1965.
[348] Fassel, V. A., Mossotti, V. G., Grossman, W. E. L., and Kniseley, R. N.: Spectrochim. Acta **22,** 347 (1966).
[349] Fassel, V. A., Myers, R. B., and Kniseley, R. N.: Spectrochim. Acta **19,** 1187 (1963).
[350] Fassel, V. A., Rasmuson, J. O., Kniseley, R. N., and Cowley, T. G.: Spectrochim. Acta **25 B,** 559 (1970).
[351] Fassel, V. A., Slack, R. W., and Kniseley, R. N.: Anal. Chem. **43,** 186 (1971).
[352] Favretto, L., Pertoldi Marletta, G., and Favretto Gabrielli, L.: At. Absorption Newslett. **12,** 101 (1973).
[353] Feldman, F. J.: Anal. Chem. **42,** 719 (1970).
[354] Feldman, F. J., Bosshart, R. E., and Christian, G. D.: Anal. Chem. **39,** 1175 (1967).
[355] Feldman, F. J., Knoblock, E. C., and Purdy, W. C.: Anal. Chim. Acta **38,** 489 (1967).
[356] Fernandez, F. J.: At. Absorption Newslett. **8,** 90 (1969).
[357] Fernandez, F. J.: At. Absorption Newslett. **12,** 70 (1973).
[358] Fernandez, F. J.: At. Absorption Newslett. **12,** 93 (1973).
[359] Fernandez, F. J., and Kahn, H. L.: At. Absorption Newslett. **10,** 1 (1971).
[360] Fernandez, F. J., and Manning, D. C.: At. Absorption Newslett. **7,** 57 (1968).
[361] Fernandez, F. J.: and Manning, D. C.: At. Absorption Newslett. **10,** 65 (1971).
[362] Fernandez, F. J., and Manning, D. C.: At. Absorption Newslett. **10,** 86 (1971).
[363] Fernandez, F. J., and Manning, D. C.: At. Absorption Newslett. **11,** 67 (1972).
[364] Fernandez, F. J., Manning, D. C., and Vollmer, J.: At. Absorption Newslett. **8,** 117 (1969).
[365] Ferris, A. P., Jepson, W. B., and Shapland, R. C.: Analyst **95,** 574 (1970).
[366] Fielding, J., and Ryall, R. G.: Clin. Chim. Acta **33,** 235 (1971).
[367] Fierens, P., and Degre, J. P.: CSI XVII, Florence 1973, p. 558.
[368] Fiorino, J., Kniseley, R. N., and Fassel, V. A.: Spectrochim. Acta **23 B,** 413 (1968).
[369] Fiorino, J. A., Moffitt, R. A., Woodson, A. L., Gajan, R. J., Huskey, G. E., and Scholz, R. G.: J. Ass. Offic. Agr. Chem. **56,** 1246 (1973).
[370] Firman, R. J.: Spectrochim. Acta **21,** 341 (1965).
[371] Fishman, M. J.: At. Absorption Newslett. **5,** 102 (1966).
[372] Fishman, M. J.: Anal. Chem. **42,** 1462 (1970).
[373] Fishman, M. J.: At. Absorption Newslett. **11,** 46 (1972).
[374] Fishman, M. J., and Downs, S. C.: US Geol. Surv. Water-Supply Papers 1540-C. US Gov. Print. Office, Washington, D.C. 1966.
[375] Fishman, M. J., and Midgett, M. R.: in Trace Inorganics in Water. Amer. Chem. Soc. Publ., Advan. Chem. Ser. **73,** 230 (1968).
[376] Fixman, M., and Boughton, L.: At. Absorption Newslett. **5,** 33 (1966).
[377] Fleet, B., Liberty, K. V., and West, T. S.: Anal. Chim. Acta. **45,** 205 (1969).
[378] Fleming, H. D.: Spectrochim. Acta. **23 B,** 207 (1967).
[379] Fleming, L. W., and Stewart, W. K.: Clin. Chim. Acta **14,** 134 (1966).
[380] Förstner, U., and Müller, G.: Schwermetalle in Flüssen und Seen. Springer-Verlag, Berlin/Heidelberg/New York 1974.
[381] Forrester, A. T., Gudmundsen, R. A., and Johnson, P. O.: J. Opt. Soc. Am. **46,** 339 (1956).

[382] Foss, R. A., and Houston, D. M.: At. Absorption Newslett. **8,** 82 (1969).
[383] Foster, R. R.: At. Absorption Newslett. **7,** 110 (1968).
[384] Fratta, M.: 4th ICAS, Toronto 1973.
[385] Frazer, A., Seconda, S. K., and Mendels, J.: Clin. Chim. Acta **36,** 499 (1972).
[386] Frey, S. W.: At. Absorption Newslett. **3,** 127 (1964).
[387] Frey, S. W., de Witt, W. G., and Bellomy, B. R.: Prc. Am. Soc. Brew. Chem. **1966,** 172.
[388] Fritze, K., and Stuart, C.: 4th ICAS, Toronto 1973.
[389] Fuchs, Ch., Brasche, M., Paschen, K., Nordbeck, H., and Quellhorst, E.: Clin. Chim. Acta **52,** 71 (1974).
[390] Fuhrman, D. L.: At. Absorption Newslett. **8,** 105 (1969).
[391] Fukushima, S.: Microchim. Acta **1959,** 596.
[392] Fuller, C. W.: Anal. Chim. Acta **62,** 261 (1972).
[393] Fuller, C. W.: Anal. Chim. Acta **62,** 442 (1972).
[394] Fuller, C. W.: At. Absorption Newslett. **11,** 65 (1972).
[395] Fuller, C. W.: At. Absorption Newslett. **12,** 40 (1973).
[396] Fuller, C. W., and Whiteheard, J.: Anal. Chim. Acta **68,** 407 (1974).
[397] Fulton, A., and Butler, L. R. P.: Spectry. Letters **1,** 317 (1968).
[398] Fuwa, K., Pulido, P., McKay, R., and Vallee, B. L.: Anal. Chem. **36,** 2407 (1964).
[399] Fuwa, K., and Vallee, B. L.: Anal. Chem. **35,** 942 (1963).
[400] Fuwa, K., and Vallee, B. L.: Anal. Chem. **41,** 188 (1969).
[401] Galindo, G. G., Appelt, H., and Schalscha, E. B.: Soil Sci. Soc. Amer. Proceed. **33,** 974 (1969).
[402] Galle, O. K.: Appl. Spectry. **22,** 404 (1968).
[403] Gambrell, J. W.: At. Absorption Newslett. **10,** 81 (1971).
[404] Gambrell, J. W.: At. Absorption Newslett. **11,** 125 (1972).
[405] Garrido, M. D., Llaguno, C., and Garrido, J.: Am. J. Enology Viticult. **1971,** 44.
[406] Gatehouse, B. M., and Walsh, A.: Spectrochim. Acta **16,** 602 (1960).
[407] Gatehouse, B. M., and Willis, J. B.: Spectrochim. Acta **17,** 710 (1961).
[408] Gaumer, M. W., Sprague, S., and Slavin, W.: At. Absorption Newslett. **5,** 58 (1966).
[409] Gelder, Z. V.: Spectrochim. Acta **25 B,** 669 (1970).
[410] Giammarise, A.: At. Absorption Newslett. **5,** 113 (1966).
[411] Gibson, J. H., Grossman, W. E. L., and Cooke, W. D.: in Analytical Chemistry. Elsevier Publishers, Amsterdam 1962.
[412] Gibson, J. H., Grossman, W. E. L., and Cooke, W. D.: Anal. Chem. **35,** 266 (1963).
[413] Gidley, J. A. F.: CSI IX, Paris 1962, p. 263.
[414] Gidley, J. A. F., and Jones, J. T.: Analyst **85,** 249 (1960).
[415] Gidley, J. A. F., and Jones, J. T.: Analyst **86,** 271 (1961).
[416] Gimblet, E. G., Marney, A. F., and Bonsnes, R. W.: Clin. Chem. **13,** 204 (1967).
[417] Ginzburg, V. L., Livshits, D. M., and Satariana, G. I.: Zh. Analit. Khim. **19,** 1089 (1964).
[418] Giron, H. C.: At. Absorption Newslett. **12,** 28 (1973).
[419] Glenn, M. T., Savory, J., Fein, S. A., Reevers, R. D., Molnar, C. J., and Winefordner, J. D.: Anal. Chem. **45,** 203 (1973).
[420] Gochman, N., and Givelber, H.: Clin. Chem. **16,** 229 (1970).
[421] Goebgen, H. G., and Brockmann, J.: Wasser, Luft, Betrieb **12,** 11 (1968).
[422] Goecke, R.: Talanta **15,** 871 (1968).
[423] Goguel, R.: Spectrochim. Acta **26 B,** 313 (1971).
[424] Goleb, J. A.: Anal. Chem. **35,** 1978 (1963).
[425] Goleb, J. A.: Anal. Chim. Acta **34,** 135 (1966).
[426] Goleb, J. A.: Anal. Chim. Acta **36,** 130 (1966).
[427] Goleb, J. A., and Yokoyama, Y.: Anal. Chim. Acta **30,** 213 (1964).
[428] Gonzalez, J. G., and Ross, R. T.: Anal. Letters **5,** 683 (1972).
[429] Gonzalez, J. G., Ross, R. T., and Segar, D. A.: 4th ICAS, Toronto 1973.
[430] Goodfellow, G. I.: Anal. Chim. Acta **36,** 1491 (1967).
[431] Goodfellow, G. I.: Appl. Spectry. **21,** 39 (1967).
[432] Goodwin, E.: At. Absorption Newsletter **9,** 95 (1970).

[433] Gough, D. S., Hannaford, P., and Walsh, A.: Spectrochim. Acta **28 B,** 197 (1973).
[434] Goulden, P. D., Brooksbank, P., and Ryan, J. F.: Int. Lab. **1973,** No. 5, 31.
[435] Govindaraju, K.: 4th ICAS, Toronto 1973.
[436] Govindaraju, K., Hermann, R., Mevelle, G., and Chouard, C.: At. Absorption Newslett. **12,** 73 (1973).
[437] Govindaraju, K., and L'homel, N.: At. Absorption Newslett. **11,** 115 (1972).
[438] Gray, R., and Pruden, E. L.: Am. J. Med. Technol. **33,** 349 (1967).
[439] Greaves, M. C.: Nature **199,** 552 (1963).
[440] Greenfield, S., Jones, I. L., and Berry, C. T.: Analyst **89,** 713 (1964).
[441] Griffin, G. F.: Soil. Sci. Soc. Amer. Proc. **32,** 803 (1968).
[442] Griffith, F. D., Parker, H. E., and Rogler, J. C.: J. Nutr. **83,** 15 (1964).
[443] Grimaldi, F. S., and Schnepfe, M. M.: Talanta **17,** 617 (1970).
[444] Grobenski, Z.: Analysentechn. Berichte **31** (1974).
[445] Groenewald, T.: Anal. Chem. **40,** 863 (1968).
[446] Grunbaum, B. W., and Pace, N.: Microchem. J. **15,** 666 (1970).
[447] Güçer, S., and Massmann, H.: CSI XVII, Florence 1973, p. 51.
[448] Guillaumin, R.: At. Absorption Newslett. **5,** 19 (1966).
[449] Gupta, H. K. L., Amore, F. J., and Boltz, D. F.: At. Absorption Newslett. **7,** 107 (1968).
[450] Gutsche, B., and Herrmann, R.: Analyst **95,** 805 (1970).
[451] Gutsche, B., and Herrmann, R.: Naunyn-Schmiedebergs Arch. Pharm. **270,** 94 (1971).
[452] Gutsche, B., and Herrmann, R.: Z. Anal. Chem. **253,** 257 (1971).
[453] Gutsche, B., and Herrmann, R.: Z. Anal. Chem. **258,** 277 (1972).
[454] Haarsma, J. P. S., DeJong, G. J., and Agterdenbos, J.: Spectrochim. Acta **29 B,** 1 (1974).
[455] Haas, T., Lehnert, G., and Schaller, K. H.: Z. Klin. Chem. Klin. Biochem. **5,** 27 (1967).
[456] Haelen, P., Cooper, G., and Pampel, C.: At. Absorption Newslett. **13,** 1 (1974).
[457] Hall, G., Cochrane, I. G., and Dorman, R. W.: IAASC, Sheffield, 1969.
[458] Hall, J. M., and Woodward, C.: Spectry. Letters **2,** 113 (1969).
[459] Halls, D. J., and Townshend, A.: Anal. Chim. Acta **36,** 278 (1966).
[460] Hambly, A. N., and Rann, C. S.: in [5], p. 241.
[461] Hammar, H. E., and Page, N. R.: At. Absorption Newslett. **6,** 33 (1967).
[462] Hanig, R. C., and Aprison, M. H.: Anal. Biochem. **21,** 169 (1967).
[463] Hankiewicz, J.: Polish Med. J. **VIII,** 779 (1969).
[464] Hanna, W. J.: Ag. Chem. **III,** 23 (1967).
[465] Hansen, A. C.: At. Absorption Newslett. **12,** 125 (1973).
[466] Hansen, J. L.: Am. J. Med. Technol. **34,** 625 (1968).
[467] Hansen, J. L., and Freier, E. F.: Am. J. Med. Technol. **33,** 217 (1967).
[468] Hareland, W. A., Ebersole, E. R., and Ramachandran, T. P.: Anal. Chem. **44,** 520 (1972).
[469] Harrington, D. E.: At. Absorption Newslett. **9,** 106 (1970).
[470] Harrington, D. E.: At. Absorption Newslett. **11,** 107 (1972).
[471] Harrison, W. W.: Anal. Chem. **37,** 1168 (1965).
[472] Harrison, W. W., and Juliano, P. O.: Anal. Chem. **41,** 1016 (1969).
[473] Harrison, W. W., and Juliano, P. O.: Anal. Chem. **43,** 248 (1971).
[474] Harrison, W. W., and Tyree, A. B.: Clin. Chim. Acta **31,** 63 (1971).
[475] Harrison, W. W., and Wadlin, W. H.: Anal. Chem. **41,** 374 (1969).
[476] Harrison, W. W., Yurachek, J. P., and Benson, C. A.: Clin. Chim. Acta **23,** 83 (1969).
[477] Harth, M., Haines, D. S. M., and Bondy, D. C.: Am. J. Clin. Pathol. **59,** 423 (1973).
[478] Hartley, F. R., and Inglis, A. S.: Analyst **92,** 622 (1967).
[479] Hartstein, A. M., Freedman, R. W., and Platter, D. W.: Anal. Chem. **45,** 611 (1973).
[480] Hasegawa, N., Hirai, A., Sugino, H., and Kashiwagi, T.: 3rd CISAFA, Paris 1971, p. 665.
[481] Hatch, W. R., and Ott, W. L.: Anal. Chem. **40,** 2085 (1968).
[482] Hauck, G.: Z. Anal. Chem. **267,** 337 (1973).
[483] Hauser, T. R., Hinners, T. A., and Kent, J. L.: Anal. Chem. **44,** 1819 (1972).
[484] Hazebroucq, G. F.: 3rd CISAFA, Paris 1971, p. 577.
[485] Headridge, J. B., and Ashy, M. A.: 4th ICAS, Toronto 1973.

[486] Headridge, J. B., and Hubbard, D. P.: Anal. Chim. Acta **37,** 151 (1967).
[487] Headridge, J. B., and Richardson, J.: Analyst **94,** 968 (1969).
[488] Headridge, J. B., and Sowerbutts, A.: Analyst **98,** 57 (1973).
[489] Heckman, M.: J. Ass. Offic. Agr. Chem. **50,** 45 (1967).
[490] Heckman, M.: J. Ass. Offic. Agr. Chem. **53,** 923 (1970).
[491] Heckman, M.: J. Ass. Offic. Agr. Chem. **54,** 666 (1971).
[492] Hein, H.: Analysentechn. Berichte **29** (1973).
[493] Heinemann, G.: Z. Klin. Chem. Klin. Biochem. **10,** 467 (1972).
[494] Heinrich, G., Spitzer, H., and Tesik, G.: Z. Anal. Chem. **226,** 124 (1967).
[495] Heinrichs, H., and Lange, J.: Fortschr. Miner. **50,** Bh. 1, 35 (1972).
[496] Heinrichs, H., and Lange, J.: Z. Anal. Chem. **265,** 256 (1973).
[497] Hell, A., Ulrich, W. F., Shifrin, N., and Ramirez-Muñoz, J.: Appl. Opt. **7,** 1317 (1968).
[498] Helsby, C. A.: Talanta **20,** 779 (1973).
[499] Helsby, C. A.: Anal. Chim. Acta **69,** 259 (1974).
[500] Heneage, P.: At. Absorption Newslett. **5,** 64 (1966).
[501] Heneage, P.: At. Absorption Newslett. **5,** 67 (1966).
[502] Henn, E. L.: At. Absorption Newslett. **12,** 109 (1973).
[503] Henning, S., and Jackson, T. L.: At. Absorption Newslett. **12,** 100 (1973).
[504] Herrmann, R., and Lang, W.: Z. Ges. Exptl. Med. **134,** 268 (1961).
[505] Herrmann, R., and Lang, W.: Z. Ges. Exptl. Med. **135,** 569 (1962).
[506] Herrmann, R., and Lang, W.: Z. Klin. Chem. **1,** 182 (1963).
[507] Herrmann, R., Lang, W., and Stamm, D.: Blut **11,** 135 (1965).
[508] Hessel, D. W.: At. Absorption Newslett. **7,** 55 (1968).
[509] Hicks, J. M., Gutierrez, A. N., and Worthy, B. E.: Clin. Chem. **19,** 322 (1973).
[510] Hicks, J. E., McPherson, R. T., and Salyer, J. W.: Anal. Chim. Acta **61,** 441 (1972).
[511] Hieftje, G. M., and Malmstadt, H. V.: Anal. Chem. **40,** 1860 (1968).
[512] Hill, G. L., and Caputi, A.: Am. J. Enology Viticult. **20,** 227 (1969).
[513] Hill, U. T.: in [3], 83 (1969).
[514] Hinkle, M. E., and Learned, R. E.: US Geol. Survey Prof. Paper **650-D,** D 251 (1969).
[515] Hoare, H. C., Mostyn, R. A., and Newland, B. T. N.: Anal. Chim. Acta **40,** 181 (1968).
[516] Hoare, H. C., Mostyn, R. A., and Newland, B. T. N.: IAASC, Sheffield 169.
[517] Hobbs, R. S., Kirkbright, G. F., Sargent, M., and West, T. S.: Talanta **15,** 997 (1968).
[518] Hobbs, R. S., Kirkbright, G. F., and West, T. S.: Analyst **94,** 554 (1969).
[519] Höhn, R., Jackwerth, E., and Koos, K.: CSI XVII, Florence 1973, p. 568.
[520] Höhn, R., and Umland, F.: Z. Anal. Chem. **258,** 100 (1972).
[521] Hoffmann, H-D., and Fiedler, H.: Z. Ges. Inn. Med. Ihre Grenzgeb. **25,** 1065 (1970).
[522] Hofton, M. E., and Hubbard, D. P.: 3rd CISAFA, Paris 1971, p. 743.
[523] Hohnadel, D. C., Sunderman, F. W., Nechay, M. W., and McNeely, M. D.: Clin. Chem. **19,** 1288 (1973).
[524] Holak, W.: Anal. Chem. **41,** 1712 (1969).
[525] Holak, W.: J. Ass. Offic. Agr. Chem. **53,** 877 (1970).
[526] Holak, W.: J. Ass. Offic. Agr. Chem. **54,** 1138 (1971).
[527] Holak, W.: At. Absorption Newslett. **12,** 63 (1973).
[528] Holak, W., Krinitz, B., and Williams, J. C.: J. Ass. Offic. Agr. Chem. **55,** 741 (1972).
[529] Holtzman, N. A., Elliott, D. A., and Heller, R. H.: New England J. Med. **275,** 347 (1966).
[530] Hoover, W. L., and Duren, S. C.: J. Ass. Offic. Agr. Chem. **52,** 708 (1969).
[531] Hoover, W. L., Melton, J. R., and Howard, P. A.: J. Ass. Offic. Agr. Chem. **54,** 860 (1971).
[532] Hoover, W. L., Reagor, J. C., and Garner, J. C.: J. Ass. Offic. Agr. Chem. **52,** 708 (1969).
[533] Horn, D. B., and Latner, A. L.: Clin. Chim. Acta **8,** 974 (1963).
[534] Hoschler, M. E., Kanabrocki, E. L., Moore, C. E., and Hattori, D. M.: Appl. Spectry. **27,** 185 (1973).
[535] Hossner, L. R., and Ferrara, L. W.: At. Absorption Newslett. **6,** 71 (1967).
[536] Hossner, L. R., Weger, S. J., and Ferrara, L. W.: ACS Meet., Atlantic City 1968.
[537] Hubbard, D. P., and Monks, H. H.: Anal. Chim. Acta **47,** 197 (1969).

[538] Huffman, C., Mensik, J. D., and Rader, L. F.: US Geol. Survey Prof. Paper **550-B,** B 189 (1966).
[539] Human, H. G. C., Butler, L. R. P., and Strasheim, A.: Analyst **94,** 81 (1969).
[540] Human, H. G. C., Strasheim, A., and Butler, L. R. P.: IAASC, Sheffield 1969.
[541] Humphrey, J. R.: Anal. Chem. **37,** 1604 (1965).
[542] Hunt, B. J.: Clin. Chem. **15,** 979 (1969).
[543] Hunter, R. E., Kelsall, M. A., Bishop, W. J., and Woodworth, P. F.: 4th ICAS, Toronto 1973.
[544] Hurford, T. R., and Boltz, D. F.: Anal. Chem. **40,** 379 (1968).
[545] Husdan, H., and Rapoport, A.: Clin. Chem. **15,** 669 (1969).
[546] Husler, J. W.: At. Absorption Newslett. **9,** 31 (1970).
[547] Husler, J. W.: At. Absorption Newslett. **10,** 60 (1971).
[548] Hwang, J. Y.: Canadian Spectry. March **1971,** 43.
[549] Hwang, J. Y., Mokeler, Ch. J., and Ullucci, P. A.: Anal. Chem. **44,** 2018 (1972).
[550] Hwang, J. Y., and Sandonato, L. M.: Anal. Chem. **42,** 744 (1970).
[551] Hwang, J. Y., Ullucci, P. A., and Mokeler, Ch. J.: Anal. Chem. **45,** 795 (1973).
[552] Hyatt, K. H., Levy, L., Nichaman, N., and Oscherwitz, M.: Appl. Spectry. **20,** 142 (1966).
[553] Ihnat, M.: 4th ICAS, Toronto 1973.
[554] Iida, C., Fuwa, K., and Wacker, W. E. C.: Anal. Biochem. **18,** 18 (1967).
[555] Ingamells, C. O.: Anal. Chem. **38,** 1228 (1966).
[556] Ingamells, C. O.: Anal. Chim. Acta **52,** 323 (1970).
[557] Intonti, R., and Stacchini, A.: Spectrochim. Acta **23 B,** 437 (1968).
[558] Jackson, A. S., Michael, L. M., and Schumacher, H. S.: Anal. Chem. **44,** 1064 (1972).
[559] Jackson, B., and Myrick, H. N.: Internat. Labor. May/June **1971,** 41.
[560] Jackwerth, E.: Z. Anal. Chem. **256,** 128 (1971).
[561] Jackwerth, E., Höhn, R., and Koos, K.: Z. Anal. Chem. **264,** 1 (1973).
[562] Jackwerth, E., Lohmar, J., and Wittler, G.: Z. Anal. Chem. **266,** 1 (1973).
[563] Jackwerth, E., Lohmar, J., and Wittler, G.: CSI XVII, Florence 1973, p. 571.
[564] Jacobs, M. B., Goldwater, L. J., and Gilbert, H.: Am. Ind. Hyg. Assoc. J. **22,** 276 (1961).
[565] Jacobs, M. B., and Singerman, A.: J. Lab. Clin. Med. **59,** 871 (1962).
[566] Jacobs, M. B., Yamaguchi, S., Goldwater, L. J., and Gilbert, H.: Am. Ind. Hyg. Assoc. J. **21,** 475 (1960).
[567] Jacobsen, E., and Harrison, G. R.: J. Opt. Soc. Amer. **39,** 1054 (1969).
[568] James, C. H., and Webb, J. S.: Trans. Instn. Min. Metall. **73,** 633 (1964).
[569] Janauer, G. E., Smith, F. E., and Mangan, J.: At. Absorption Newslett. **6,** 3 (1967).
[570] Janoušková, J., Nehasilova, M., and Sychra, V.: At. Absorption Newslett. **12,** 161 (1973).
[571] Janssen, A., and Umland, F.: Z. Anal. Chem. **251,** 101 (1970).
[572] Janssens, M., and Dams, R.: Anal. Chim. Acta **65,** 41 (1973).
[573] Janssens, M., and Dams, R.: Anal. Chim. Acta **70,** 25 (1974).
[574] Jaworowski, R. J., and Weberling, R. P.: At. Absorption Newslett. **5,** 125 (1966).
[575] Jaworowski, R. J., Weberling, R. P., and Bracco, D. J.: Anal. Chim. Acta **37,** 285 (1967).
[576] Jeffus, M. T., Elkins, J. S., and Kenner, C. T.: J. Ass. Offic. Agr. Chem. **53,** 1172 (1970).
[577] Jenkins, D. R.: Spectrochim. Acta **23 B,** 167 (1967).
[578] Jenkins, D. R.: IAASC, Sheffield 1969.
[579] Jenkins, D. R.: Spectrochim. Acta **25 B,** 47 (1970).
[580] Jenkins, D. R., and Sudgen, T. M.: in [5], p. 151.
[581] Johnson, F. J., Woodis, T. C., and Cummings, J. M.: At. Absorption Newslett. **11,** 118 (1972).
[582] Johnson, H. N., Kirkbright, G. F., and Whitehouse, R. J.: Anal. Chem. **45,** 1603 (1973).
[583] Johnson, J. R. K., and Riechman, G. C.: Clin. Chem. **14,** 1218 (1968).
[584] Jones, A. H.: Anal. Chem. **37,** 1761 (1965).
[585] Jones, A. H.: At Absorption Newslett. **9,** 1 (1970).
[586] Jones, D. I. H., and Thomas, T. A.: Hilger J. **9,** 39 (1965).
[587] Jones, G. W., Lewis, B., and Seaman, H.: J. Amer. Chem. Soc. **53,** 3992 (1931).
[588] Jones, J. L., and Eddy, R. D.: Anal. Chim. Acta **43,** 165 (1968).
[589] Jones, W. G., and Walsh, A.: Spectrochim. Acta **16,** 249 (1960).

[590] Jordan, J.: At. Absorption Newslett. **7,** 48 (1968).
[591] Joselow, M. M., and Bogden, J. D.: At. Absorption Newslett. **11,** 99 (1972).
[592] Joselow, M. M., and Bogden, J. D.: At. Absorption Newslett. **11,** 127 (1972).
[593] Joselow, M. M., and Singh, N. P.: At. Absorption Newslett. **12,** 128 (1973).
[594] Joseph, K. T., Panday, V. K., Raut, S. J., and Soman, S. D.: At. Absorption Newslett. **7,** 25 (1968).
[595] Joseph, K. T., Parameswaran, M., and Soman, S. D.: At. Absorption Newslett. **8,** 127 (1969).
[596] Joyner, T., and Finley, J. S.: At. Absorption Newslett. **5,** 4 (1966).
[597] Joyner, T., Healy, M. L., Chakrabarti, D., and Koyanagi, T.: Environm. Sci. Technol. **1,** 417 (1967).
[598] Julshamn, K., and Braekkan, O. R.: At. Absorption Newslett. **12,** 139 (1973).
[599] Jung, P. D., and Clarke, D.: J. Ass. Offic. Agr. Chem. **57,** 379 (1974).
[600] Jursik, M. L.: At. Absorption Newslett. **6,** 21 (1967).
[601] Kägler, S. H.: Analysentechn. Berichte **6** (1965).
[602] Kägler, S. H.: Erdöl, Kohle, Erdgas, Petrochemie **19,** 879 (1966).
[603] Kägler, S. H.: Erdöl, Kohle, Erdgas, Petrochemie **24,** 13 (1971).
[604] Kahn, H. L.: At. Absorption Newslett. **2,** 35 (1963).
[605] Kahn, H. L.: J. Chem. Education **43,** A 7 & A 103 (1966).
[606] Kahn, H. L.: J. Metals, **1966,** 1101.
[607] Kahn, H. L.: At. Absorption Newslett. **6,** 51 (1967).
[608] Kahn, H. L.: At. Absorption Newslett. **7,** 40 (1968).
[609] Kahn, H. L.: in Trace Inorganics in Water. Advan. Chem. Ser. **73,** 183 (1968).
[610] Kahn, H. L.: Amer. Laboratory **1969,** 52.
[611] Kahn, H. L.: At. Absorption Newslett. **10,** 58 (1971).
[612] Kahn, H. L., Fernandez, F. J., and Slavin, S.: At. Absorption Newslett. **11,** 42 (1972).
[613] Kahn, H. L., and Manning, D. C.: At. Absorption Newslett. **4,** 264 (1965).
[614] Kahn, H. L., and Manning, D. C.: Amer. Laboratory **1972,** 8.
[615] Kahn, H. L., Peterson, G. E., and Manning, D. C.: At. Absorption Newslett. **9,** 79 (1970).
[616] Kahn, H. L., Peterson, G. E., and Schallis, J. E.: At. Absorption Newslett. **7,** 35 (1968).
[617] Kahn, H. L., and Schallis, J. E.: At. Absorption Newslett. **7,** 5 (1968).
[618] Kahn, H. L., and Sebestyen, J. E.: At. Absorption Newslett. **9,** 33 (1970).
[619] Kahn, H. L., and Slavin, S.: At. Absorption Newslett. **10,** 125 (1971).
[620] Kaiser, G., Tschöpel, P., and Tölg, G.: Z. Anal. Chem. **253,** 177 (1971).
[621] Kalb, G. W.: At. Absorption Newslett. **9,** 84 (1970).
[622] Kallmann, S., and Hobart, E. W.: Anal. Chim. Acta **51,** 120 (1970).
[623] Kallmann, S., and Hobart, E. W.: Talanta **17,** 845 (1970).
[624] Kapetan, J. P.: in [3] 78 (1969).
[625] Karakhanis, P. P., and Anfinsen, J. R.: J. Ass. Offic. Agr. Chem. **56,** 358 (1973).
[626] Kashiki, M., and Oshima, S.: Anal. Chim. Acta **51,** 387 (1970).
[627] Kashiki, M., Yamazoe, S., Ikeda, N., and Oshima, S.: Anal. Letters **7,** 53 (1974).
[628] Kashiki, M., Yamazoe, S., and Oshima, S.: Anal. Chim. Acta **53,** 95 (1971).
[629] Kawamura, H., Tanaka, G., and Ohyagi, Y.: Spectrochim. Acta **28 B,** 309 (1973).
[630] Keats, G. H.: At. Absorption Newslett. **4,** 319 (1965).
[631] Keller, E., and Parsons, M. L.: At. Absorption Newslett. **9,** 92 (1970).
[632] Kelly, W. R., and Moore, C. B.: Anal. Chem. **45,** 1274 (1973).
[633] Kerber, J. D.: Appl. Spectry. **20,** 212 (1966).
[634] Kerber, J. D.: At. Absorption Newslett. **10,** 104 (1971).
[635] Kerber, J. D., and Barnett, W. B.: At. Absorption Newslett. **8,** 113 (1969).
[636] Kerber, J. D., Barnett, W. B., and Kahn, H. L.: At. Absorption Newslett. **9,** 39 (1970).
[637] Kerber, J. D., Koch, A., and Peterson, G. E.: At. Absorption Newslett. **12,** 104 (1973).
[638] Kerber, J. D., Russo, A. J., Peterson, G. E., and Ediger, R. D.: At. Absorption Newslett. **12,** 106 (1973).
[639] Kerbyson, J. D., and Ratzkowski, C.: Canad. Spectry. **13,** 102 (1968).
[640] Kettner, H.: Schriftenr. V. Wasser-, Boden-, Lufthyg. **29,** 55 (1969).

[641] King, A. S.: Astrophys. **28,** 300 (1908).
[642] Kinnunen, J., and Lindsjö, O.: Chemist-Analyst **56,** 25 (1967).
[643] Kinnunen, J., and Lindsjö, O.: Chemist-Analyst **56,** 76 (1967).
[644] Kinser, R. E.: Am. Ind. Hyg. Ass. J. **27,** 260 (1966).
[645] Kinson, K., and Belcher, C. B.: Anal. Chim. Acta **30,** 64 (1964).
[646] Kinson, K., and Belcher, C. B.: Anal. Chim. Acta **31,** 180 (1964).
[647] Kirchhof, H.: Spectrochim. Acta **24 B,** 235 (1969).
[648] Kirchhoff, G.: Pogg. Annalen, **109,** 275 (1880).
[649] Kirchhoff, G.: Phil. Mag. (4), **20,** 1 (1860).
[650] Kirchhoff, G., and Bunsen, R.: Phil. Mag. (4), **20,** 89 (1860).
[651] Kirchhoff, G., and Bunsen, R.: Phil. Mag. (4), **22,** 329 (1861).
[652] Kirkbright, G. F.: IAASC, Sheffield 1969.
[653] Kirkbright, G. F.: Analyst **96,** 609 (1971).
[654] Kirkbright, G. F.: II ČS. Conf. Flame Spectry., Zvikov 1973.
[655] Kirkbright, G. F., Peters, M. K., and West, T. S.: Analyst **91,** 705 (1966).
[656] Kirkbright, G. F., Rao, A., and West, T. S.: Anal. Letters **2,** 465 (1969).
[657] Kirkbright, G. F., Sargent, M., and West, T. S.: At. Absorption Newslett. **8,** 34 (1969).
[658] Kirkbright, G. F., Sargent, M., and West, T. S.: Talanta **16,** 245 (1969).
[659] Kirkbright, G. F., Sargent, M., and West, T. S.: Talanta **16,** 1467 (1969).
[660] Kirkbright, G. F., Semb, A., and West, T. S.: Talanta **14,** 1011 (1967).
[661] Kirkbright, G. F., Semb, A., and West, T. S.: Talanta **15,** 441 (1968).
[662] Kirkbright, G. F., Semb, A., and West, T. S.: Spectry. Letters **1,** 7 (1968).
[663] Kirkbright, G. F., Smith, A. M., and West, T. S.: Analyst **91,** 700 (1966).
[664] Kirkbright, G. F., and Troccoli, O. E.: Spectrochim. Acta **28 B,** 33 (1973).
[665] Kirkbright, G. F., Troccoli, O. E., and Vetter, S.: Spectrochim. Acta **28 B,** 1 (1973).
[666] Kirkbright, G. F., and Vetter, S.: Spectrochim. Acta **26 B,** 505 (1971).
[667] Kirkbright, G. F., Ward, A. F., and West, T. S.: Anal. Chim. Acta **64,** 353 (1973).
[668] Kirkbright, G. F., and West, T. S.: Appl. Optics **7,** 1305 (1968).
[669] Kirkbright, G. F., West, T. S., and Wilson, P. J.: At. Absorption Newslett. **11,** 53 (1972).
[670] Kirkbright, G. F., West, T. S., and Wilson, P. J.: At. Absorption Newslett. **11,** 113 (1972).
[671] Klein, B., Kaufman, J. H., and Morgenstern, S.: Clin, Chem. **13,** 388 (1967).
[672] Klein, B., Kaufman, J. H., and Oklander, M.: Clin. Chem. **13,** 788 (1967).
[673] Klein, B., Kaufman, J. H., and Oklander, M.: Clin. Chem. **13,** 79 (1967).
[674] Knauer, G. A.: Analyst **95,** 476 (1970).
[675] Knight, D. M., and Pyzyna, M. K.: At. Absorption Newslett. **8,** 129 (1969).
[676] Kniseley, R. N.: in [5], p. 189.
[677] Kniseley, R. N., D'Silva, A. P., and Fassel, V. A.: Anal. Chem. **35,** 911 (1963).
[678] Knudson, E. J., and Christian, G. D.: Anal. Letters **6,** 1039 (1973).
[679] König, P., Schmitz, K. H., and Thiemann, E.: Z. Anal. Chem. **244,** 232 (1969).
[680] König, P., Schmitz, K. H., and Thiemann, E.: Arch. Eisenhüttenwesen **40,** 53 (1969).
[681] Koirtyohann, S. R.: Anal. Chem. **37,** 601 (1965).
[682] Koirtyohann, S. R.: At. Absorption Newslett. **6,** 77 (1967).
[683] Koirtyohann, S. R.: in [5], p. 295.
[684] Koirtyohann, S. R.: 4th ICAS, Toronto 1973.
[685] Koirtyohann, S. R., and Feldman, C.: in Developments in Applied Spectroscopy, Vol. 3. Plenum Press, New York 1964, p. 180.
[686] Koirtyohann, S. R., and Pickett, E. E.: Anal. Chem. **37,** 601 (1965).
[687] Koirtyohann, S. R., and Pickett, E. E.: Anal. Chem. **38,** 585 (1966).
[688] Koirtyohann, S. R., and Pickett, E. E.: CSI XIII, Ottawa 1967.
[689] Koirtyohann, S. R., and Pickett, E. E.: Spectrochim. Acta **23 B,** 673 (1968).
[690] Koirtyohann, S. R., and Pickett, E. E.: Spectrochim. Acta **26 B,** 349 (1971).
[691] Koirtyohann, S. R., and Wen, J. W.: Anal. Chem. **45,** 1986 (1973).
[692] Kolb, B., Kemmner, G., Schleser, F. H., and Wiedeking, E.: Z. Anal. Chem. **221,** 166 (1966).
[693] Kolihová, D., and Sychra, V.: Anal. Chim. Acta **59,** 477 (1972).

[694] Kometain, T. Y.: Plating **56,** 1251 (1969).
[695] Kopito, L., Byers, R. K., and Schwachman, H.: New England J. Med. **276,** 949 (1967).
[696] Kopito, L., Davis, M. A., and Schwachman, H.: Clin. Chem. **20,** 205 (1974).
[697] Kopito, L., and Schwachman, H.: J. Lab. & Clin. Med. **70,** 326 (1967).
[698] Kornblum, G. R., and DeGalan, L.: Spectrochim. Acta **28 B,** 139 (1973).
[699] Kopp, J. F.: 4th ICAS, Toronto 1973.
[700] Kotz, L., Kaiser, G., Tschöpel, P., and Tölg, G.: Z. Anal. Chem. **260,** 207 (1972).
[701] Kraft, E. A.: Amer. Laboratory, August **1969,** 8.
[702] Kranz, E.: Emissionsspektroskopie. Akademie-Verlag, Berlin 1964, p. 160.
[703] Krauser, L. A., Henderson, R., Shotwell, H. P., and Culp, D. A.: Amer. Ind. Hyg. Assn. J. **32,** 331 (1971).
[704] Kremling, K., and Petersen, H.: Anal. Chim. Acta **70,** 35 (1974).
[705] Kriege, O. H., and Welcher, G. G.: Talanta **15,** 781 (1968).
[706] Krinitz, B., and Franco, V.: J. Ass. Offic. Agr. Chem. **56,** 869 (1973).
[707] Krishnan, S. S., and Crapper, D. R.: 4th ICAS, Toronto 1973.
[708] Krishnan, S. S., Gillespie, K. A., and Crapper, D. R.: Anal. Chem. **44,** 1469 (1972).
[709] Kriss, R. H., and Bartels, T. T.: At. Absorption Newslett. **9,** 78 (1970).
[710] Kriss, R. H., and Bartels, T. T.: At. Absorption Newslett. **11,** 110 (1972).
[711] Kubasik, N. P., Sine, H. E., and Volosin, M. T.: Clin. Chem. **18,** 1326 (1972).
[712] Kubasik, N. P., Volosin, M. T., and Tuzzeo, T.: 4th ICAS, Toronto 1973.
[713] Kumamaru, T.: Anal. Chim. Acta **43,** 19 (1968).
[714] Kumamaru, T., Hayashi, Y., Okamoto, N., Tao, E., and Yamamoto, Y.: Anal. Chim. Acta **35,** 524 (1966).
[715] Kumamaru, T., Otani, Y., and Yuroka, Y.: Bull. Chem. Soc. Jap. **40,** 429 (1967).
[716] Kumamaru, T., Tao, E., Okamoto, N., and Yamamoto, Y.: Bull. Chem. Soc. Jap. **38,** 2204 (1966).
[717] Kunishi, M., and Ohno, S.: At. Absorption Newslett. **13,** 29 (1974).
[718] Kuwata, K., Hisatomi, K., and Hasegawa, T.: At. Absorption Newslett. **10,** 111 (1971).
[719] Kuzovlev, I. A., Kuznetsov, Y. N., and Sverdlina, O. A.: Zaw. Lab. **39,** 428 (1973).
[720] Laflamme, Y.: At. Absorption Newslett. **6,** 70 (1967).
[721] Laflamme, Y.: At. Absorption Newslett. **7,** 101 (1968).
[722] Lagathu, J., and Desirant, J.: Rev. Franc. Corps Gras **19,** 169 (1972).
[723] Lambert, H.: Photometria, sive de mesura et gradibus luminis colorum et umbrae. 1760.
[724] Lamm, S., Cole, B., Glynn, K., and Ullmann, W.: New England J. Med. **289,** 574 (1973).
[725] Langmyhr, F. J., and Paus, P. E.: Anal. Chim. Acta **43,** 397 (1968).
[726] Langmyhr, F. J., and Paus, P. E.: Anal. Chim. Acta **43,** 506 (1968).
[727] Langmyhr, F. J., and Paus, P. E.: Anal. Chim. Acta **43,** 508 (1968).
[728] Langmyhr, F. J., and Paus, P. E.: At. Absorption Newslett. **7,** 103 (1968).
[729] Langmyhr, F. J., and Paus, P. E.: At. Absorption Newslett. **8,** 131 (1969).
[730] Langmyhr, F. J., Solberg, R., and Wold, L. T.: Anal. Chim. Acta **69,** 267 (1974).
[731] Langmyhr, F. J., and Thomassen, Y.: Z. Anal. Chem. **264,** 122 (1973).
[732] Langmyhr, F. J., Thomassen, Y., and Massoumi, A.: Anal. Chim. Acta **68,** 305 (1974).
[733] Larkins, P. L.: Spectrochim. Acta **26 B,** 477 (1971).
[734] Larkins, P. L., and Willis, J. B.: Spectrochim. Acta **26 B,** 491 (1971).
[735] Least, C. J., Rejent, T. A., and Lees, H.: At. Absorption Newslett. **13,** 4 (1974).
[736] Leaton, J. R.: J. Ass. Offic. Agr. Chem. **53,** 237 (1970).
[737] Lebedev, V. I., and Dolidze, L. D.: IAASC, Sheffield 1969.
[738] Lehnert, G., Klavis, G., Schaller, K. H., and Haas, T.: Brit. J. Industr. Med. **26,** 156 (1969).
[739] Lehnert, G., Schaller, K. H., and Haas, T.: Z. Klin. Chem. **6,** 174 (1968).
[740] Lemonds, A. J., and McClellan, B. E.: Anal. Chem. **45,** 1455 (1973).
[741] Leonard, E. N.: At. Absorption Newslett. **10,** 84 (1971).
[742] Levine, J. R., Moore, S. G., and Levine, S. L.: Anal. Chem. **42,** 412 (1970).
[743] Levine, S. L.: Anal. Chem. **40,** 1376 (1968).
[744] Levy, A. L., and Katz, E. M.: Clin. Chem. **15,** 787 (1969).

[745] Levy, A. L., and Katz, E. M.: Clin. Chem. **16,** 840 (1970).
[746] Lewis, B., Seaman, H., and Jones, G. W.: J. Franklin Inst. **215,** 149 (1933).
[747] Lewis, C., Ott, W. L., and Sine, N. M.: The Analysis of Nickel. Pergamon Press, Oxford 1966, p. 192.
[748] Lewis, R. R.: At. Absorption Newslett. **7,** 61 (1968).
[749] Lieber, E. R.: At. Absorption Newslett. **9,** 51 (1970).
[750] Lieberman, K. W.: Clin. Chim. Acta **46,** 217 (1973).
[751] Lind, B., Kjellström, T., Linnman, L., and Nordberg, G.: 4th ICAS, Toronto 1973.
[752] Lindner, J.: Gesundheits-Ingenieur **95,** 39 (1974).
[753] Lindstedt, G.: Analyst **95,** 264 (1970).
[754] Lindstedt, G., and Skare, I.: 3rd CISAFA, Paris 1971, p. 581.
[755] Lindstrom, O.: Anal. Chem. **31,** 461 (1959).
[756] Ling, C.: Anal. Chem. **39,** 798 (1967).
[757] Ling, C.: Anal. Chem. **40,** 1876 (1968).
[758] Lloyd, P. D., and Lowe, R. M.: Spectrochim. Acta **27 B,** 23 (1972).
[759] Lockyer, R., and Hames, G. E.: Analyst **84,** 385 (1959).
[760] Loftin, H. P., Christian, C. M., and Robinson, J. W.: Spectry. Letters **3,** 161 (1970).
[761] Loken, H. F., Teal, J. S., and Eisenberg, E.: Anal. Chem. **35,** 875 (1963).
[762] Lorber, A., Cohen, R. L., Chang, C. C., and Anderson, H. E.: Arthritis and Rheumatism **11,** 170 (1968).
[763] Lowe, R. M.: Spectrochim. Acta **26 B,** 201 (1971).
[764] Luecke, W.: N. Jb. Miner. Mh. **1971,** 263.
[765] Luecke, W.: N. Jb. Miner. Mh. **1971,** 469.
[766] Luecke, W., and Emmermann, R.: Analysentechn. Berichte **19** (1970).
[767] Luecke, W., and Emmermann, R.: At. Absorption Newslett. **10,** 45 (1971).
[768] Luecke, W., Eschermann, F., Lennartz, U., and Papastamataki, A. J.: N. Jb. Miner. Abh. **120,** 178 (1974).
[769] Luecke, W., and Zielke, H-J.: Z. Anal. Chem. **253,** 20 (1971).
[770] Lundgren, G.: 4th ICAS, Toronto 1973.
[771] Lundgren, G., Lundmark, L., and Johansson, G.: Anal. Chem. **46,** 1028 (1974).
[772] Lundy, R. G., and Watje, W. F.: At. Absorption Newslett. **8,** 124 (1969).
[773] Lurie, H. H., and Sherman, G. W.: Ind. Eng. Chem. **25,** 404 (1933).
[774] Luyten, S., Smeyers-Verbeke, J., and Massart, D. L.: At. Absorption Newslett. **12,** 131 (1973).
[775] L'vov, B. V.: Ing. Fiz. Zhur. **11,** 44 (1959).
[776] L'vov, B. V.: Ing. Fiz. Zhur. **11,** 56 (1959).
[777] L'vov, B. V.: Spectrochim. Acta **17,** 761 (1961).
[778] L'vov, B. V.: Spectrochim. Acta **24 B,** 53 (1969).
[779] L'vov, B. V., and Khartsyzov, A. D.: Zh. Prikl. Spektrosk. **11,** 413 (1969).
[780] L'vov, B. V., and Khartsyzov, A. D.: Zh. Analit. Khim. **24,** 799 (1969).
[781] L'vov, B. V., and Khartsyzov, A. D.: Zh. Analit. Khim. **25,** 1824 (1970).
[782] MacDonald, M. A., and Watson, L.: Clin. Chim. Acta **14,** 233 (1966).
[783] Machata, G.: Wiener Klin. Wochenschr. **85,** 216 (1973).
[784] Machata, G., and Binder, R.: Z. Rechtsmed. **73,** 29 (1973).
[785] Maes, D., Adiwinata, Y., Egglestone, D., Fyles, T., Tilgner, F., and Pate, B. D.: 4th ICAS, Toronto 1973.
[786] Magos, L., and Cernik, A. A.: British J. Ind. Med. **26,** 144 (1969).
[787] Mahoney, J. R., Sargent, K., Greland, M., and Small, W.: Clin. Chem. **15,** 312 (1969).
[788] Makarov, D. F., Kukushkin, Y. N., and Eroshevich, T. A.: Zh. Analit. Khim. **24,** 1436 (1969).
[789] Manahan, S. E., and Jones, D. R.: Anal. Letters **6,** 745 (1973).
[790] Manahan, S. E., and Kunkel, R.: Anal. Letters **6,** 547 (1973).
[791] Manning, D. C.: At. Absorption Newslett. **3,** 84 (1964).
[792] Manning, D. C.: At. Absorption Newslett. **4,** 267 (1965).
[793] Manning, D. C.: At. Absorption Newslett. **5,** 63 (1966).
[794] Manning, D. C.: At. Absorption Newslett. **5,** 127 (1966).

[795] Manning, D. C.: At. Absorption Newslett. **6,** 35 (1967).
[796] Manning, D. C.: At. Absorption Newslett. **6,** 75 (1967).
[797] Manning, D. C.: At. Absorption Newslett. **9,** 97 (1970).
[798] Manning, D. C.: At. Absorption Newslett. **9,** 109 (1970).
[799] Manning, D. C.: At. Absorption Newslett. **10,** 123 (1971).
[800] Manning, D. C., and Capacho-Delgado, L.: Anal. Chim. Acta **36,** 312 (1966).
[801] Manning, D. C., and Fernandez, F. J.: At. Absorption Newslett. **6,** 15 (1967).
[802] Manning, D. C., and Fernandez, F. J.: At. Absorption Newslett. **9,** 65 (1970).
[803] Manning, D. C., and Fernandez, F. J.: Pittsburgh Conf. Anal. Chem. Appl. Spectry., Cleveland 1973.
[804] Manning, D. C., and Fernandez, F. J.: Pittsburgh Conf. Anal. Chem. Appl. Spectry., Cleveland 1974.
[805] Manning, D. C., and Heneage, P.: At. Absorption Newslett. **6,** 124 (1967).
[806] Manning, D. C., and Heneage, P.: At. Absorption Newslett. **7,** 80 (1968).
[807] Manning, D. C., and Kahn, H. L.: At. Absorption Newslett. **4,** 224 (1965).
[808] Manning, D. C., and Slavin, W.: At. Absorption Newslett. **1,** 39 (1962).
[809] Manning, D. C., and Slavin, S.: At. Absorption Newslett. **8,** 132 (1969).
[810] Manning, D. C., Trent, D. J., Sprague, S., and Slavin, W.: At. Absorption Newslett. **4,** 255 (1965).
[811] Manning, D. C., Trent, D. J., and Vollmer, J.: At. Absorption Newslett. **4,** 234 (1965).
[812] Manning, D. C., and Vollmer, J.: At. Absorption Newslett. **6,** 38 (1967).
[813] Manning, D. C., Vollmer, J., and Fernandez, F. J.: At. Absorption Newslett. **6,** 17 (1967).
[814] Mansell, R. E.: At. Absorption Newslett. **4,** 276 (1965).
[815] Mansell, R. E., and Emmel, H. W.: At. Absorption Newslett. **4,** 365 (1965).
[816] Mansell, R. E., Emmel, H. W., and McLaughlin, E. L.: Appl. Spectry. **20,** 231 (1966).
[817] Mansfield, J. M., Bratzel, M. P., Norgordon, H. O., Knapp, D. C., Zacha, K. E., and Winefordner, J. D.: Spectrochim. Acta **23 B,** 389 (1968).
[818] Mansfield, J. M., Winefordner, J. D., and Veillon, C.: Anal. Chem. **37,** 1049 (1965).
[819] Marček, M. V., Kinson, K., and Belcher, C. B.: Anal. Chim. Acta **41,** 447 (1968).
[820] Margoshes, M., and Darr, M. M.: NBS Techn. Note **272,** 18 (1965).
[821] Mariée, M., and Pinta, M.: Méth. Pys. d'Anal. (GAMS) **6,** 361 (1970).
[822] Marinkovic, M., and Vickers, T. J.: Appl. Spectry. **25,** 319 (1971).
[823] Marks, J. Y., and Welcher, G. G.: Anal. Chem. **42,** 1033 (1970).
[824] Marshal, G. B., and West, T. S.: Talanta **14,** 823 (1967).
[825] Marshal, G. B., and West, T. S.: Analyst **95,** 343 (1970).
[826] Martin, M.: Chem. Ind. **1971,** 514.
[827] Maruta, T., and Takeuchi, T.: Anal. Chim. Acta **62,** 253 (1972).
[828] Massmann, H.: Z. Instrumentenkunde **71,** 225 (1963).
[829] Massmann, H.: XII CSI, Exeter 1965, p. 275.
[830] Massmann, H.: 2nd Int. Sympos. Reinststoffe in Wissenschaft und Technik, Dresden 1965. II, p. 297.
[831] Massmann, H.: Z. Anal. Chem. **225,** 203 (1967).
[832] Massmann, H.: Spectrochim. Acta **23 B,** 215 (1968).
[833] Massmann, H.: Méthodes Physiques d'Analyse **4,** 193 (1968).
[834] Massmann, H.: CSI XVI, Heidelberg 1971, p. 285.
[835] Matoušek, J. P.: CSI XVII, Florence 1973, p. 57.
[836] Matoušek, J. P., and Stevens, B. J.: Clin. Chem. **17,** 363 (1971).
[837] Matoušek, J. P., and Sychra, V.: IAASC, Sheffield 1969.
[838] Matoušek, J. P., and Sychra, V.: Anal. Chem. **41,** 518 (1969).
[839] Matoušek, J. P., and Sychra, V.: Anal. Chim. Acta **49,** 175 (1970).
[840] Matsumoto, C., Taniquchi, S., Suzusho, K., and Sakaguchi, T.: J. Spect. Soc. Japan **2,** 61 (1968).
[841] Matz, A. R.: Bull. Parenteral Drug Ass. **20,** 130 (1966).
[842] Mavrodineanu, R., and Boiteux, H.: Flame Spectroscopy. John Wiley 1965.
[843] Mavrodineanu, R., and Hughes, R. C.: Spectrochim. Acta **19,** 1309 (1963).

[844] Mavrodineanu, R., and Hughes, R. C.: Appl. Optics **7,** 1281 (1968).
[845] McAuliffe, J. J.: At. Absorption Newslett. **6,** 69 (1967).
[846] McBride, C. H.: At. Absorption Newslett. **3,** 144 (1964).
[847] McBryde, W. T., and Williams, F.: US AEC Rept. Y-1178 (1957).
[848] McCrackan, M. L., Webb, H. J., Hammar, H. E., and Loadholt, C. B.: J. Ass. Offic. Agr. Chem. **50,** 5 (1967).
[849] McFarren, E. F., and Lishka, R. J.: in Trace Inorganics in Water. Am. Chem. Soc. Publ., Advan. Chem. Ser. **73,** 253 (1968).
[850] McGee, W. W., and Winefordner, J. D.: Anal. Chim. Acta **37,** 429 (1967).
[851] McPherson, G. L.: At. Absorption Newslett. **4,** 186 (1965).
[852] McPherson, G. L., Price, J. W., and Scaife, P. H.: Nature **199,** 371 (1963).
[853] Means, E. A., and Ratcliff, D.: At. Absorption Newslett. **4,** 174 (1965).
[854] Meddings, B., and Kaiser, H.: At. Absorption Newslett. **6,** 28 (1967).
[855] Medlin, J. H., Suhr, N. H., and Bodkin, J. B.: At. Absorption Newslett. **8,** 25 (1969).
[856] Meggers, W. F., Corliss, C. H., and Scribner, B. F.: Tables of Spectral Line Intensities. NBS Monograph 32-I (1961).
[857] Meggers, W. F., and Westfall, F. O.: J. Res. Nat. Bur. Stand. **44,** 447 (1950).
[858] Melton, J. R., Hoover, W. L., and Howard, P. A.: J. Ass. Offic. Agr. Chem. **52,** 950 (1969).
[859] Melton, J. R., Hoover, W. L., Howard, P. A., and Ayers, J. L.: J. Ass. Offic. Agr. Chem. **53,** 682 (1970).
[860] Menis, O., and Rains, T. C.: Anal. Chem. **32,** 1837 (1960).
[861] Mensik, J. D., and Seidemann, H. J.: At. Absorption Newslett. **13,** 8 (1974).
[862] Menzies, A. C.: Anal. Chem. **32,** 898 (1960).
[863] Mesman, B. B., and Smith, B. S.: At. Absorption Newslett. **9,** 81 (1970).
[864] Mesman, B. B., Smith, B. S., and Pierce, J. O.: Am. Ind. Hyg. Assn. J. **31,** 701 (1970).
[865] Midgett, M. R., and Fishman, M. J.: At. Absorption Newslett. **6,** 128 (1967).
[866] Miller, J. R., Helprin, J. J., and Finlayson, J. S.: J. Pharm. Sci. **58,** 455 (1969).
[867] Mitchell, A. C. G., and Zemansky, M. W.: Resonance Radiation and Excited Atoms. Cambridge Univ. Press, New York 1961.
[868] Mitchell, D. G., Jackson, K. W., and Aldous, K. M.: Anal. Chem. **45,** 1215A (1973).
[869] Mitchell, D. G., Ryan, F. J., and Aldous, K. M.: At. Absorption Newslett. **11,** 120 (1972).
[870] Mitchell, K. B.: J. Opt. Soc. Amer. **51,** 846 (1961).
[871] Moffitt, A. E., and Kupel, R. E.: At. Absorption Newslett. **9,** 113 (1970).
[872] Moldan, B.: IAASC, Sheffield 1969.
[873] Monder, C., and Sells, N.: Anal. Biochem. **20,** 215 (1967).
[874] Monkman, J. L., Maffet, P. A., and Doherty, T. F.: Ind. Hyg. Foundation Am. Quart. **17,** 418 (1956).
[875] Montagut-Buscas, M., Obiols, J., and Rodriquez, E.: At. Absorption Newslett. **6,** 61 (1967).
[876] Montford, B., and Cribbs, S. C.: At. Absorption Newslett. **8,** 77 (1969).
[877] Moody, R. R., and Taylor, R. B.: J. Pharm. Pharmac. **24,** 848 (1972).
[878] Moore, E. J., Milner, O. I., and Glass, J. R.: Microchem. J. **10,** 148 (1966).
[879] Morgan, M. E.: At. Absorption Newslett. **3,** 43 (1964).
[880] Morrison, G. H., and Freiser, H.: Solvent Extraction in Analytical Chemistry. John Wiley & Sons, New York 1957.
[881] Morrison, G. H., and Talmi, Y.: Anal. Chem. **42,** 809 (1970).
[882] Morrow, R. W., and McElhaney, R. J.: Appl. Spectry. **27,** 387 (1973).
[883] Morrow, R. W., and McElhaney, R. J.: At. Absorption Newslett. **13,** 45 (1974).
[884] Mossotti, V. G., and Duggan, M.: Appl. Optics **7,** 1325 (1968).
[885] Mossotti, V. G., and Fassel, V. A.: Spectrochim. Acta **20,** 1117 (1964).
[886] Mossotti, V. G., Laqua, K., and Hagenah, W. D.: Spectrochim. Acta **23B,** 197 (1967).
[887] Mostyn, R. A., and Cunningham, A. F.: Anal. Chem. **38,** 121 (1966).
[888] Mostyn, R. A., and Cunningham, A. F.: Anal. Chem. **39,** 433 (1967).
[889] Mostyn, R. A., and Cunningham, A. F.: J. Inst. Petrol. **53,** 101 (1967).
[890] Moten, L.: J. Ass. Offic. Agr. Chem. **53,** 916 (1970).

[891] Mrozowski, S.: Z. Physik **112,** 223 (1939).
[892] Mulford, C. E.: At. Absorption Newslett. **5,** 28 (1966).
[893] Mulford, C. E.: At. Absorption Newslett. **5,** 63 (1966).
[894] Mulford, C. E.: At. Absorption Newslett. **5,** 88 (1966).
[895] Munns, R. K., and Holland, D. C.: J. Ass. Offic. Agr. Chem. **54,** 202 (1971).
[896] Muntz, J. H.: At. Absorption Newslett. **10,** 9 (1971).
[897] Murthy, L., Menden, E. E., Eller, P. M., and Petering, H. G.: Anal. Biochem. **53,** 365 (1973).
[898] Muter, R., and Cockrell, C.: Appl. Spectry. **23,** 493 (1969).
[899] Muzzarelli, R. A. A., and Rocchetti, R.: Anal. Chim. Acta **64,** 371 (1973).
[900] Muzzarelli, R. A. A., and Rocchetti, R.: Anal. Chim. Acta **69,** 35 (1974).
[901] Muzzarelli, R. A. A., and Rocchetti, R.: Anal. Chim. Acta **70,** 283 (1974).
[902] Myers, D.: At. Absorption Newslett. **6,** 89 (1967).
[903] Nakahara, T., Munemori, M., and Musha, S.: Anal. Chim. Acta **62,** 267 (1972).
[904] Nakahara, T., Munemori, M., and Musha, S.: Bull. Chem. Soc. Japan **46,** 1162 (1973).
[905] Nakahara, T., Munemori, M., and Musha, S.: Bull. Chem. Soc. Japan **46,** 1172 (1973).
[906] Nesbitt, R. W.: Anal. Chim. Acta **35,** 413 (1966).
[907] Neumann, G. M.: Z. Anal. Chem. **258,** 180 (1972).
[908] Neumann, G. M.: Z. Anal. Chem. **259,** 337 (1972).
[909] Neumann, G. M.: Z. Anal. Chem. **261,** 108 (1972).
[910] Newbrun, E.: Nature **192,** 1182 (1961).
[911] Newland, B. T. N., and Mostyn, R. A.: At. Absorption Newslett. **10,** 89 (1971).
[912] Neybon, R., and Rey-Coquais, B.: At. Absorption Newslett. **6,** 92 (1967).
[913] Ng, W. K.: Anal. Chim. Acta **63,** 469 (1973).
[914] Ng, W. K.: Anal. Chim. Acta **64,** 292 (1973).
[915] Nitis, G. J., Svoboda, V., and Winefordner, J. D.: Spectrochim. Acta **27 B,** 345 (1972).
[916] Nix, J., and Goodwin, T.: At. Absorption Newslett. **9,** 119 (1970).
[917] Nixon, D. E., Fassel, V. A., and Kniseley, R. N.: Anal. Chem. **46,** 210 (1974).
[918] Nomoto, S., and Sunderman, F. W.: Clin. Chem. **16,** 477 (1970).
[919] Nonnenmacher, G., and Schleser, F. H.: Z. Anal. Chem. **209,** 284 (1965).
[920] Nord, P. J., Kadaba, M. P., and Sorenson, J. R. J.: Arch. Environ. Health. **27,** 40 (1973).
[921] Norwitz, G., and Gordon, H.: Talanta **20,** 905 (1973).
[922] Obermiller, E. L., and Freedman, R. W.: Fuel **44,** 199 (1965).
[923] Oddo, N.: Riv. Combustibili **XXV,** 153 (1971).
[924] Olivier, M.: Z. Anal. Chem. **248,** 145 (1969).
[925] Olivier, M.: Z. Anal. Chem. **257,** 135 (1971).
[926] Olivier, M.: Z. Anal. Chem. **257,** 187 (1971).
[927] Olsen, E. D., Jatlow, P. I., Fernandez, F. J., and Kahn, H. L.: Clin. Chem. **19,** 326 (1973).
[928] Olsen, R. D., and Sommerfeld, M. R.: At. Absorption Newslett. **12,** 165 (1973).
[929] Olson, A. D., and Hamlin, W. B.: At. Absorption Newslett. **7,** 69 (1968).
[930] Olson, A. D., and Hamlin, W. B.: Clin. Chem. **15,** 438 (1969).
[931] Olson, A. M.: At. Absorption Newslett. **4,** 278 (1965).
[932] Omang, S. H.: Anal. Chim. Acta **55,** 439 (1971).
[933] Omang, S. H.: Anal. Chim. Acta **56,** 470 (1971).
[934] Omang, S. H.: Anal. Chim. Acta **63,** 247 (1973).
[935] Omenetto, N., Benetti, P., Hart, L. P., Winefordner, J. D., and Alkemade, C. Th. J.: Spectrochim. Acta **28 B,** 289 (1973).
[936] Omenetto, N., Benetti, P., and Rossi, G.: Spectrochim. Acta **27 B,** 453 (1972).
[937] Omenetto, N., Hart, L. P., Benetti, P., and Winefordner, J. D.: Spectrochim. Acta **28 B,** 301 (1973).
[938] Omenetto, N., Hatch, N. N., Fraser, L. M., and Winefordner, J. D.: Spectrochim. Acta **28 B,** 65 (1973).
[939] Omenetto, N., and Rossi, G.: Spectrochim. Acta **24 B,** 95 (1969).
[940] Omenetto, N., and Rossi, G.: IAASC, Sheffield 1969.
[941] Ormer, D. G., and Purdy, W. C.: Anal. Chim. Acta **64,** 93 (1973).

[942] Osborn, K. R., and Gunning, H. E.: J. Opt. Soc. Amer. **45,** 552 (1955).
[943] Osolinski, T. W., and Knight, N. H.: Appl. Spectry. **22,** 532 (1968).
[944] Ottaway, J. M., Coker, D. T., Rowston, W. B., and Bhattarai, D. R.: Analyst **95,** 567 (1970).
[945] Ottaway, J. M., and Pradhan, N. K.: Talanta **20,** 927 (1973).
[946] Pagliai, V., and Pozzi, F.: 3rd CISAFA, Paris 1971, p. 907.
[947] Panday, V. K., and Ganguly, A. K.: At. Absorption Newslett. **7,** 50 (1968).
[948] Pappas, E. G., and Rosenberg, L. A.: J. Ass. Offic. Agr. Chem. **49,** 782 (1966).
[949] Pappas, E. G., and Rosenberg, L. A.: J. Ass. Offic. Agr. Chem. **49,** 792 (1966).
[950] Paralusz, C. M.: Appl. Spectry. **22,** 520 (1968).
[951] Parker, H. E.: At. Absorption Newslett. **2,** 23 (1963).
[952] Parker, M. W., Humoller, F. L., and Mahler, D. J.: Clin. Chem. **13,** 40 (1967).
[953] Parson, M. L., McCarthy, W. J., and Winefordner, D. J.: Appl. Spectry. **20,** 223 (1966).
[954] Parsons, J. A., Dawson, B., Callahan, E., and Pots, J. T.: Biochem. J. **119,** 791 (1970).
[955] Paschen, A.: Physik **50,** 901 (1916).
[956] Paschen, K.: Deut. Med. Wochenschr. **95,** 2570 (1970).
[957] Paschen, K.: Private communication.
[958] Paschen, K., and Fritz, G.: Ärztl. Forsch. **24,** 202 (1970).
[959] Passmore, W. O., and Adams, P. B.: At. Absorption Newslett. **4,** 237 (1965).
[960] Passmore, W. O., and Adams, P. B.: At. Absorption Newslett. **5,** 77 (1966).
[961] Patassy, F. Z.: Plant Soil **22,** 395 (1965).
[962] Patel, B. M., and Winefordner, J. D.: Anal. Chim. Acta **64,** 135 (1973).
[963] Pau, J. C-M., Pickett, E. E., and Koirtyohann, S. R.: Analyst **97,** 860 (1972).
[964] Paus, P. E.: At. Absorption Newslett. **10,** 44 (1971).
[965] Paus, P. E.: At. Absorption Newslett. **10,** 69 (1971).
[966] Paus, P. E.: At. Absorption Newslett. **11,** 129 (1972).
[967] Paus, P. E.: Z. Anal. Chem. **264,** 118 (1973).
[968] Pawluk, S.: At. Absorption Newslett. **6,** 53 (1967).
[969] Payne, C. E., and Combs, H. F.: Appl. Spectry. **22,** 786 (1968).
[970] Pekarek, R. S., and Hauer, E. C.: Fed. Proc. **31,** 700 (1972).
[971] Perdins, J.: Analyst **88,** 324 (1963).
[972] Peterson, E. A.: At. Absorption Newslett. **9,** 129 (1970).
[973] Peterson, G. E.: At. Absorption Newslett. **5,** 177 (1966).
[974] Peterson, G. E., and Kahn, H. L.: At. Absorption Newslett **9,** 71 (1970).
[975] Pichotin, B., and Chasseur, P.: Private communication.
[976] Pickett, E. E., and Koirtyohann, S. R.: Spectrochim. Acta **23 B,** 235 (1968).
[977] Pickett, E. E., and Koirtyohann, S. R.: Spectrochim. Acta **24 B,** 325 (1969).
[978] Pickett, E. E., and Koirtyohann, S. R.: Anal. Chem. **41,** 28 A (1969).
[979] Pickett, E. E., and Pau, J. C-M.: J. Ass. Offic. Agr. Chem. **56,** 151 (1973).
[980] Pickett, E. E., Pau, J. C-M., and Koirtyohann, S. R.: J. Ass. Offic. Agr. Chem. **54,** 796 (1971).
[981] Pickford, C. J., and Rossi, G.: Analyst **97,** 647 (1972).
[982] Pickford, C. J., and Rossi, G.: Analyst **98,** 329 (1973).
[983] Pierce, J. O., and Cholak, J.: Arch. Environ. Health **13,** 208 (1966).
[984] Pinta, M., and Riandey, C.: CSI XVII, Florence 1973, p. 71.
[985] Piper, K. G., and Higgins, G.: Proc. Assoc. Clin. Biochem. **4,** 190 (1967).
[986] Pitts, A. E., VanLoon, J. C., and Beamish, F. E.: Anal. Chim. Acta **50,** 181 (1970).
[987] Pitts, A. E., VanLoon, J. C., and Beamish, F. E.: Anal. Chim. Acta **50,** 195 (1970).
[988] Platte, J. A.: in Trace Inorganics in Water. Amer. Chem. Soc. Publ., Advan. Chem. Ser. **73,** 247 (1968).
[989] Platte, J. A., and Marcy, V. M.: At. Absorption Newslett. **4,** 289 (1965).
[990] Pollock, E. N.: At. Absorption Newslett. **9,** 47 (1970).
[991] Pollock, E. N.: At. Absorption Newslett. **10,** 77 (1971).
[992] Pollock, E. N., and Anderson, S. I.: Anal. Chim. Acta **41,** 441 (1968).
[993] Pollock, E. N., and West, S. J.: At. Absorption Newslett. **11,** 104 (1972).
[994] Pollock, E. N., and West, S. J.: At. Absorption Newslett. **12,** 6 (1973).

[995] Poluektov, N. S., and Vitkun, R. A.: Zh. Anal. Khim. **18,** 33 (1963).
[996] Poluektov, N. S., Vitkun, R. A., and Zelyukova, Y. V.: Zh. Anal. Khim. **19,** 873 (1964).
[997] Popham, R. E., and Schrenk, W. G.: Spectrochim. Acta **23 B,** 543 (1968).
[998] Porter, C.: At. Absorption Newslett. **8,** 112 (1969).
[999] Posener, D. W.: Austr. J. Physics **12,** 184 (1959).
[1000] Prasad, A. S., Oberleas, D., and Halsted, J. A.: J. Lab. Clin. Med. **66,** 508 (1965).
[1001] Prévôt, A.: At. Absorption Newslett. **5,** 13 (1966).
[1002] Price, J. P.: At. Absorption Newslett. **11,** 1 (1972).
[1003] Price, W. J., and Roos, J. T. H.: J. Sci. Fd. Agric. **20,** 437 (1969).
[1004] Price, W. J., and Roos, J. T. H.: Met. Ital. **61,** 423 (1969).
[1005] Price, W. J., Roos, J. T. H., and Clay, A. F.: Analyst **95,** 760 (1970).
[1006] Prugger, H.: Optik **21,** 320 (1964).
[1007] Prugger, H., Grosskopf, R., and Torge, R.: Spectrochim. Acta **26 B,** 191 (1971).
[1008] Purushottam, A., Naidu, P. P., and Lal, S. S.: Talanta **20,** 631 (1973).
[1009] Pybus, J., and Bowers, G. N.: Clin. Chem. **16,** 139 (1970).
[1010] Pyrih, R. S., and Bisque, R. E.: Econ. Geol. **64,** 825 (1969).
[1011] Quarrell, T. M., Powell, R. J. W., and Cluley, H. J.: Analyst **98,** 443 (1973).
[1012] Rains, T. C.: in [5], p. 349.
[1013] Rains, T. C., Epstein, M. S., and Menis, O.: Anal. Chem. **46,** 207 (1974).
[1014] Ramakrishna, T. V., Robinson, J. W., and West, P. W.: Anal. Chim. Acta **36,** 57 (1966).
[1015] Ramakrishna, T. V., Robinson, J. W., and West, P. W.: Anal. Chim. Acta **37,** 20 (1967).
[1016] Ramakrishna, T. V., Robinson, J. W., and West, P. W.: Anal. Chim. Acta **39,** 81 (1969).
[1017] Rann, C. S., and Hambly, A. N.: Anal. Chim. Acta **32,** 346 (1965).
[1018] Rann, C. S., and Hambly, A. N.: Anal. Chem. **37,** 879 (1965).
[1019] Rantala, R. T. T., and Loring, D. H.: At. Absorption Newslett. **12,** 97 (1973).
[1020] Rao, P. D.: At. Absorption Newslett. **9,** 131 (1970).
[1021] Rao, P. D.: At. Absorption Newslett. **10,** 118 (1971).
[1022] Rao, P. D.: At. Absorption Newslett. **11,** 45 (1972).
[1023] Rasmuson, J. O., Fassel, V. A., and Kniseley, R. N.: Spectrochim. Acta **28 B,** 365 (1973).
[1024] Rathje, A. O.: Amer. Ind. Hyg. Assoc. J. **30,** 126 (1969).
[1025] Rawling, B. S., Amos, M. D., and Greaves, M. C.: Aust. Inst. Mining Met. Proc. **199,**(1961).
[1026] Rawling, B. S., Greaves, M. C., and Amos, M. D.: Nature **188,** 137 (1960).
[1027] Rawson, R. A. G.: 4th ICAS, Toronto 1973.
[1028] Reevers, J. R.: Econ. Geol. **62,** 426 (1967).
[1029] Reif, I., Fassel, V. A., and Kniseley, R. N.: Spectrochim. Acta **28 B,** 105 (1973).
[1030] Reinhold, J. G., Pascoe, E., and Kfoury, G. A.: Anal. Biochem. **25,** 557 (1968).
[1031] Renshaw, G. D.: At. Absorption Newslett. **12,** 158 (1973).
[1032] Renshaw, G. D., Pounds, C. A., and Pearson, E. F.: At. Absorption Newslett. **12,** 55 (1973).
[1033] Reversat, G.: 3rd CISAFA, Paris 1971, p. 769.
[1034] Riandey, C., and Pinta, M.: 3rd CISAFA, Paris 1971, p. 321.
[1035] Ringhardtz, I., and Welz, B.: Z. Anal. Chem. **243,** 190 (1968).
[1036] Rivalenti, G., and Sighinolfi, G. P.: Contr. Mineral. Petr. **23,** 173 (1969).
[1037] Robinson, J. W.: Anal. Chim. Acta **23,** 458 (1960).
[1038] Robinson, J. W.: Anal. Chim. Acta **24,** 451 (1961).
[1039] Robinson, J. W.: Anal. Chem. **33,** 1067 (1961).
[1040] Robles, J., and Lachica, M.: 3rd CISAFA, Paris 1971, p. 453.
[1041] Rodgerson, D. O., and Moran, I. K.: Clin. Chem. **14,** 1206 (1968).
[1042] Roe, D. A., Miller, P. S., and Lutwak, L.: Anal. Biochem. **15,** 313 (1966).
[1043] Roelandts, I.: At. Absorption Newslett. **11,** 48 (1972).
[1044] Rohleder, H. A., Dietl, F., and Sansoni, B.: Spectrochim. Acta **29 B,** 19 (1974).
[1045] Roos, J. T. H.: IAASC, Sheffield 1969.
[1046] Roos, J. T. H.: Spectrochim. Acta **24 B,** 255 (1969).
[1047] Roos, J. T. H.: Spectrochim. Acta **25 B,** 539 (1970).
[1048] Roos, J. T. H.: Spectrochim. Acta **26 B,** 285 (1971).

[1049] Roos, J. T. H.: Spectrochim. Acta **27 B,** 473 (1972).
[1050] Roos, J. T. H.: Spectrochim. Acta **28 B,** 407 (1973).
[1051] Roos, J. T. H., and Price, W. J.: Spectrochim. Acta **26 B,** 279 (1971).
[1052] Roosels, D., and Vanderkeel, J. V.: At. Absorption Newslett. **7,** 9 (1968).
[1053] Roosney, R. C., and Pratt, C. G.: IAASC, Sheffield 1969.
[1054] Roschnik, R. K.: 4th ICAS, Toronto 1973.
[1055] Roschnik, R. K.: Analyst **98,** 596 (1973).
[1056] Rose, G. A., and Willden, E. G.: Analyst **98,** 243 (1973).
[1057] Rose, S. A., and Boltz, D. F.: Anal. Chim. Acta **44,** 239 (1969).
[1058] Ross, R. T., and Gonzalez, J. G.: Anal. Chim. Acta **70,** 443 (1974).
[1059] Ross, R. T., Gonzalez, J. G., and Segar, D. A.: Anal. Chim. Acta **63,** 205 (1973).
[1060] Rossi, G., and Omenetto, N.: Appl. Spectry. **21,** 329 (1967).
[1061] Roth, D. J., Bohl, D. R., and Sellers, D. E.: At. Absorption Newslett. **7,** 87 (1968).
[1062] Rousselet, F., Antonetti, A., Englander, J., and Amiel, C.: 3rd CISAFA, Paris 1971, p. 175.
[1063] Rowston, W. B., and Ottaway, J. M.: Anal. Letters **3,** 411 (1970).
[1064] Rubeška, I.: Anal. Chim. Acta **40,** 187 (1968).
[1065] Rubeška, I.: in [5], p. 317.
[1066] Rubeška, I.: At. Absorption Newslett. **12,** 33 (1973).
[1067] Rubeška, I.: II CŠ. Conf. Flame Spectry., Zvikov 1973.
[1068] Rubeška, I., and Mikšovsky, M.: At. Absorption Newslett. **11,** 57 (1972).
[1069] Rubeška, I., and Moldan, B.: Anal. Chim. Acta **37,** 421 (1967).
[1070] Rubeška, I., and Moldan, B.: Appl. Optics **7,** 1341 (1968).
[1071] Rubeška, I., and Štupar, J.: At. Absorption Newslett. **5,** 69 (1966).
[1072] Rubeška, I., Šulcek, Z., and Moldan, B.: Anal. Chim. Acta **37,** 27 (1967).
[1073] Rubeška, I., and Svoboda, V.: Anal. Chim. Acta **32,** 253 (1965).
[1074] Russell, B. J., Shelton, J. P., and Walsh, A.: Spectrochim. Acta **8,** 317 (1957).
[1075] Russell, B. J., and Walsh, A.: Spectrochim. Acta **10,** 883 (1959).
[1076] Russos, G. F., and Morrow, B. H.: Appl. Spectry. **22,** 769 (1968).
[1077] Sachdev, S. L., Robinson, J. W., and West, P. W.: Anal. Chim. Acta **38,** 499 (1967).
[1078] Salama, C.: At. Absorption Newslett. **10,** 72 (1971).
[1079] Sattur, T. W.: At. Absorption Newslett. **5,** 37 (1966).
[1080] Savory, J., Rozel, N. O., Mushak, P., and Sunderman, F. W.: Am. J. Clin. Pathol. **50,** 505 (1968).
[1081] Savory, J., Wiggins, J. W., and Heintges, M. G.: Am. J. Clin. Pathol. **51,** 720 (1969).
[1082] Scarborough, J. M.: Anal. Chem. **41,** 250 (1969).
[1083] Scarborough, J. M., Bingham, C. D., and de Vries, P. F.: Anal. Chem. **39,** 1394 (1967).
[1084] Schaefer, C., and Vomhof, D. W.: At. Absorption Newslett. **12,** 133 (1973).
[1085] Schaller, K. H., Essing, H. G., Valentin, H., and Schäcke, G.: Z. Klin. Chem. Klin. Biochem. **10,** 434 (1972).
[1086] Schaller, K. H., Kühnert, A., and Lehnert, G.: Blut **17,** 155 (1968).
[1087] Schallis, J. E., and Kahn, H. L.: At. Absorption Newslett. **7,** 75 (1968).
[1088] Schiller, R.: At. Absorption Newslett. **9,** 111 (1970).
[1089] Schlebusch, H., and Niehoff, B.: Biochem. Anal. 74, Munich 1974.
[1090] Schlewitz, J. H., and Shields, M. G.: At. Absorption Newslett. **10,** 39 (1971).
[1091] Schlewitz, J. H., and Shields, M. G.: At. Absorption Newslett. **10,** 43 (1971).
[1092] Schmidt, F. J., and Royer, J. L.: Anal. Letters **6,** 17 (1973).
[1093] Schmidt, W., and Sansoni, B.: Biochem. Anal. 74, Munich 1974.
[1094] Schneider, W., and Matter, L.: Lebensmittelchem. Gerichtl. Chem. **28,** 3 (1974).
[1095] Schnepfe, M. M., and Grimaldi, F. S.: Talanta **16,** 1461 (1969).
[1096] Schramel, P.: Anal. Chim. Acta **67,** 69 (1973).
[1097] Schrenk, W. G., Lehman, D. A., and Neufeld, L.: Appl. Spectry. **20,** 389 (1966).
[1098] Schroeder, H. A., and Nason, A. P.: Clin. Chem. **17,** 461 (1971).
[1099] Schulz-Baldes, M.: Marine Biology **16,** 226 (1972).
[1100] Schwab, M. R., and Hembree, N. H.: At. Absorption Newslett. **10,** 15 (1971).

[1101] Schweizer, V. B.: Analysentechn. Berichte **30,** (1973).
[1102] Scott, R. L.: At. Absorption Newslett. **9,** 46 (1970).
[1103] Searle, B., Chan, W., Jensen, C., and Davidow, B.: At. Absorption Newslett. **8,** 126 (1969).
[1104] Sebens, C., Vollmer, J., and Slavin, W.: At. Absorption Newslett. **3,** 165 (1964).
[1105] Sebestyen, N. A.: Spectrochim. Acta **25 B,** 261 (1970).
[1106] Segal, R. J.: Clin. Chem. **15,** 1124 (1969).
[1107] Segar, D. A.: 3rd CISAFA, Paris 1971, p. 523.
[1108] Segar, D. A.: Anal. Letters **7,** 89 (1974).
[1109] Segar, D. A., and Gonzalez, J. G.: Anal. Chim. Acta **58,** 7 (1972).
[1110] Segar, D. A., and Gonzalez, J. G.: 4th ICAS, Toronto 1973.
[1111] Segar, D. A., Gonzalez, J. G., Gilio, J. L., and Pellenbarg, R. E.: 4th ICAS, Toronto 1973.
[1112] Selander, S., and Cramer, K.: British J. Ind. Med. **25,** 139 (1968).
[1113] Seller, R. H., Ramirez-Muxo, O., Brest, A. N., and Moyer, J. H.: J. Amer. Med. Ass. **191,** 118 (1965).
[1114] SenGupta, J. G.: Anal. Chim. Acta **58,** 23 (1972).
[1115] SenGupta, J. G.: Anal. Chim. Acta **63,** 19 (1973).
[1116] SenGupta, J. G.: Miner. Sci. Engng. **5,** 207 (1973).
[1117] Severne, B. C., and Brooks, R. R.: Talanta **19,** 1467 (1972).
[1118] Severne, B. C., and Brooks, R. R.: Anal. Chim. Acta **58,** 216 (1972).
[1119] Shafto, R. G.: At. Absorption Newslett. **3,** 115 (1964).
[1120] Shkolnik, G. M., and Bevill, R. F.: At. Absorption Newslett. **12,** 112 (1973).
[1121] Sideman, L., Murphy, J. J., and Wilson, D. T.: Clin. Chem. **16,** 597 (1970).
[1122] Siemer, D., Lech, J. F., and Woodriff, R.: Spectrochim. Acta **28 B,** 469 (1973).
[1123] Sighinolfi, G. P.: At. Absorption Newslett. **11,** 96 (1972).
[1124] Sighinolfi, G. P.: At. Absorption Newslett. **12,** 136 (1973).
[1125] Silvester, M. D., Koop, D. J., and Barringer, A. R.: 4th ICAS, Toronto 1973.
[1126] Simmons, E. C.: At. Absorption Newslett. **4,** 281 (1965).
[1127] Simmons, W. J.: Anal. Chem. **45,** 1947 (1973).
[1128] Simonian, J. V.: At. Absorption Newslett. **7,** 63 (1968).
[1129] Simpson, G. R., and Blay, R. A.: Food Trade Rev. **36,** No. 8, 35 (1966).
[1130] Siren, M. J.: Sci. Tools **11,** 37 (1964).
[1131] Skerry, P. J., and Chapman, W.: 4th ICAS, Toronto 1973.
[1132] Skewes, H. R.: Aust. Inst. Mining Met. Proc. **211,** 217 (1964).
[1133] Skogerboe, R. K.: in [5], p. 381.
[1134] Skogerboe, R. K., and Woodriff, R. A.: Anal. Chem. **35,** 1977 (1963).
[1135] Slavin, S., Barnett, W. B., and Kahn, H. L.: At. Absorption Newslett. **11,** 37 (1972).
[1136] Slavin, S., Fernandez, F. J., and Manning, D. C.: 25th Pittsburgh Conf. Anal. Chem. Appl. Spectry., Cleveland 1974.
[1137] Slavin, S., and Sattur, T. W.: At. Absorption Newslett. **7,** 99 (1968).
[1138] Slavin, S., and Slavin, W.: At. Absorption Newslett. **5,** 106 (1966).
[1139] Slavin, W.: At. Absorption Newslett. **2,** 1 (1963).
[1140] Slavin, W.: At. Absorption Newslett. **3,** 93 (1964).
[1141] Slavin, W.: At. Absorption Newslett. **4,** 192 (1965).
[1142] Slavin, W.: At. Absorption Newslett. **4,** 243 (1965).
[1143] Slavin, W.: At. Absorption Newslett. **4,** 330 (1965).
[1144] Slavin, W.: Occupational Health Rev. **17,** 9 (1965).
[1145] Slavin, W.: At. Absorption Newslett. **6,** 9 (1967).
[1146] Slavin, W., and Manning, D. C.: Anal. Chem. **35,** 253 (1963).
[1147] Slavin, W., and Manning, D. C.: Appl. Spectry. **19,** 65 (1965).
[1148] Slavin, W., Sebens, C., and Sprague, S.: At. Absorption Newslett. **4,** 341 (1965).
[1149] Slavin, W., and Slavin, S.: Appl. Spectry. **23,** 421 (1969).
[1150] Slavin, W., and Sprague, S.: At. Absorption Newslett. **3,** 1 (1965).
[1151] Slavin, W., Trent, D. J., and Sprague, S.: At. Absorption Newslett. **4,** 180 (1965).
[1152] Slavin, W., Venghiattis, A., and Manning, D. C.: At. Absorption Newslett. **5,** 84 (1966).

[1153] Smith, D., and McLain, M. E.: Radiochem. Radioanal. Letters **16,** 89 (1974).
[1154] Smith, K. E., and Frank, C. W.: Appl. Spectry. **22,** 765 (1968).
[1155] Smith, R., and Winefordner, J. D.: Spectry. Letters **1,** 157 (1968).
[1156] Smith, R., and Winefordner, J. D.: IAASC, Sheffield 1969.
[1157] Smith, R. V.: Int. Laboratory **1973,** No. 2, 39.
[1158] Smyly, D. S., Townsend, W. P., Zeegers, P. J. Th., and Winefordner, J. D.: Spectrochim. Acta **26 B,** 531 (1971).
[1159] Snelleman, W.: in [5], p. 213.
[1160] Sobolev, N. N.: Spectrochim. Acta **11,** 310 (1956).
[1161] Spector, H., Clusman, S., Jatlow, P., and Seligson, D.: Clin. Chim. Acta **31,** 5 (1971).
[1162] Spielholtz, G. I., and Toralballa, G. C.: Analyst **94,** 1072 (1969).
[1163] Spitz, J., and Uny, G.: Appl. Optics **7,** 1345 (1968).
[1164] Spitzer, H.: Z. Erzbergbau Metallhüttenwes. **19,** 567 (1966).
[1165] Spitzer, H., and Tesik, G.: Z. Anal. Chem. **232,** 40 (1967).
[1166] Spooner, C. M., and Crassweller, P. O.: At. Absorption Newslett. **11,** 72 (1972).
[1167] Sprague, S., Manning, D. C., and Slavin, W.: At. Absorption Newslett. **3,** 27 (1964).
[1168] Sprague, S., and Slavin, W.: At. Absorption Newslett. **3,** 37 (1964).
[1169] Sprague, S., and Slavin, W.: At. Absorption Newslett. **3,** 72 (1964).
[1170] Sprague, S., and Slavin, W.: At. Absorption Newslett. **3,** 160 (1964).
[1171] Sprague, S., and Slavin, W.: At. Absorption Newslett. **4,** 228 (1965).
[1172] Sprague, S., and Slavin, W.: At. Absorption Newslett. **4,** 293 (1965).
[1173] Sprague, S., and Slavin, W.: At. Absorption Newslett. **4,** 367 (1965).
[1174] Stevens, B. J.: Clin. Chem. **18,** 1379 (1972).
[1175] Stewart, W. K., Hutchinson, F., and Fleming, L. W.: J. Lab. Clin. Med. **61,** 858 (1963).
[1176] Stone, M., and Chesher, S. E.: Analyst **94,** 1063 (1969).
[1177] Strasheim, A., and Butler, L. R. P.: Appl. Spectry. **16,** 109 (1962).
[1178] Strasheim, A., and Human, H. G. C.: Spectrochim. Acta **23 B,** 265 (1968).
[1179] Strasheim, A., Norval, E., and Butler, L. R. P.: J. S. Afr. Chem. Inst. **17,** 55 (1964).
[1180] Strasheim, A., Strelow, F. W. E., and Butler, L. R. P.: J. S. Afr. Chem. Inst. **13,** 73 (1960).
[1181] Strasheim, A., and Wessels, G. J.: Appl. Spectry. **17,** 65 (1963).
[1182] Strelow, F. W. E., Feast, E. C., Mathews, P. M., Bothma, C. J. C., and van Zyl, C. R.: Anal. Chem. **38,** 115 (1966).
[1183] Stresko, V., and Martiny, E.: At. Absorption Newslett. **11,** 4 (1972).
[1184] Strunk, D. H., and Andreasen, A. A.: At. Absorption Newslett. **6,** 111 (1967).
[1185] Stupar, J.: Z. Anal. Chem. **203,** 401 (1964).
[1186] Stupar, J.: Microchim. Acta **1966,** 722.
[1187] Stupar, J., and Dawson, J. B.: Appl. Optics **7,** 1351 (1968).
[1188] Stupar, J., and Dawson, J. B.: At. Absorption Newslett. **8,** 38 (1969).
[1189] Stupar, J., Podobnik, B., and Korosin, J.: Croat. Chem. Acta **37,** 141 (1965).
[1190] Suhr, N. H.: XIII CSI, Ottawa 1967.
[1191] Suhr, N. H., and Ingamells, C. O.: Anal. Chem. **38,** 730 (1966).
[1192] Sullivan, J. V., Parker, M., and Carson, S. B.: J. Lab. Clin. Med. **71,** 893 (1968).
[1193] Sullivan, J. V., and Walsh, A.: Spectrochim. Acta **21,** 721 (1965).
[1194] Sullivan, J. V., and Walsh, A.: Spectrochim. Acta **21,** 727 (1965).
[1195] Sullivan, J. V., and Walsh, A.: Spectrochim. Acta **22,** 1843 (1966).
[1196] Sullivan, J. V., and Walsh, A.: VI Aust. Spectry. Conf., Brisbane 1967.
[1197] Sullivan, J. V., and Walsh, A.: XIII CSI, Ottawa 1967.
[1198] Sullivan, J. V., and Walsh, A.: Appl. Optics **7,** 1271 (1968).
[1199] Sundberg, L. L.: Anal. Chem. **45,** 1460 (1973).
[1200] Sunderman, F. W.: Am. J. Clin. Pathol. **44,** 182 (1965).
[1201] Sunderman, F. W.: Human Pathol. **4,** 549 (1973).
[1202] Sunderman, F. W., and Carroll, J. E.: Am. J. Clin. Pathol. **43,** 302 (1965).
[1203] Sunderman, F. W., and Nechay, M. W.: Biochem. Anal. 74, Munich 1974.
[1204] Sunderman, F. W., and Roszel, N. O.: Am. J. Clin. Pathol. **48,** 286 (1967).

[1205] Supp, G. R.: At. Absorption Newslett. **11,** 122 (1972).
[1206] Supp, G. R., Gibbs, I., and Juszli, M.: At. Absorption Newslett. **12,** 66 (1973).
[1207] Sutter, E., Platman, S. R., and Fieve, R. R.: Clin. Chem. **16,** 602 (1970).
[1208] Swider, R. T.: At. Absorption Newslett. **7,** 111 (1968).
[1209] Sychra, V., and Kolihová, D.: 3rd CISAFA, Paris 1971, p. 265.
[1210] Sychra, V., and Matoušek, J.: Anal. Chim. Acta **52,** 376 (1970).
[1211] Sychra, V., and Matoušek, J.: Talanta **17,** 363 (1970).
[1212] Sychra, V., Slevin, P. J., Matoušek, J., and Bek, F.: Anal. Chim. Acta **52,** 259 (1970).
[1213] Syty, A.: Anal. Letters **4,** 531 (1971).
[1214] Syty, A.: At. Absorption Newslett. **12,** 1 (1973).
[1215] Takahashi, M., and Uruno, Y.: Bunko Kenkyu **10,** 110 (1962).
[1216] Tarlin, I. H., and Batchelder, M.: J. Pharmac. Sci. **59,** 1328 (1970).
[1217] Taulli, T. A., and Kaelble, E. F.: At. Absorption Newslett. **9,** 100 (1970).
[1218] Tavenier, P., and Hellendoorn, H. B. A.: Clin. Chim. Acta **23,** 47 (1969).
[1219] Taylor, J. H.: At. Absorption Newslett. **8,** 95 (1969).
[1220] Taylor, M. L., and Belcher, C. B.: Anal. Chim. Acta **45,** 219 (1969).
[1221] Temperli, A. T., and Misteli, H.: Anal. Biochem. **27,** 361 (1969).
[1222] Tenoutasse, N.: 3rd CISAFA, Paris 1971, p. 817.
[1223] Terashima, S.: Japan Analyst **18,** 1259 (1969).
[1224] Thilliez, G.: Anal. Chem. **39,** 427 (1967).
[1225] Thilliez, G.: Chimie Analytique **50,** 226 (1968).
[1226] Thompson, M. H.: J. Ass. Offic. Agr. Chem. **52,** 55 (1969).
[1227] Thompson, R. J., Morgan, G. B., and Purdue, L. J.: At. Absorption Newslett. **9,** 53 (1970).
[1228] Thormahlen, D. J., and Frank, E. H.: At. Absorption Newslett. **10,** 63 (1971).
[1229] Thorpe, V. A.: J. Ass. Offic. Agr. Chem. **54,** 206 (1971).
[1230] Thorpe, V. A.: J. Ass. Offic. Agr. Chem. **56,** 147 (1973).
[1231] Tindall, F. M.: At. Absorption Newslett. **4,** 339 (1965).
[1232] Tindall, F. M.: At. Absorption Newslett. **5,** 140 (1966).
[1233] Tindall, F. M.: At. Absorption Newslett. **6,** 104 (1967).
[1234] Toffoli, P., and Pannetier, G.: 3rd CISAFA, Paris 1971, p. 707.
[1235] Topping, G., and Pirie, J. M.: Anal. Chim. Acta **62,** 200 (1972).
[1236] Toshimitsu, M.: J. Physic. Soc. Japan. **17,** 1440 (1962).
[1237] Toth, S. J., and Reimer, D. N.: Indust. Water Eng. Sept. **1967,** 42.
[1238] Townsend, W. P., Smyly, D. S., Zeegers, P. J. T., Svoboda, V., and Winefordner, J. D.: Spectrochim. Acta **26 B,** 595 (1971).
[1239] Trent, D. J.: At. Absorption Newslett. **4,** 348 (1965).
[1240] Trent, D. J., Manning, D. C., and Slavin, W.: At. Absorption Newslett. **4,** 335 (1965).
[1241] Trent, D. J., and Slavin, W.: At. Absorption Newslett. **3,** 17 (1964).
[1242] Trent, D. J., and Slavin, W.: At. Absorption Newslett. **3,** 53 (1964).
[1243] Trent, D. J., and Slavin, W.: At. Absorption Newslett. **3,** 118 (1964).
[1244] Trent, D. J., and Slavin, W.: At. Absorption Newslett. **4,** 300 (1965).
[1245] Trudeau, D. L., and Freier, E. F.: Clin. Chem. **13,** 101 (1967).
[1246] Tušl, J.: J. Ass. Offic. Agr. Chem. **53,** 1190 (1970).
[1247] Tyler, J. B.: At. Absorption Newslett. **6,** 14 (1967).
[1248] Tyndall, J.: Six Lectures on Light. D. Appleton & Co, New York 1898.
[1249] Ulfvarson, U.: Acta Chem. Scand. **21,** 641 (1967).
[1250] Ure, A. M.: IAASC, Sheffield 1969.
[1251] Uthe, J. F., Armstrong, F. A. J., and Stainton, M. P.: J. Fish. Res. Bd. Canada **27,** 805 (1970).
[1252] VanLoon, J. C.: At. Absorption Newslett. **7,** 3 (1968).
[1253] VanLoon, J. C.: At. Absorption Newslett. **8,** 6 (1969).
[1254] VanLoon, J. C.: At. Absorption Newslett. **11,** 60 (1972).
[1255] VanLoon, J. C., Aarden, D., and Galbraith, J.: IAASC, Sheffield 1969.
[1256] VanLoon, J. C., and Parissis, C. M.: Anal. Letters **1,** 249 (1968).
[1257] VanLoon, J. C., and Parissis, C. M.: Analyst **94,** 1057 (1969).

[1258] Varju, M. E.: At. Absorption Newslett. **11,** 45 (1972).
[1259] Varju, M. E.: II ČS. Conf. Flame Spectry., Zvikov 1973.
[1260] Varju, M. E., and Elek, E.: At. Absorption Newslett. **10,** 128 (1971).
[1261] Varley, J. A., and Chin, P. Y.: Analyst **95,** 592 (1970).
[1262] Vaughn, W. W., and McCarthy, J. H.: US Geol. Survey Prof. Paper **501-D,** D 123 (1964).
[1263] Veall, N.: Medical uses of Ca-47. Int. Atom. Energ. Ag., Vienna, Techn. Report **32,** (1964).
[1264] Veillon, C., Mansfield, J. M., Parsons, M. L., and Winefordner, J. D.: Anal. Chem. **38,** 204 (1966).
[1265] Veillon, C., and Margoshes, M.: Spectrochim. Acta **23 B,** 503 (1968).
[1266] Venghiattis, A. A.: Spectrochim. Acta **23 B.** 67 (1967).
[1267] Venghiattis, A. A.: Appl. Optics **7,** 1313 (1968).
[1268] Vickers, T. J., and Vaughn, R. M.: Anal. Chem. **41,** 1476 (1969).
[1269] Vidale, G. L.: General Electric T.I.S. Report R60SD330 (1961).
[1270] Vidale, G. L.: General Electric T.I.S. Report R60SD331 (1962).
[1271] Vijan, P. N., and Wood, G. R.: At. Absorption Newslett. **13,** 33 (1974).
[1272] Vogliotti, F. L.: At. Absorption Newslett. **9,** 123 (1970).
[1273] Voinovitch, I., Legrand, G., and Louvrier, J.: 3rd CISAFA, Paris 1971, p. 843.
[1274] Volhard, J.: J. Prakt. Chem. **2,** 217 (1874).
[1275] Vollmer, J.: At. Absorption Newslett. **5,** 12 (1966).
[1276] Vollmer, J.: At. Absorption Newslett. **5,** 35 (1966).
[1277] Vollmer, J., Sebens, C., and Slavin, W.: At. Absorption Newslett. **4,** 306 (1965).
[1278] Vulfson, E. K., Karyakin, A. V., and Shidlovsky, A. I.: Zh. Anal. Khim. **28,** 1253 (1973).
[1279] Wadman, B. W.: Diesel and Gas Turbine Progress, July **1971,** 16.
[1280] Walker, C. R., and Vita, O. A.: Anal. Chim. Acta **43,** 27 (1968).
[1281] Wallace, F. J.: Analyst **88,** 259 (1963).
[1282] Walsh, A.: Spectrochim. Acta **7,** 108 (1955).
[1283] Walsh, A.: X CSI, Maryland 1962.
[1284] Walsh, A.: IAASC, Sheffield 1969.
[1285] Walsh, A.: Int. Congr. Anal. Chem., Kyoto 1972.
[1286] Walsh, A.: Appl. Spectry. **27,** 335 (1973).
[1287] Warren, J. M., and Spencer, H.: Clin. Chim. Acta **38,** 435 (1972).
[1288] Weger, S. J., Hossner, L. R., and Ferrara, L. W.: J. Agr. Food Chem. **17,** 1276 (1969).
[1289] Welcher, G. G., and Kriege, O. H.: At. Absorption Newslett. **8,** 97 (1969).
[1290] Welcher, G. G., and Kriege, O. H.: At. Absorption Newslett. **9,** 61 (1970).
[1291] Welcher, G. G., Kriege, O. H., and Marks, J. Y.: 4th ICAS, Toronto 1973.
[1292] Welz, B.: CZ **95,** T99 (1971).
[1293] Welz, B.: 3rd CISAFA, Paris 1971, p. 655.
[1294] Welz, B.: CZ-Chemie-Tech. **1,** 455 (1972).
[1295] Welz, B.: Fortschr. Miner. **50** (Bh. 1), 106 (1972).
[1296] Welz, B.: Z. Werkstofftechnik **4,** 285 (1973).
[1297] Welz, B.: CSI XVII, Florence 1973, p. 67.
[1298] Welz, B.: in Vom Wasser **42.** Verlag Chemie, Weinheim 1974, p. 119.
[1299] Welz, B., and Sebestyen, J. E.: Joint Sympos. Accurate Meth. Anal. Maj. Const., London 1970.
[1300] Welz, B., and Wiedeking, E.: Int. Sympos. Mikrochem., Graz 1970.
[1301] Welz, B., and Wiedeking, E.: Z. Anal. Chem. **252,** 111 (1970).
[1302] Welz, B., and Wiedeking, E.: Z. Anal. Chem. **264,** 110 (1973).
[1303] Welz, B., and Witte, W.: 4th ICAS, Toronto 1973.
[1304] Wendt, R. H., and Fassel, V. A.: Anal. Chem. **37,** 920 (1965).
[1305] Wendt, R. H., and Fassel, V. A.: Anal. Chem. **38,** 337 (1966).
[1306] West, A. C., Fassel, V. A., and Kniseley, R. N.: Anal. Chem. **45,** 1586 (1973).
[1307] West, C. D., and Hume, D. N.: Anal. Chem. **36,** 412 (1964).
[1308] West, F. K., West, P. W., and Ramakrishna, T. V.: Environ. Sci. Technol. **1,** 717 (1967).
[1309] West, T. S.: Endeavour **26,** 44 (1967).

[1310] West, T. S.: Int. Sympos. Mikrochem., Graz 1970.
[1311] West, T. S., and Williams, X. K.: Anal. Chim. Acta **45,** 27 (1969).
[1312] Westerlund-Helmerson, U.: At. Absorption Newslett. **5,** 97 (1966).
[1313] Wheat, J. A.: XI Conf. Anal. Chem. Nucl. Techn., Gatlinburg, Tenn., 1967.
[1314] Wheat, J. A.: USAEC-Report DP-1164, Dupont Savannah River Lab., 1968.
[1315] Wheat, J. A.: Appl. Spectry. **25,** 3 (1971).
[1316] White, A. D.: J. Appl. Phys. **30,** 711 (1959).
[1317] White, R. A.: J. Sci. Instrum. **44,** 678 (1967).
[1318] Whittington, C. M., and Willis, J. B.: Plating **51,** 767 (1964).
[1319] Williams, A. I.: Analyst **98,** 233 (1973).
[1320] Williams, C. H., David, D. J., and Iismaa, O.: J. Agric. Sci. **59,** 381 (1962).
[1321] Williams, T. R., Wilkinson, B., Wadsworth, G. A., Barther, D. H., and Beer, W. J.: J. Sci. Fd. Agric. **17,** 344 (1967).
[1322] Willis, J. B.: Nature **184,** 186 (1959).
[1323] Willis, J. B.: Nature **186,** 249 (1960).
[1324] Willis, J. B.: Spectrochim. Acta **16,** 259 (1960).
[1325] Willis, J. B.: Spectrochim. Acta **16,** 273 (1960).
[1326] Willis, J. B.: Spectrochim. Acta **16,** 551 (1960).
[1327] Willis, J. B.: Nature **191,** 381 (1961).
[1328] Willis, J. B.: Anal. Chem. **33,** 556 (1961).
[1329] Willis, J. B.: Anal. Chem. **34,** 614 (1962).
[1330] Willis, J. B.: in Methods of Biochemical Analysis, Vol. XI. Interscience, New York/London/Sydney 1963.
[1331] Willis, J. B.: Aust. J. Dairy Technol., June **1964,** 70.
[1332] Willis, J. B.: Clin. Chem. **11,** 251 (1965).
[1333] Willis, J. B.: Nature **207,** 715 (1965).
[1334] Willis, J. B.: Appl. Optics **7,** 1295 (1968).
[1335] Willis, J. B.: Spectrochim. Acta **26 B,** 177 (1971).
[1336] Willis, J. B., Fassel, V. A., and Fiorino, J. A.: Spectrochim. Acta **24 B,** 157 (1969).
[1337] Willis, J. B., Rasmuson, J. O., Kniseley, R. N., and Fassel, V. A.: Spectrochim. Acta **23 B,** 725 (1968).
[1338] Wilson, J.: Unpublished.
[1339] Wilson, L.: Anal. Chim. Acta **30,** 377 (1964).
[1340] Wilson, L.: Anal. Chim. Acta **40,** 503 (1968).
[1341] Winefordner, J. D.: IAASC, Sheffield 1969.
[1342] Winefordner, J. D.: CSI XVII, Florence 1973, p. 58.
[1343] Winefordner, J. D., and Mansfield, J. M.: Appl. Spectry. Rev. **1,** 1 (1967).
[1344] Winefordner, J. D., and Parsons, M. P.: Anal. Chem. **39,** 1593 (1966).
[1345] Winefordner, J. D., and Staab, R. A.: Anal. Chem. **36,** 165 (1964).
[1346] Winefordner, J. D., and Staab, R. A.: Anal. Chem. **36,** 1367 (1964).
[1347] Winefordner, J. D., and Vickers, T. J.: Anal. Chem. **36,** 161 (1964).
[1348] Winefordner, J. D., and Vickers, T. J.: Anal. Chem. **36,** 1947 (1964).
[1349] Witkind, I. J.: Am. Mineral. **54,** 1118 (1969).
[1350] Wollin, A.: At. Absorption Newslett. **9,** 43 (1970).
[1351] Woodriff, R., and Ramelow, G.: Spectrochim. Acta **23 B,** 665 (1968).
[1352] Woodriff, R., and Stone, R. W.: Appl. Optics **7,** 1337 (1968).
[1353] Woodriff, R., Stone, R. W., and Held, A. M.: Appl. Spectry. **22,** 408 (1968).
[1354] Woodson, T. T.: Rev. Sci. Instrum. **10,** 308 (1939).
[1355] Woodward, C.: At. Absorption Newslett. **8,** 121 (1969).
[1356] Woolley, J. F.: Spectrovision **1969,** 7.
[1357] Woolley, J. F.: 4th ICAS, Toronto 1973.
[1358] Wu, J. Y. L., Droll, H. A., and Lott, P. F.: At. Absorption Newslett. **7,** 90 (1968).
[1359] Yamamoto, Y., Kumamaru, T., and Hayashi, Y.: Talanta **14,** 611 (1967).
[1360] Yanagisawa, M., Suzuki, M., and Takeuchi, T.: Talanta **14,** 933 (1967).

[1361] Yanagisawa, M., Suzuki, M., and Takeuchi, T.: Microchim. Acta **1973,** 475.
[1362] Yeh, Y-Y., and Zee, P.: Clin. Chem. **20,** 360 (1974).
[1363] Youden, W. J.: J. Ass. Offic. Agr. Chem. **46,** 55 (1963).
[1364] Yoza, N., and Ohashi, S.: Anal. Letters **6,** 595 (1973).
[1365] Yuan, T. L., and Breland, H. L.: Soil Sci. Soc. Amer. Proc. **33,** 868 (1969).
[1366] Yule, J. W., and Swanson, G. A.: At. Absorption Newslett. **8,** 30 (1969).
[1367] Zacha, K. E., Bratzel, M. P., Winefordner, J. D., and Mansfield, J. M.: Anal. Chem. **40,** 1733 (1968).
[1368] Zacha, K. E., and Winefordner, J. D.: Anal. Chem. **38,** 1537 (1966).
[1369] Zaidel, A. N., and Korennoi, E. P.: Opt. Spectry. **10,** 299 (1961).
[1370] Zaino, E. C.: At. Absorption Newslett. **6,** 93 (1967).
[1371] Zaugg, W. S.: At. Absorption Newslett. **6,** 63 (1967).
[1372] Zaugg, W. S., and Knox, R. J.: Anal. Chem. **38,** 1759 (1966).
[1373] Zaugg, W. S., and Knox, R. J.: Anal. Biochem. **20,** 282 (1967).
[1374] Zeegers, P. J. T., Smith, R., and Winefordner, J. D.: Anal. Chem. **40,** 26 (1968).
[1375] Zeegers, P. J. T., and Winefordner, J. D.: Spectrochim. Acta **26 B,** 161 (1971).
[1376] Zeeman, P. B., and Brink, J. A.: Analyst **93,** 388 (1968).
[1377] Zeeman, P. B., and Butler, L. R. P.: Tegnicon **13,** 96 (1960).
[1378] Zeeman, P. B., and Butler, L. R. P.: Appl. Spectry. **16,** 120 (1962).
[1379] Zettner, A.: in Adv. in Clin. Chem., Vol. 7. Academic Press, New York 1965.
[1380] Zettner, A., and Mansbach, L.: Am. J. Clin. Pathol. **44,** 517 (1965).
[1381] Zettner, A., and Mensch, A. H.: Am. J. Clin. Pathol. **48,** 225 (1967).
[1382] Zettner, A., and Mensch, A. H.: Am. J. Clin. Pathol. **49,** 196 (1968).
[1383] Zettner, A., Rafferty, K., and Jarecky, H. J.: At. Absorption Newslett. **7,** 32 (1968).
[1384] Zettner, A., and Seligson, D.: Clin. Chem. **10,** 869 (1964).
[1385] Zettner, A., Sylvia, L. C., and Capacho-Delgado, L.: Am. J. Clin. Pathol. **45,** 533 (1966).
[1386] Zinterhofer, L. J. M., Jatlow, P. I., and Fappiano, A.: J. Lab. Clin. Med. **78,** 664 (1971).
[1387] Zlalkis, A., Bruewing, W., and Bayler, E.: Anal. Chem. **41,** 1692 (1969).
[1388] Zuehlke, C. W., and Ballard, A. E.: Anal. Chem. **22,** 953 (1950).

Index